Dedication to Professor Michael Tribelsky: 50 Years in Physics

Dedication to Professor Michael Tribelsky: 50 Years in Physics

Editors

Andrey Miroshnichenko
Boris Malomed
Fernando Moreno

MDPI • Basel • Beijing • Wuhan • Barcelona • Belgrade • Manchester • Tokyo • Cluj • Tianjin

Editors
Andrey Miroshnichenko
University of New South Wales Canberra
Australia

Boris Malomed
Tel Aviv University
Israel

Fernando Moreno
University of Cantabria
Spain

Editorial Office
MDPI
St. Alban-Anlage 66
4052 Basel, Switzerland

This is a reprint of articles from the Special Issue published online in the open access journal *Physics* (ISSN 2624-8174) (available at: https://www.mdpi.com/journal/physics/special_issues/DPMTP).

For citation purposes, cite each article independently as indicated on the article page online and as indicated below:

LastName, A.A.; LastName, B.B.; LastName, C.C. Article Title. *Journal Name* **Year**, *Volume Number*, Page Range.

ISBN 978-3-0365-5637-6 (Hbk)
ISBN 978-3-0365-5638-3 (PDF)

Contents

About the Editors

Andrey Miroshnichenko

Prof. Miroshnichenko obtained his Ph.D. in 2003 from the Max-Planck Institute for Physics of Complex Systems in Dresden, Germany. In 2004, he moved to Australia to join the Nonlinear Physics Centre at ANU. During that time, he made fundamentally important contributions to the field of photonic crystals and brought the concept of the Fano resonances to nanophotonics. In 2007, he was awarded by Australian Postdoctoral and, in 2011, by Future Fellowships from the Australian Research Council. In 2017, he moved to the University of New South Wales Canberra and obtained a very prestigious UNSW Scientia Fellowship. During his career, Prof Miroshnichenko published more than 250 journal publications and authored a book, in addition to writing seven book chapters in other publications. He was recognized as a 'Highly Cited Researcher' for three consecutive years (2019–2021) by the Web of Science Group, which is awarded to fewer than 0.1% of the world's most influential researchers. Recently, he was also recognized as a top-40 researcher in Australia (and top-5 in physics) by The Australian Lifetime Achievers Leaderboard (2020, 2021). The topics of his research are nonlinear nanophotonics, nonlinear optics, and resonant interaction of light with nanoclusters, including optical nanoantennas and metamaterials.

Boris Malomed

Boris Malomed received his MS in physics from Belorussian State University (Minsk) in 1977, a PhD in theoretical physics from the Moscow Institute of Physics and Technology in 1981, and a Doctor of Sciences degree (habilitation) in theoretical physics from the N. N. Bogoliubov Institute for Theoretical Physics, Ukrainian Academy of Sciences (Kiev) in 1989. Until 1991, he was a senior researcher at the Institute for Oceanology of the Russian Academy of Sciences (Moscow). Since 1991, he has been an Associate Professor, and, since 1998, a Full Professor at the Tel Aviv University. Since 2015, he has held a position as a research chair for "Optical solitons", funded by Tel Aviv University. He has published 1200+ papers and four books. His h-index is 82 (Web of Science)/95 (Google Scholar). He is an editor of three major international journals: Physics Letters A, Chaos, Solitons & Fractals, and Frontiers in Physics, and an Editorial Board member of the Journal of Optics, Symmetry, Photonics, Scientific Reports, Optics Communications, and Chaos. He is a Senior Member of the Optical Society of America.

Fernando Moreno

Fernando Moreno has been Professor of Optics at the University of Cantabria since 1993, where he heads the Optics group. He has a Ph.D. in physics (University of Zaragoza, 1982). Assistant Professor and Associate Professor at the University of Cantabria with research in photon statistics applied to light scattering methods (1982–1993). Guest researcher at the Royal Signals and Radar Establishment (UK, 1990–1991) and visiting professor at the Department of Physics at the University of California San Diego (USA, 1995–1996), working in plasmonics at the nanometer range. Published over 200 articles in high impact journals, edited a book in Springer and co-authored several book chapters. Co-author of four patents with members of the Army Research Laboratory in Maryland (Washington, USA) and also of five other patents. Responsible for different research projects (including tutoring Thesis, Dissertations and Thesis work), both public (European Commission, Spanish Research Council, NATO, The Royal Society, USAITCA, etc.) and with private companies. He maintains permanent collaborations with other research groups, both in Spain and abroad (mainly

Italy, USA, France, Russia and UK). Co-founder of the spin-off FOTOGLASS (www.fotoglass.es), dedicated to research and development in optical materials and awarded by the Cantabria Gob. As being one of the best Technological-Based Local Companies. Senior Member of Optical Society of America, member of the European Optical Society, the Electromagnetics Academy (USA) and the Spanish Society of Optics. Of the latter, he is the former Chairman of the Education Committee. Current research is on nanooptics, plasmonics, light scattering and phase change materials.

Preface to "Dedication to Professor Michael Tribelsky: 50 Years in Physics"

Michael Tribelsky graduated from Lomonosov Moscow State University, USSR in 1973, defended his Ph.D. thesis at the Moscow Institute of Physics and Technologies in 1976, and receive a second Doctorate (habilitation, known as "Doctor of Physical-Mathematical Sciences") from the Landau Institute for Theoretical Physics in 1983. In 1979, at the age of 28, he received the highest USSR national prize for junior scientists, the Lenin Komsomol Prize in Science and Technologies, for his outstanding achievements in the study of optical damage to glass.

His first scientific paper (of which he was a single author) was published in 1971, when he still was a student. The paper is devoted to the Gunn effect in semiconductors. It is worth mentioning that, in this paper, he developed an original approximate method to solve the Schrödinger equation in the vicinity of the ground state with complicated potential (the limit opposite to the WKB). Since then, Prof. Tribelsky has made numerous fundamental contributions to an extremely broad area of physics and mathematics, including (but not limited to) quantum solid-state physics, various problems of light–matter interaction, liquid crystals, physical hydrodynamics, nonlinear waves, pattern formation in nonequilibrium systems and transition to chaos, bifurcation and probability theory, and even a prediction of the dynamics of actual market prices. He has published several fundamental papers, which included the following results:

- The pioneering detailed study of the structure and stability of domain walls between various nonequilibrium (dissipative) structures, such as grains of roll with different orientations in Rayleigh–Bénard–Marangoni convection.
- The prediction of a new type of stable dissipative structure with quasicrystal symmetry and a determination of the criteria for their stability.
- The discovery of the drift bifurcation transforming a steady dissipative pattern into a traveling wave.
- The formulation of an approach to determine the height of the barrier separating various locally stable solutions of the Ginzburg–Landau equation in the corresponding functional space and calculating this barrier in the explicit form. This made it possible to determine the "stability margin" of the corresponding solutions for finite-amplitude perturbations.
- A simple topological explanation of the nature of the violation of weak conservation laws, which led to a "slip" in the phase of the complex-order parameter when its modulus vanishes, and the prediction of the universality of the dynamics of the order parameter in the vicinity of the phase-slip-points.
- In collaboration with I.M. Lifshitz, the formulation and solution of the problem of the propagation of nonlinear elastic waves in metals near the point of the electron-topological phase transition, which required the derivation of the nonlinear quantum elasticity theory equations.
- The prediction of a new and very unusual type of transition to turbulence analogous to second-order phase transitions, when the turbulent state smoothly and directly detaches from the rest state of a fluid (experimentally observed in the electroconvection of liquid crystals).

In the course of his work on light–matter interactions, Prof. Tribelsky's contribution to the optical breakdown of glass acquired special importance, as well as his explanation of the deep laser melting of metals made shortly after the experimental discovery of the effect. Another noteworthy study direction is represented by his work on the optical-thermodynamic phenomena in liquids, where a light beam is used as a tool to transfer liquid to a given point in the phase diagram. Finally, it is worht mentioning the pioneering research of Prof. Tribelsky on the dynamics of thermochemical instability in polymers. In his recent work, he focused on fundamentals of resonant light scattering by subwavelength particles and made a significant contribution to the understanding of the physics of the Fano resonances, anomalous scattering, and absorption, as well as the excitation of longitudinal modes in subwavelength particles made of materials with a spatial dispersion. Last but not least, his most recent results, devoted to the dynamic resonant scattering, should be mentioned, which opened door to a new subfield in subwavelength optics.

We would like to stress that all his important (and often counterintuitive) theoretical predictions have found solid experimental evidence. Most of these results remain highly relevant to the current research in this vast area.

Professor Tribelsky's accomplishments are highly appreciated by the international community. The best indications of this are the high citation rates of his publications, and the numerous awards and titles he has received. In particular, in addition to the aforementioned Lenin Komsomol Prize, he received the Max Planck Society Fellowship many times to carry out research in Germany; the JSPS Fellowship for Senior Scientists, to carry out research in Japan, Center of Excellency Professorship from the University of Tokyo and Kyushu University, Japan; Honorary Doctor of Philosophy from the Yamaguchi University, Japan; and numerous invitations for Visiting Professorships from the best universities around the world.

We congratulate Michael on his double anniversary and wish him many happy, fruitful years ahead, new fundamental discoveries, and talented disciples. We believe that this Special Issue constitutes a timely celebration of our respected scholar, researcher, and friend. Furthermore, we hope that this Special Issue inspires scholars, especially junior researchers, to continue these advancements in physics.

Andrey Miroshnichenko, Boris Malomed, and Fernando Moreno
Editors

Brief Report

Spontaneous Curvature Induced Stretching-Bending Mode Coupling in Membranes

Efim I. Kats

Landau Institute for Theoretical Physics, RAS, 142432 Chernogolovka, Moscow Region, Russia; kats@itp.ac.ru

Abstract: In this paper, a simple example to illustrate what is basically known from the Gauss' times interplay between geometry and mechanics in thin shells is presented. Specifically, the eigen-mode spectrum in spontaneously curved (i.e., up-down asymmetric) extensible polymerized or elastic membranes is studied. It is found that in the spontaneously curved crystalline membrane, the flexural mode is coupled to the acoustic longitudinal mode, even in the harmonic approximation. If the coupling (proportional to the membrane spontaneous curvature) is strong enough, the coupled modes dispersions acquire the imaginary part, i.e., effective damping. The damping is not related to the entropy production (dissipation); it comes from the redistribution of the energy between the modes. The curvature-induced mode coupling makes the flexural mode more rigid, and the acoustic mode becomes softer. As it concerns the transverse acoustical mode, it remains uncoupled in the harmonic approximation, keeping its standard dispersion law. We anticipate that the basic ideas inspiring this study can be applied to a large variety of interesting systems, ranging from still fashionable graphene films, both in the freely suspended and on a substrate states, to the not yet fully understood lipid membranes in the so-called gel and rippled phases.

Keywords: membranes; vibration modes

Citation: Kats, E.I. Spontaneous Curvature Induced Stretching-Bending Mode Coupling in Membranes. *Physics* **2021**, *3*, 367–371. https://doi.org/10.3390/physics3020025

Received: 15 April 2021
Accepted: 10 May 2021
Published: 14 May 2021

1. Introduction

It is my pleasure and honor to present my work in the Special Issue of journal of *Physics*, dedicated to the 70th birthday of Prof. Mikhail Tribelsky. Many years ago, we collaborated with Misha (as I became used to calling him), investigating phase transitions in biaxial liquid crystals. Misha came up with the idea of our joint article and guided me through the long process of writing. The paper I am presenting now (also from the realm of soft matter physics) is a small token of my gratitude and respect to Misha Tribelsky.

The study is motivated by a recent influential paper [1], where the authors found that even in the harmonic approximation, in-plane and out-of-plane vibrations of a low-dimension ($D = 1, 2$) elastic system can be coupled. The harmonic coupling in Ref. [1] occurs if the stress-free state of the membrane is curved. Provided that the "up-down" symmetry of the system is not broken, the only way to couple the modes is related to the Gaussian curvature, which is symmetric (even) with respect to "up-down" inversion; for more details, see monographs and review [2–6] or original papers [7–10]. However, in many realistic and experimentally relevant situations, the "up-down" symmetry is spontaneously broken [3,5]. Moreover, it can be nonuniformly broken over the membrane surface, e.g., due to asymmetric adsorption of different molecular species. In such a situation, the up-down non-symmetric free energy in the harmonic approximation can be written in terms of the scalar out-of-plane displacement f and the $2D$ vector $\mathbf{u}$ of the in-plane displacements: [2,3,5]

$$F_{\mathrm{elastic}} = \int d^2r \left[\frac{1}{2}\kappa(\nabla^2 f)^2 + C_0(r)\nabla^2 f + \frac{1}{2}\lambda u_{ii}^2 + \mu\left(\frac{\partial u_i}{\partial x_j} + \frac{\partial u_j}{\partial x_i}\right)^2\right], \quad (1)$$

where κ is the bending (curvature) elastic modulus, the subscripts i and j take on the values $1, 2$ for the Cartesian axis within the membrane plane (in the limit of small displacements, the topological Gaussian curvature contribution is neglected here), μ and λ are the stretching Lame coefficients, and $C_0(r)$ is non-uniform over the membrane spontaneous curvature. In the linear approximation, the only term allowed by symmetry is

$$C_0 = \gamma_{\text{int}} \text{div}\mathbf{u}\,, \tag{2}$$

where γ_{int} is the curvature-induced mode coupling coefficient.

2. Basic Derivation

With the free energy expansion (1) one can find the eigen-modes of the system. Since the intent of this study is to investigate dynamic behavior of spontaneously curved membranes, first of all, the free energy (1) has to be supplemented by the kinetic energy terms related to the in-plane and out-of-plane displacements dynamics. As a result, one gets the dynamic action,

$$S = \frac{1}{2}\left(\frac{\partial f}{\partial t}\right)^2 + \frac{1}{2}\left(\frac{\partial u_i}{\partial t}\right)^2 + F_{\text{elastic}}\,, \tag{3}$$

where, for the sake of simplicity, the units with the mass density $\rho = 1$ are used, and t is the time. Then, in the Fourier space, the corresponding Euler–Lagrange equations for the coupled eigen-modes read:

$$\begin{aligned} \omega^2 \tilde{f} - \kappa q^4 \tilde{f} - i\gamma_{\text{int}} q^2 q_i \tilde{u}_i = 0\,, \\ \omega^2 \tilde{u}_i - i\gamma_{\text{int}} q_i q^2 \tilde{f} - \mu q^2 \tilde{u}_i - \mu q_i (q_j \tilde{u}_j) - \lambda q_i (q_m \tilde{u}_m) = 0\,, \end{aligned} \tag{4}$$

where $\tilde{f}$ and $\tilde{\mathbf{u}}$ are the Fourier transforms of the displacements

$$f(\mathbf{r}, t) = \int \frac{d^2 q}{(2\pi)^2} \frac{d\omega}{2\pi} \exp(-i\omega t + i\mathbf{q}\mathbf{r}) \tilde{f}(\mathbf{q}, \omega)\,,$$

and similarly for the in-plane displacements u_i.

From Equation (4), one can see that the out-of-plane displacement is coupled to the only longitudinal component (with respect to the wave-vector) of the in-plane displacements, similarly to the case of up-down symmetric crystalline membranes considered in Ref. [11]. Therefore, it is convenient to express the in-plane displacements in terms of the longitudinal, u_l, and transverse, u_t, components, namely,

$$u_i = u_l \frac{q_i}{q} + u_t \epsilon_{ij} \frac{q_j}{q}\,, \tag{5}$$

where $\epsilon_{ij} = -\epsilon_{ji}$ is the antisymmetric second-rank tensor. Then, the equations of motion can be rewritten as:

$$\begin{aligned} \omega^2 \tilde{f} - \kappa q^4 \tilde{f} - i\gamma_{\text{int}} q^3 \tilde{u}_l = 0\,, \\ \omega^2 \tilde{u}_l - i\gamma_{\text{int}} q_i q^2 \tilde{f} - (2\mu + \lambda) q^2 \tilde{u}_l = 0\,, \\ \omega^2 \tilde{u}_t - \mu q^2 \tilde{u}_t = 0\,. \end{aligned} \tag{6}$$

3. Eigen-Modes in Spontaneously Curved Membranes

From Equation (6), one can see that the transverse acoustic mode, which is decoupled from flexural mode in the harmonic approximation, has the standard dispersion law:

$$\omega_t^2 = \mu q^2\,. \tag{7}$$

In the harmonic approximation, the dispersion laws for the two coupled (in the spontaneously curved membrane) longitudinal acoustic and flexural modes are:

$$\omega_{1,2}^2 = \frac{\kappa q^4 + (\lambda + 2\mu)q^2}{2} \pm \left[\left(\frac{\kappa q^4 - (\lambda + 2\mu)q^2}{2} \right)^2 - \gamma_{\rm int}^2 \kappa^2 q^5 \right]^{1/2} . \tag{8}$$

This expression is the main result of this paper and is ready for further inspection. From Equation (8), one arrives at the two following conclusions.

- For the sufficiently strong spontaneous curvature-induced coupling, there is an interval of the wave vectors,

$$\frac{(\lambda + 2\mu)^2}{\kappa^2 \gamma_{\rm int}^2} < q < \sqrt{\frac{\lambda + 2\mu}{\kappa}} , \tag{9}$$

  where the coupled-modes dispersions acquire the imaginary part, i.e., effective damping. The damping is not related to the entropy production (dissipation), since there is no any dissipative term in the action (3). The damping occurs from the redistribution of the energy of the modes.
- In the limit of weak coupling,

$$\gamma_{\rm int}^2 \kappa^2 q^5 < \frac{(\kappa q^4 - \mu q^2)^2}{4} , \tag{10}$$

  the coupled modes remain purely propagating, and their dispersion laws read:

$$\omega_1^2 = \kappa q^4 + \frac{\gamma_{\rm int}^2 \kappa^2}{\lambda + 2\mu} q^3 , \tag{11}$$

$$\omega_2^2 = (2\mu + \lambda) q^2 - \frac{\gamma_{\rm int}^2 \kappa^2}{(2\mu + \lambda)} q^3 . \tag{12}$$

  Therefore, the curvature-induced mode coupling makes the flexural mode more rigid, and the acoustic mode becomes softer.

4. Outlook and Conclusions

To summarize, in this paper, the eigen-mode spectrum in spontaneously curved, i.e., up-down asymmetric, extensible polymerized or elastic membranes is calculated. It is found that in the spontaneously-curved crystalline membrane, the flexural mode is coupled to the acoustic longitudinal mode, even in the harmonic approximation. If the coupling, being proportional to the membrane spontaneous curvature, is strong enough, the coupled-mode's dispersions acquire the imaginary part, i.e., effective damping. The damping is not related to the entropy production (dissipation): it comes from the redistribution of the energy of the modes. The curvature-induced mode coupling makes the flexural mode more rigid, and the acoustic mode becomes softer. What concerns the transverse acoustical mode, it remains uncoupled in the harmonic approximation, keeping its standard dispersion law.

Let us close with some conclusions about where the results presented here can be applied. First, one may think about the famous graphene films [12–16] either on substrate or freely suspended. In both cases, experimental observations suggest that the graphene film becomes spontaneously corrugated. Although the main physical phenomena leading to the graphene film corrugations are not yet fully understood (see, e.g., Refs. [17–19]), the conclusion of the very existence of the corrugated state seems inescapable. The second system one might have in mind is a rippled state of lipid membranes, below so-called main phase transition [4–6]. The macroscopic structure and physical properties of the rippled gel state are still debated in the literature; see, e.g., Refs. [20,21]. The simple calculations, presented in this paper, are a step forward to check whether the state is liquid, liquid-crystalline (liquid layers), or crystalline (with a positional order within the layer). Analysis of eigen-mode spectra provides a hint to disentangle both states.

It is worth noting that only the mean-field approximation was considered in this paper. However, the membranes are effectively two-dimensional objects (on scales larger than the membrane thickness). Therefore, thermal fluctuations can affect behavior, first of all, renormalizing the membrane elastic moduli. This renormalization of the bending modulus κ is well known for liquid-like membranes. It turns out that thermal fluctuations make the liquid-like membrane softer, i.e., reduce the bending modulus [5,6]. For crystalline membranes (see Refs. [9,10] and recent studies [22,23]), the bending modulus increases, i.e., the membrane becomes harder, whereas the stretching Lame coefficients μ and λ decrease. Similarly, one can also find the fluctuation renormalization of the spontaneous curvature or, in terms of this study, the curvature-induced mode coupling coefficient γ_{int}. The one-loop approximation calculations [5,6] yield the following scaling law:

$$\gamma_{\text{int}} \propto (3T/4\pi\kappa)^{-1/3},$$

with T being the temperature, i.e., thermal fluctuations decrease the curvature-induced mode coupling. In this paper, only the surface of this reach subject is stratched. This study deliberately focused on the most limited possible question, which can be answered by calculations "on a back of the envelope". There is a plenty of work ahead.

Acknowledgments: I thank V.V. Lebedev for useful discussions.

Conflicts of Interest: The author declares no conflict of interest.

References

1. Kernes, J.; Levine, A.J. Effects of curvature on the propagation of undulatory waves in lower dimensional elastic materials. *Phys. Rev. E* **2021**, *103*, 013002. [CrossRef] [PubMed]
2. Nelson, D.; Piran, T.; Weinberg, S. (Eds.) *Statistical Mechanics of Membranes and Surfaces*; World Scientific: Singapore, 1989. [CrossRef]
3. Chaikin, P.M.; Lubensky, T.C. *Principles of Condensed Matter Physics*; Cambridge University Press: Cambridge, UK, 2000. [CrossRef]
4. Bowick, M.J.; Travesset, A. The statistical mechanics of membranes. *Phys. Rep.* **2001**, *344*, 255–308. [CrossRef]
5. Safran, S.A. *Statistical Thermodynamics of Surfaces, Interfaces, and Membranes*; CRC Press: Boca Raton, FL, USA, 2003. [CrossRef]
6. Kats, E.I.; Lebedev, V.V. *Fluctuational Effects in the Dynamics of Liquid Crystals*; Springer: New York, NY, USA, 1994. [CrossRef]
7. Helfrich, W. Elastic properties of lipid bilayers: Theory and possible experiment. *Z. Naturforsch.* **1973**, *28*, 693–703. [CrossRef] [PubMed]
8. Seifert, U. Configurations of fluid membranes and vesicles. *Adv. Phys.* **1997**, *46*, 13–137. [CrossRef]
9. Nelson, D.R.; Peliti, L. Fluctuations in membranes with crystalline and hexatic order. *J. Phys. France* **1987**, *48*, 1085–1092. [CrossRef]
10. Aronovitz, J.A.; Lubensky, T.C. Fluctuations of solid membranes. *Phys. Rev. Lett.* **1988**, *60*, 2634–2638. [CrossRef] [PubMed]
11. Kernes, J.; Levine, A.J. Geometrically-induced localization of flexural waves on thin warped physical membranes. *arXiv* **2020**, arXiv:2011.07152v1.
12. Novoselov, K.S.; Geim, A.K.; Morozov, S.V.; Jiang, D.; Zhang, Y.; Dubonos, S.V.; Grigorieva, I.V.; Firsov, A.A. Electric field effect in atomically thin carbon films. *Science* **2004**, *306*, 666–669. [CrossRef] [PubMed]
13. Novoselov, K.S.; Jiang, D.; Schedin, F.; Booth, T.J.; Khotkevich, V.V.; Morozov, V.V.; Geim, A.K. Two-dimensional atomic crystals. *Proc. Natl. Acad. Sci. USA* **2005**, *102*, 10451–10453. [CrossRef] [PubMed]
14. Neto, A.H.C.; Guinea, F.; Peres, N.M.R.; Novoselov, K.S.; Geim, A.K. The electronic properties of graphene. *Rev. Mod. Phys.* **2009**, *81*, 109–161. [CrossRef]
15. Vozmediano, M.A.H.; Katsnelson, M.I.; Guinea, F. Gauge fields in graphene. *Phys. Rep.* **2010**, *496*, 109–152. [CrossRef]
16. Geim, A.K.; Grigorieva, I.V. Van der Waals heterostructures. *Nature* **2013**, *499*, 419–425. [CrossRef] [PubMed]
17. Fasolino, A.; Los, J.H.; Katsnelson, M.I. Intrinsic ripples in graphene. *Nat. Mater.* **2007**, *6*, 858–861. [CrossRef]
18. Bao, W.; Miao, F.; Chen, Z.; Zhang, H.; Jang, W.; Dames, C.; Lau, C.N. Controlled ripple texturing of suspended graphene and ultrathin graphite membranes. *Nat. Nanotechnol.* **2009**, *4*, 562–566. [CrossRef] [PubMed]
19. Deng, S.; Berry, V. Wrinkled, rippled and crumpled graphene: an overview of formation mechanism, electronic properties, and applications. *Mater. Today* **2016**, *19*, 197–213. [CrossRef]
20. Raghunathan, V.A.; Katsaras, J. $L'_\beta \rightarrow L_c$ phase transition in phosphatidylcholine lipid bilayers: A disorder-order transition in two dimensions. *Phys. Rev. E* **1996**, *54*, 4446–4449. [CrossRef] [PubMed]
21. Akabori, K.; Nagle, J.F. Structure of the DMPC lipid bilayer ripple phase. *Soft Matter* **2015**, *11*, 918–926. [CrossRef] [PubMed]

22. Burmistrov, I.S.; Gornyi, I.V.; Kachorovskii, V.Y.; Katsnelson, M.I.; Los, J.H.; Mirlin, A.D. Stress-controlled Poisson ratio of a crystalline membrane: Application to graphene. *Phys. Rev. B* **2018**, *97*, 125402. [CrossRef]
23. Saykin, D.R.; Kachorovskii, V.Y.; Burmistrov, I.S. Phase diagram of a flexible two-dimensional material. *Phys. Rev. Res.* **2020**, *2*, 043099. [CrossRef]

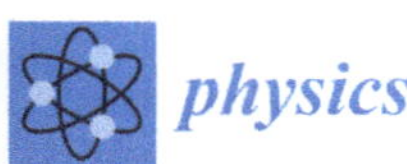

MDPI

Perspective

Laws of Spatially Structured Population Dynamics on a Lattice

Natalia L. Komarova [1,*], Ignacio A. Rodriguez-Brenes [1] and Dominik Wodarz [1,2]

1 Department of Mathematics, University of California Irvine, Irvine, CA 92617, USA; ignacio.rodriguezbrenes@uci.edu (I.A.R.-B.); dwodarz@uci.edu (D.W.)
2 Department of Population Health and Disease Prevention, Program in Public Health Susan and Henry Samueli College of Health Sciences, University of California Irvine, Irvine, CA 92697, USA
* Correspondence: komarova@uci.edu

Abstract: We consider spatial population dynamics on a lattice, following a type of a contact (birth–death) stochastic process. We show that simple mathematical approximations for the density of cells can be obtained in a variety of scenarios. In the case of a homogeneous cell population, we derive the cellular density for a two-dimensional (2D) spatial lattice with an arbitrary number of neighbors, including the von Neumann, Moore, and hexagonal lattice. We then turn our attention to evolutionary dynamics, where mutant cells of different properties can be generated. For disadvantageous mutants, we derive an approximation for the equilibrium density representing the selection–mutation balance. For neutral and advantageous mutants, we show that simple scaling (power) laws for the numbers of mutants in expanding populations hold in 2D and 3D, under both flat (planar) and range population expansion. These models have relevance for studies in ecology and evolutionary biology, as well as biomedical applications including the dynamics of drug-resistant mutants in cancer and bacterial biofilms.

Keywords: evolutionary dynamics; mutations; agent-based modeling; somatic evolution; computational methods; mathematical modeling

Citation: Komarova, N.L.; Rodriguez-Brenes, I.A.; Wodarz, D. Laws of Spatially Structured Population Dynamics on a Lattice. *Physics* **2022**, *4*, 812–832. https://doi.org/10.3390/physics4030052

Received: 14 April 2022
Accepted: 14 June 2022
Published: 22 July 2022

1. Introduction

Before we begin describing some of our attempts to derive a number of mathematical laws for biological population dynamics, one of the authors (N.L.K.) would like to express her eternal gratitude to Michael Tribelsky, who was her teacher in the Physics Department, Moscow State University, at the beginning of the 1990s. Without his advice and guidance, N.L.K. would not be what she is now. Tribelsky's relentless optimism (often disguised) has taught her to overcome whatever difficulties life has posed. N.L.K. always follows Tribelsky's principle that problems (in life, as well as in science) must be addressed at the same pace, and not faster, than they are thrown at you ("Проблемы нужно решать по мере их поступления"). Tribelsky's "special course", delivered to the theoreticians from the Low Temperature Division, opened up a universe of diverse phenomena that are, at the same time, fascinating and amenable to understanding. He helped instill the sense of wonder at the beauty of the surrounding world, and this has stayed with N.L.K. no matter what subject matter she happened to focus on, from language evolution to virus dynamics.

Much has been said about the existence of laws in biology. Physicists in particular often feel that some beautiful (and, hopefully, simple) formal relations must exist to help us navigate the complexities of life. In this paper, we look for universal biological laws in the area of population dynamics, which is an area relevant for studies in ecology and evolutionary biology. A rich literature exists about the evolutionary dynamics of mutant spread and invasion, investigating aspects such as fixation probabilities and fixation times of different kinds of mutants in constant populations [1,2], as well as mutant dynamics in growing populations [3–5], where mathematical approaches were motivated by the famous

Luria–Delbruck experiments [6]. Traditionally, much of this work has focused on well-mixed (non-spatial), homogeneous populations. In a wide array of biological scenarios, however, cells and organisms evolve in more complex settings, such as in spatially structured habitats (bacterial biofilms, cells in tissues and tumors) and within heterogeneous population structures (such as stem and more differentiated cells in tissues [7] or bacteria with different degrees of specialization in biofilms [8,9]). In the last 15 years, theoretical work has extended our understanding of mutant emergence in spatially structured populations [10–18]. An excess of mutational jackpot events was observed in spatial compared to well-mixed systems; such events result from mutations arising at the surface of expanding, spatially structured populations, surfing at the edge of range expansions, and appearing as mutant "bubbles" or "slices". These jackpot events have been implicated in the finding that the average number of neutral mutants when the total population reaches a given threshold size is significantly larger in spatial compared to non-spatial settings [18]. Further work by our group [19] showed that the evolutionary dynamics of mutants in spatially structured, expanding populations are characterized by additional complexities.

In general, spatial population dynamics are significantly more complex compared to the mass-action dynamics, and analytical approximations are not easily derived. In this paper, we describe two examples where our group has been successful in obtaining analytical approximations for laws of spatial dynamics: (a) steady-state density in space; (b) scaling laws for the number of mutants in expanding populations.

(a) Analytical approximations for steady-state density. In [20], we considered the phenomenon of range expansion, the process in biology by which a species spreads to new areas. We derived a numerical methodology that allowed for efficient computations of the number of individuals as the species expands its range in space. In addition, we found approximations for the steady-state density of populations, or the core density of expanding colonies, in several cases, such as the von Neumann and Moore neighborhoods on a square lattice and for a honeycomb neighborhood on a hexagonal lattice; see Table 1. For the same death-to-birth ratio, the grid is more packed under the Moore neighborhood, because of the availability of more neighbors per site. As a consequence, the equilibrium density corresponding to the Moore neighborhood is higher and closer to that of mass-action. A general formula that provides an approximation for the steady-state density as a function of the number of neighbors is also given (Table 1). In this paper, we provide a novel mathematical derivation of these laws, which generalizes to any neighborhood size (Section 3).

Table 1. Steady-state densities for different types of lattices, as well as the mass-action system and a generalized, N_b-neighbor system; see [20]. Here, $\delta = D/L$, the division-to-death rate ratio. Red dots denote the focal cell and the blue ones stand for its neighbors.

Lattice	Geometry	Density
von Neumann		$\frac{3-4\delta}{3-\delta}$
Moore		$\frac{7-8\delta}{7-\delta}$
Hexagonal		$\frac{5-6\delta}{5-\delta}$
Mass-action		$1-\delta$
N_b neighbors		$\frac{N_b-1-N_b\delta}{N_b-1-\delta}$

(b) Laws of disadvantageous, neutral, and advantageous mutant spread in different geometries. In [19], we studied the evolutionary dynamics of disadvantageous, neutral, and advantageous mutants in spatially expanding populations, where the growing front was characterized by different symmetries in two dimensions (2D) and 3D; see Table 2. In particular, while disadvantageous mutants grow linearly with N (and differences among different cases are more subtle), neutral mutants grow as a decreasing power of N as we go from a 2D flat front (which is essentially a 1D growth), to a 2D and 3D range. There are always fewer neutral or advantageous mutants in the mass-action (exponentially growing) case compared to any spatially restricted growth. Here (Sections 4 and 5), we provide a derivation of these laws, which follows [19].

Table 2. Mutant scaling laws for expanding populations; see [19]. Here, N is the population size, u is the mutation rate, and $\alpha > 1$. See text for details.

	2D Flat	2D Range	3D Flat	3D Range	Exponential
Mutant property					
Disadvantageous	uN	uN	uN	uN	uN
Neutral	uN^2	$uN^{3/2}$	uN^2	$uN^{4/3}$	$uN\ln N$
Advantageous	uN^3	uN^2	uN^4	uN^2	$uN^{(2\alpha-1)/\alpha}$

In the case of disadvantageous mutants, the number of mutants always scales with the first power of N. In addition to this scaling law, for the case of disadvantageous mutants, we also derive an approximate expression for the number of mutants in quasi-steady-state, given that the number of wild-type cells is N:

$$\#\,\text{mutants} = \frac{D_w L_w (4D_m^2 + 3D_m L_w + D_w L_m) uN}{D_m (4D_m + 3L_w)(D_m L_w - D_w L_m)}; \tag{1}$$

see Section 4.1 for details and Section 2 for definitions (this expression is valid for a 2D von Neumann grid, and can be generalized to other cases). We show that the proportion of mutants at selection–mutation balance is higher for spatially distributed systems compared to well-mixed systems at equilibrium.

2. General Setup and Agent-Based Modeling of Population Dynamics

In order to describe the spatial growth and turnover of cells, as well as the population dynamics of species, we use a continuous-time Markov process, which is a generalization of the usual birth–death process and a type of a contact process. We assume that individuals of one or two different types exist, which we call "wild-type" and "mutant" individuals (to reflect the versatility of these models, we use both "individuals" and "cells" to refer to the biological agents under consideration). At the core of the description is a lattice that specifies possible locations of cells. This can be viewed as a geometric network, where each node is connected with its neighbors. The state space consists of different locations of the cells of the two types on the lattice. Cells are characterized by division and death rate parameters, which are denoted by L_w and D_w for the division and death rates of the wild-type cells and L_m and D_m for division and death rates of mutants. In the case where only one type of cell is considered, the notation is simply L and D for the two rates.

Each of the nodes can either be empty or contain one cell of either type. During an infinitesimally small time increment, Δt, a given wild-type cell can attempt a division with probability L_w and death with probability D_w. If division is attempted, then an offspring is placed in one of the neighboring locations, chosen randomly and uniformly; division only happens if the chosen node is empty. The offspring cell is wild-type with probability $1-u$ and mutant with probability u, where $0 \le u < 1$ is the mutation rate. Mutant cells divide and die according to similar rules, except in the models considered here, we always assume that an offspring of a mutant cell is a mutant cell.

A number of biological phenomena can be studied by slight modifications of this process. To study the growth laws in the absence [20] or presence [19] of mutants, we assume an infinitely large grid. To study the quasi-steady-state (which describes, e.g., the turnover of cells in homeostasis), we make the grid finite and impose relevant boundary conditions.

For numerical explorations of these processes, agent-based modeling (ABM) is often used; see, e.g., [21–23]. For example, consider a two-dimensional ABM on a square grid with individuals of two types. A spot on the grid can be empty or can contain a cell, which is either wild-type or mutant. At each time step, the grid is randomly sampled N times, where N is the total number of cells currently in the system. If the sampled cell is wild-type, the cell attempts division (described below) with a probability proportional to L_w or dies with a probability proportional to D_w. When reproduction is attempted, a target spot is chosen randomly among the N_b nearest neighbors of the cell. A neighborhood may contain, e.g., $N_b = 4$ cells (the von Neumann neighborhood) or $N_b = 8$ neighbors (the Moore neighborhood). If that spot is empty, the offspring cell is placed there. If it is already filled, the division event is aborted. (This modeling choice represents the assumption that the probability of divisions is reduced under more crowded conditions. This is similar to the logistic growth term often used in deterministic models.) The offspring cell is assigned wild-type with probability $1 - u$, and it is a mutant with probability u. If the sampled spot contains a mutant cell, the same processes occur. Attempted division occurs with a probability $\propto L_m$, and the cell dies with a probability $\propto D_m$. Initial and boundary conditions are determined by the specific geometric setting investigated. For 2D spatial simulations, an $n \times n$ square or an $n \times w$ rectangular domain could be considered. At the boundaries of the domain, a spot is assumed to have fewer neighbors, i.e., more division events will fail. The process starts with a single wild-type cell, placed, e.g., in the center of the grid. Simulations always stop before the boundary of the grid is reached.

In what follows below, we derive some approximations of important observables from these types of dynamics, which have a clear biological meaning. We discuss both single-type populations and the co-dynamics of wild-type and mutant individuals.

3. Analytical Approximations for Steady-State Density (A Single Type)

In this Section, we consider populations consisting of only a single cell type with division and death rates L and $D < L$ and no mutations ($u = 0$).

Equilibrium Density in Space: An Analytical Method

The process of range expansion (colonization) is one of the basic types of biological dynamics, whereby a species grows and spreads outwards, occupying new territories. Spatial modeling of this process is naturally implemented as a stochastic ABM of the type described above, with individuals (in this case, of only a single type) occupying nodes on a rectangular grid, births and deaths occurring probabilistically, and individuals only reproducing onto un-occupied neighboring spots. This approach is known to be computationally expensive. In [20], we derived a set of efficient computational tools, which were shown to be in good agreement with the underlying stochastic process of spatial expansion. As part of the method development, we were able to obtain approximate expressions for the quasi-steady-state (core) density of the individuals for different types of grids. In [20], we provided the density formula for the contact process:

$$\rho = \frac{N_b - 1 - N_b\delta}{N_b - 1 - \delta}, \tag{2}$$

where $\delta = D/L$ and N_b is the number of neighbors. Equation (2) was derived for several cases of N_b, but no general derivation that would work for a given N_b was supplied. Here, we present such a derivation, starting with $N_b = 2$ and generalizing to any N_b.

In 1D, with only $N_b = 2$ neighbors, Equation (2) gives

$$\rho = \frac{1-2\delta}{1-\delta}, \tag{3}$$

which, although a slight overestimation of the density, provides a good approximation. To derive this formula, consider a 1D ABM of the type described above. At the equilibrium, denote by p_1 the probability that a cell is located next to an existing cell and by p_0 the probability that a cell is located next to an existing empty spot. These two quantities can be estimated numerically.

A 1D realization of the population at a fixed moment of time can be viewed as a Markov chain with states $\{0,1\}$, where state 0 corresponds to the absence of a cell at a given spot and state 1 denotes the presence of a cell at a given spot. The transition probability matrix is given by

$$P = \begin{pmatrix} p_1 & 1-p_1 \\ p_0 & 1-p_0 \end{pmatrix}$$

where P_{11} is the probability that there is a cell on the right of a cell, P_{12} is the probability that the spot on the right of a cell is empty, P_{21} is the probability that there is a cell on the right of an empty spot, and P_{22} is the probability that an empty spot is on the right of an empty spot. The steady-state probability distribution of this process is $(\rho, 1-\rho)$, where ρ is the probability that a given spot contains a cell. This can be found as the eigenvector corresponding to the unit eigenvalue and is given by

$$\rho = \frac{p_0}{1-p_1+p_0}. \tag{4}$$

This quantity (given the numerically calculated p_0 and p_1) is a very good approximation of the actual (numerical) density. Another way to derive this connection between ρ and p_0, p_1 is as follows:

$$\rho = \rho p_1 + (1-\rho)p_0, \tag{5}$$

where the left-hand side is the probability to have a cell at a given point, the first term on the right assumes that there is a cell to the left of that point (ρ) and, then, the given point contains a cell with probability p_1, and the second term on the right assumes that there is no cell to the left of that point $(1-\rho)$ and, then the given point contains a cell with probability p_0. Solving this for ρ gives expression (4).

Now, suppose that a cell is located at a given location. Then, a cell is located to the right of it (which we call the focal location) with probability p_1. We have, after a single update of the contact process:

$$p_1 = (1-p_1)\left(\frac{L}{2} + p_0\frac{L}{2}\right)\frac{1}{N} + p_1\left(1-\frac{D}{N}\right). \tag{6}$$

The first term on the right assumes that there was no cell at the focal location $(1-p_1)$, but a cell on its left was chosen (probability $1/N$) that divided to its right $(L/2)$ or that a cell on its right exists (probability p_0), was chosen $(1/N)$, and divided to its left $(L/2)$. The second term assumes that there was a cell at the focal location (p_1) and it did not die $(1-D/N)$. The equality follows from the assumption of having an equilibrium.

Similarly, we can assume that there is no cell at a given location, then to its right (the focal location), the probability to have a cell, p_0, satisfies:

$$p_0 = (1-p_0)p_0\frac{L}{2N} + p_0\left(1-\frac{D}{N}\right). \tag{7}$$

The first term on the right assumes that there was no cell at the focal location $(1-p_0)$, but a cell on its right exists (probability p_0), was chosen $(1/N)$, and divided to its left $(L/2)$.

The second term assumes that there was a cell at the focal location (p_0) and it did not die $(1 - D/N)$.

Solving Equations (6) and (7) for p_0 and p_1, we obtain:

$$p_1 = 1 - \frac{D}{L}, \quad p_0 = 1 - \frac{2D}{L}.$$

Substituting this into Equation (4), we obtain expression (3).

This analysis easily generalizes to other systems (including higher dimensionalities), where the number of neighbors is given by N_b. Instead of Equations (6) and (7), we have

$$p_1 = (1 - p_1)\left(\frac{L}{N_b} + \frac{(N_b - 1)p_0 L}{n}\right)\frac{1}{N} + p_1\left(1 - \frac{D}{N}\right), \tag{8}$$

$$p_0 = (1 - p_0)\frac{(N_b - 1)p_0 L}{N_b N} + p_0\left(1 - \frac{D}{N}\right). \tag{9}$$

Solving this system, we obtain

$$p_1 = 1 - \frac{D}{L}, \quad p_0 = 1 - \frac{N_b D}{(N_b - 1)L}.$$

Substitution into Equation (4) (which holds in these systems because Equation (5) remains valid in these systems) yields Equation (2).

Figure 1 plots the quasi-equilibrium density approximated by Equation (2) as a function of D/L for two finite values of N_b corresponding to the von Neumann and Moore neighborhoods. Figure 1a compares them with the well-known mass-action result, showing that higher connectivity of the underlying network corresponds to a higher density of the cells. Figure 1b and Figure 1c compare the results with numerical simulations, demonstrating that Equation (2) provides a very good approximation. We notice that the approximation works better at higher densities, compared to lower ones. In the derivation, we assumed that the state of a given position on the lattice only depends on its immediate neighbor, and not on other, more distant, sites. This assumption becomes less valid at low densities, because macroscopic structures with strong correlations over distances > 1 form. Therefore, the approximation becomes worse as the death rate approaches the division rate.

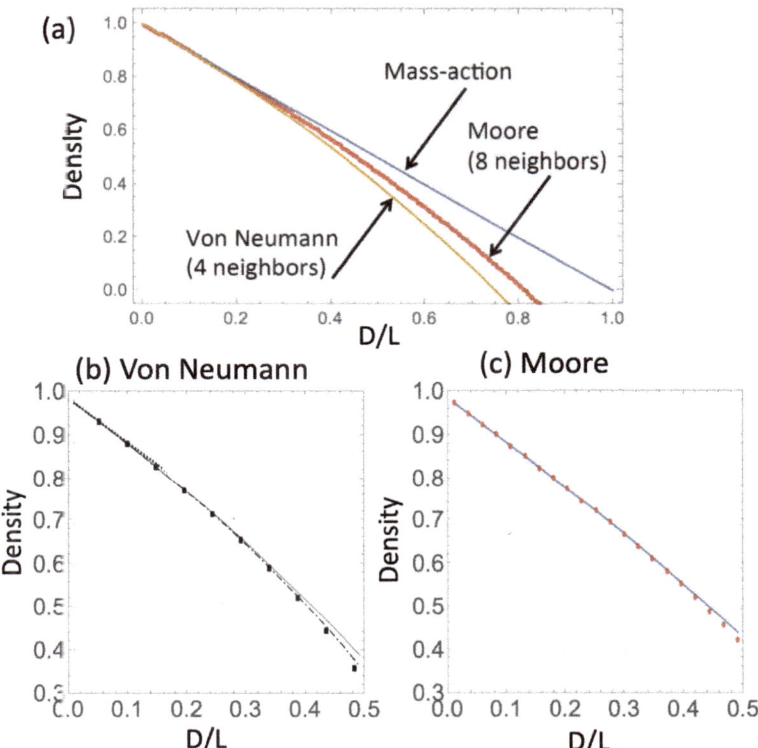

Figure 1. Quasi-equilibrium density of cells. (**a**) Equation (2), showing the equilibrium density as a function of D/L for $N_b = 4$ (von Neumann lattice), $N_b = 8$ (Moore lattice), and $N_b \to \infty$ (mass-action). (**b**) Comparison of Equation (2) (solid line) with numerically calculated density values (dots) for $N_b = 4$. The dashed-dotted line represents a higher-order approximation; see [20] for details. (**c**) Comparison of Equation (2) (solid line) with numerically calculated density values (dots) for $N_b = 8$. See text for details.

4. Disadvantageous Mutants and Selection–Mutation Balance in Spatial Models

In the remainder of the paper, we consider two types of individuals, wild-type and mutants. This Section deals with disadvantageous mutants (defined below), while Section 5 focuses on neutral and advantageous mutants. We start with an Ordinary Differential Equation (ODE) formulation.

4.1. A Basic ODE Formulation

Let us denote the wild-type population as $x(t)$ and the mutant population as $y(t)$. Suppose that mutations happen at the rate u, and $D_w < L_w$. Then, the competition dynamics of cells in a mass-action system can be formulated as follows:

$$\dot{x} = L_w x(1-u)\left(1 - \frac{x+y}{\mathcal{K}}\right) - D_w x, \quad (10)$$

$$\dot{y} = (L_w x u + L_m y)\left(1 - \frac{x+y}{\mathcal{K}}\right) - D_m y, \quad (11)$$

where $\mathcal{K}$ is the carrying capacity. This system illustrates differences in the behavior of advantageous and disadvantageous mutants. If

$$\frac{D_m}{D_w} - \frac{L_m}{L_w(1-u)} < 0,$$

then the globally stable solution is $x = 0$, $y = \mathcal{K}(1 - D_m/L_m)$, that is the mutant excludes the wild-type and takes over. In this case, we say that the mutant is advantageous. If the inequality above is reversed, the mutants are disadvantageous and the so-called selection–mutation balance is reached, where mutants and wild-type cells coexist. This happens because the (stronger) wild-type cells produce mutants at a nonzero rate u, but the latter cannot take over and remains at a fraction of the population. For simplicity, let us assume that the mutation rate is low:

$$u \ll \frac{D_m}{D_w} - \frac{L_m}{L_w} \equiv \gamma.$$

Then, we can say that the mutants are disadvantageous if

$$\frac{D_m}{D_w} - \frac{L_m}{L_w} > 0, \tag{12}$$

and then, the equilibrium solution is given by

$$\bar{x} = \mathcal{K}\left(1 - \frac{D_w}{L_w}\right), \tag{13}$$

where we neglected terms of the order u/γ, while the number of mutants is

$$\bar{y} = \bar{x}u\left(\frac{D_m}{D_w} - \frac{L_m}{L_w}\right)^{-1}, \tag{14}$$

where we neglected terms of the order $(u/\gamma)^2$.

Below, we calculate the equilibrium densities of disadvantageous mutants and (advantageous) wild-type cells in a spatially distributed system at steady-state. This will also correspond to the densities in the core of an expanding system away from the advancing front.

4.2. A Spatial Description: Equations for the Densities

We restrict our description to a 2D square grid, with the von Neumann neighborhood (that is, each location has four nearest neighbors); the methodology is generalizable to the Moore neighborhood (eight neighbors). We use a method similar to that of [20]. Two random variables describe the state of the stochastic system at each spatial location, x: ρ_x describes wild-type cells, such that

$$\rho_x = \begin{cases} 1, & \text{if a wild-type cell is at location } x, \\ 0, & \text{otherwise,} \end{cases}$$

and η_x describes mutant cells, such that

$$\eta_x = \begin{cases} 1, & \text{if a mutant cell is at location } x, \\ 0, & \text{otherwise.} \end{cases}$$

Note that ρ_x and η_x cannot be equal to one simultaneously; an empty spot corresponds to $\rho_x = \eta_x = 0$. We assume that wild-type cells have division and death rates L_w and D_w and mutant cells have division and death rates L_m and D_m. Wild-type cells mutate with probability u, and no back mutations are considered.

Denote the expectation of ρ_x and η_x by

$$\langle \rho_x \rangle = \rho, \quad \langle \eta_x \rangle = \eta,$$

where we assumed that the expected values do not depend on spatial location, since we are interested in spatially homogeneous equilibrium solutions. We have

$$\dot{\rho} = \left\langle \frac{L_w}{N_b}(1-u)(1-\rho_x)(1-\eta_x)\sum_k \rho_x^{(k)} - D_w\rho_x \right\rangle, \tag{15}$$

$$\dot{\eta} = \left\langle \frac{1}{N_b}(1-\rho_x)(1-\eta_x)\sum_k \left(L_w u\rho_x^{(k)} + L_m\eta_x^{(k)}\right) - D_m\eta_x \right\rangle, \tag{16}$$

where the product $(1-\rho_x)(1-\eta_x)$ is nonzero only if location x is empty, and the summation goes over all the neighbors of point x, which reproduce into location x at rates L_w/N_b and L_m/N_b if they are wild-type or mutant, respectively. The superscript in the notation $\rho_x^{(k)}$, $\eta_x^{(k)}$ refers to the quantity at a location, k, neighboring the focal location, x.

Let us consider the von Neumann neighborhood ($N_b = 4$). In the right-hand side of Equation (15), the terms are the summation having the form

$$\langle(1-\rho_x)(1-\eta_x)\rho_x^{(k)}\rangle = \langle\rho_x^{(k)}\rangle - \langle\rho_x\rho_x^{(k)}\rangle - \langle\rho_x^{(k)}\eta_x\rangle + \langle\rho_x^{(k)}\rho_x\eta_x\rangle = \rho - W - I, \tag{17}$$

and in Equation (16), there are also terms of the form

$$\langle(1-\rho_x)(1-\eta_x)\eta_x^{(k)}\rangle = \langle\eta_x^{(k)}\rangle - \langle\rho_x\eta_x^{(k)}\rangle - \langle\eta_x^{(k)}\eta_x\rangle + \langle\eta_x^{(k)}\rho_x\eta_x\rangle = \eta - I - M. \tag{18}$$

In the expressions above, we have $\langle\rho_x\rho_x^{(k)}\eta_x^{(k)}\rangle = 0$, because either $\eta_x^{(k)}$ or $\rho_x^{(k)}$ is zero at location $x^{(k)}$, and the three types of dyads are defined as follows:

- $W = \langle\rho_x\rho_x^{(k)}\rangle$ is the probability to have two wild-type cells at two neighboring locations;
- $I = \langle\rho_x\eta_x^{(k)}\rangle$ is the probability to have a wild-type cell and a mutant at two neighboring locations;
- $M = \langle\eta_x\eta_x^{(k)}\rangle$ is the probability to have two mutant cells at two neighboring locations.

Figure 2a illustrates these three configurations. In terms of these correlations, Equations (15) and (16) can be rewritten as

$$\dot{\rho} = L_w(1-u)(\rho - W - I) - D_w\rho, \tag{19}$$

$$\dot{\eta} = L_w u(\rho - W - I) + L_m(\eta - I - M) - D_m\eta. \tag{20}$$

The correlations for the three dyads that appear in these equations require their own equations to close the system. Let us derive an equation for W. We have

$$\dot{W} = \left\langle 2(1-\rho_x)(1-\eta_x)\rho_x^{(k)}\sum_j \rho_x^{(j)}\frac{L_w}{N_b}(1-u) - 2D_wW \right\rangle,$$

where we assume that one of the points in the dyad contains a wild-type cell (term $\rho_x^{(k)}$), while the other point is empty (term $(1-\rho_x)(1-\eta_x)$), and that one of its neighbors (location $x^{(j)}$) contains a wild-type cell, which reproduces faithfully into point x at rate $L_w(1-u)/N_b$. Note that either of the two points could be empty, which results in the multiplier of two in the first term on the right-hand side. Similarly, either of the dyad's locations can experience cell death, resulting in the negative rate $2D_w$. In order to calculate the average, we need to consider terms

$$\langle(1-\rho_x)(1-\eta_x)\rho_x^{(k)}\rho_x^{(j)}\rangle. \tag{21}$$

Note that here and below, the operation of averaging makes the expression independent of the actual location x. Further, the superscripts (k) and (j) do not refer to any specific neighbor of x, but to any neighbor of x; in particular, location $x^{(j)}$ may be the same as or different than location $x^{(k)}$. In the case when the two locations are different, the correlation (21) is presented in Figure 2b, on the left.

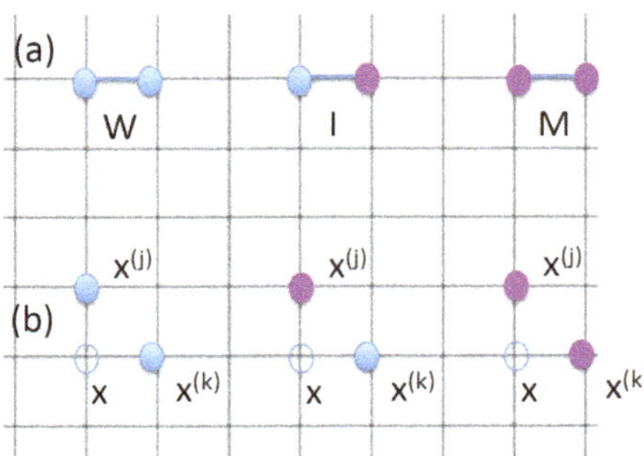

Figure 2. Steps in the derivation of equations for a two-component system of wild-type and mutant cells. Blue circles denote wild-type, and purple denote mutant cells. (**a**) Three configurations, whose correlations appear in Equations (19) and (20). (**b**) Three types of correlations needed for equations for W, I, and M.

In the equations for $\dot{M}$ and $\dot{I}$, the following expressions appear in addition to Equation (21):

$$\langle(1-\rho_x)(1-\eta_x)\rho_x^{(k)}\eta_x^{(j)}\rangle, \quad \langle(1-\rho_x)(1-\eta_x)\eta_x^{(k)}\eta_x^{(j)}\rangle.$$

These correlations are shown in Figure 2b, center and right. Therefore, denoting by a and b either ρ or η, we evaluate the average of the form

$$\langle(1-\rho_x)(1-\eta_x)a_x^{(k)}b_x^{(j)}\rangle, \tag{22}$$

which corresponds to a dyad with one of the locations (location x) empty and the other (location $x^{(k)}$), containing type "a", while a neighbor of x (location $x^{(j)}$) contains type "b". First, let us assume that location $x^{(j)}$ is different from location $x^{(k)}$. Under von Neumann neighborhoods, this implies that $x^{(j)}$ and $x^{(k)}$ are not each other's neighbors because on a square grid, there could not be a non-degenerate triangle with diameter 1 or less. expression (22) is equal to

$$\begin{aligned} &P(b_x^{(j)}=1|a_x^{(k)}=1,\rho_x=\eta_x=0)P(a_x^{(k)}=1,\rho_x=\eta_x=0)\approx \\ &P(b_x^{(j)}=1|\rho_x=\eta_x=0)P(a_x^{(k)}=1,\rho_x=\eta_x=0)= \\ &\frac{P(b_x^{(j)}=1,\rho_x=\eta_x=0)P(a_x^{(k)}=1,\rho_x=\eta_x=0)}{P(\rho_x=\eta_x=0)} \end{aligned} \tag{23}$$

The expression in the denominator is calculated as follows:

$$P(\rho_x=\eta_x=0)=\langle(1-\rho_x)(1-\eta_x)\rangle=\langle 1-\rho_x-\eta_x+\rho_x\eta_x\rangle=1-\rho-\eta.$$

Depending on the types at location x, the expressions in the numerator of Equation (23) can be of two types:

$$P(\rho_x^{(k)}=1,\rho_x=\eta_x=0) \text{ or } P(\eta_x^{(k)}=1,\rho_x=\eta_x=0),$$

and they are calculated in (17) and (18), respectively.

Next, we assume that location $x^{(j)}$ is the same as $x^{(k)}$. Then, if types "a" and "b" in expression (22) are different, then we obtain $\langle(1-\rho_x)(1-\eta_x)\rho_x^{(k)}\eta_x^{(k)}=0$. If the types are the same, then we obtain expression (17) or (18). To summarize, expressions of type (22) are given as follows:

$$\langle(1-\rho_x)(1-\eta_x)\rho_x^{(k)}\rho_x^{(j)}\rangle = \begin{cases} \frac{(\rho-W-I)^2}{1-\rho-\eta}, & x^{(j)} \neq x^{(k)}, \\ \rho - W - I, & x^{(j)} = x^{(k)}, \end{cases}$$
$$\langle(1-\rho_x)(1-\eta_x)\rho_x^{(k)}\eta_x^{(j)}\rangle = \begin{cases} \frac{(\rho-W-I)(\eta-I-M)}{1-\rho-\eta}, & x^{(j)} \neq x^{(k)}, \\ 0, & x^{(j)} = x^{(k)}, \end{cases}$$
$$\langle(1-\rho_x)(1-\eta_x)\eta_x^{(k)}\eta_x^{(j)}\rangle = \begin{cases} \frac{(\eta-I-M)^2}{1-\rho-\eta}, & x^{(j)} \neq x^{(k)}, \\ \eta - I - M, & x^{(j)} = x^{(k)}. \end{cases}$$

The equation for W is then given by

$$\dot{W} = \frac{L_w}{2}(1-u)\left(\rho - W - I + \frac{3(\rho-W-I)^2}{1-\rho-\eta}\right) - 2D_wW. \tag{24}$$

Similarly, the other two equations can be derived:

$$\begin{aligned} \dot{I} &= \frac{3}{4}[L_w(1-u)+L_m]\frac{(\rho-W-I)(\eta-I-M)}{1-\rho-\eta} + \frac{L_w u}{4}L_m\left(\rho - W - I + \frac{3(\rho-W-I)^2}{1-\rho-\eta}\right) \\ &- (D_w+D_m)I, \end{aligned} \tag{25}$$

$$\dot{M} = \frac{3L_w u}{2}\frac{(\rho-W-I)(\eta-I-M)}{1-\rho-\eta} + \frac{L_m}{2}\left(\eta - I - M + \frac{3(\eta-I-M)^2}{1-\rho-\eta}\right) - 2D_mM. \tag{26}$$

The closed system of equations for ρ, η, W, I, and M is given by Equations (19), (20), and (24)–(26).

4.3. Selection–Mutation Balance Solution

Solving these equations in the steady-state exactly is difficult, but if the mutation rate $u \ll 1$, we can find the approximate solution. We start by setting $u = 0$ and obtaining the steady-state solution. Apart from the trivial solution and a negative solution, there are two symmetric solutions where only one type survives (competitive exclusion). We use the one where the wild-type excludes mutants:

$$\rho^{(0)} = 1 + \frac{3D_w}{D_w - 3L_w}, \quad W^{(0)} = 1 - \frac{6D_w}{D_w - 3L_w} - \frac{4D_w}{L_w}, \quad \eta^{(0)} = I^{(0)} = M^{(0)} = 0, \tag{27}$$

where the superscript corresponds to the zeroth order in the expansion in terms of small u. Note that, as expected, the expression for $\rho^{(0)}$ coincides with the one given in Table 1 for von Neumann grid ($N_b = 4$). We then look for the first correction by substituting

$$\rho = \rho^{(0)} + u\rho^{(1)}, \quad \eta = u\eta^{(1)}, \quad W = W^{(0)} + uW^{(1)}, \quad I = uI^{(1)}, \quad M = uM^{(1)},$$

inserting in the system of five equations, keeping only the first order of expansion in u, and solving for $\rho^{(1)}, \ldots, M^{(1)}$. We obtain

$$\eta = \eta^{(1)}u = \frac{D_wL_w(4D_w - 3L_w)(4D_m^2 + 3D_mL_w + D_wL_m)u}{D_m(D_w - 3L_w)(4D_m + 3L_w)(D_mL_w - D_wL_m)}. \tag{28}$$

This is the equilibrium solution corresponding to mutation–selection balance in the presence of spatial interactions. This approach is valid as long as the wild-type is advantageous (inequality (12)). In the opposite scenario, this solution is unstable, and the system converges to the mutants excluding the wild-type.

Under selection–mutation balance, of interest is the equilibrium proportion of mutants in the system given by

$$\nu_{\rm vN}^{eq} = \frac{\eta^{(1)} u}{\rho^{(0)}} = \frac{D_w L_w (4D_m^2 + 3D_m L_w + D_w L_m) u}{D_m (4D_m + 3L_w)(D_m L_w - D_w L_m)}. \quad (29)$$

It is instructive to compare this quantity with the equilibrium proportion of mutants in a mass-action system, $\nu_{\rm m-a}^{eq} = \bar{y}/\bar{x}$, Equation (14):

$$\nu_{\rm m-a}^{eq} = \frac{D_w L_w u}{D_m L_w - D_w L_m}.$$

We have

$$\frac{\nu_{\rm vN}^{eq}}{\nu_{\rm m-a}^{eq}} = 1 + \frac{D_w L_w}{D_m (4D_m + 3L_w)} > 1.$$

In other words, the relative contents of mutants (in proportion to the wild-types) are higher in spatially distributed systems compared to mass-action systems at equilibrium. This result resembles our recent analysis presented in [24], where we showed that the mean number of disadvantageous mutants is higher in fragmented populations compared to well-mixed populations. A comparison with non-equilibrium, growing, well-mixed populations is presented in [19] and not discussed here.

4.4. Applications and Comparison with Computations

The result in Equation (29) is applicable to two relevant scenarios.

- Quasi-equilibrium density in finite populations. If simulations are continued until a finite grid is filled, the population reaches a quasi-equilibrium state where wild-type and disadvantageous mutant cells coexist. For a 2D square grid under a von Neumann neighborhood, the density of the mutants is approximated by η in Equation (28), while the density of wild-types is given by $\rho^{(0)}$, Equation (27). The total numbers of mutant and wild-type cells are obtained by multiplying the quantities η and $\rho^{(0)}$ by the total number of grid points, respectively. Note that this scenario is not interesting in the case of advantageous or neutral mutants, as the entire population will eventually consist of mutant cells.
- Number of mutants in spatially expanding populations. Simulating a growing population (on a large grid where the boundaries are not reached), we can ask how the number of disadvantageous mutants scales with the total number of cells, N. Since the core of the expanding colony is in quasi-equilibrium, Equation (29) shows that the number of mutants grows as the first power of N. Results for the scaling laws for neutral and advantageous mutants are discussed in the next Section.

The expected number of mutants predicted theoretically was compared with the results of numerical simulations. This was done in the following way. At size N, the number of mutants (in the von Neumann case) is predicted to be $Nu\nu_{\rm vN}^{eq}$; see Equation (29). Solving the equation $Nu\nu_{\rm vN}^{eq}$=const, we can obtain the pairs (L_m, D_m) of mutant kinetic rates corresponding to a predicted given number of mutants in a system of size N. Figure 3a shows the predicted number of mutant as a contour plot. The closer to the "neutrality" line (see inequality (12)), the larger the predicted number of mutants. The solution of equation $Nu\nu_{\rm vN}^{eq} = const$ is shown in Figure 3b and for five points from the solution set, the predicted number of mutants (given by $Nu\nu_{\rm vN}^{\rm eq} = 10$) is compared with the numerically obtained mean (plotted together with the standard deviation, Figure 3c. We can see that for larger mutant division rates, the deviation from the theory becomes significant. Figure 3d shows a histogram of the numbers of mutants for the parameters corresponding to the fifth point. One can see that the distribution has a long tail and a very large standard deviation. This is the consequence of macroscopic structures ("slices") that cannot be handled by the present, local, method.

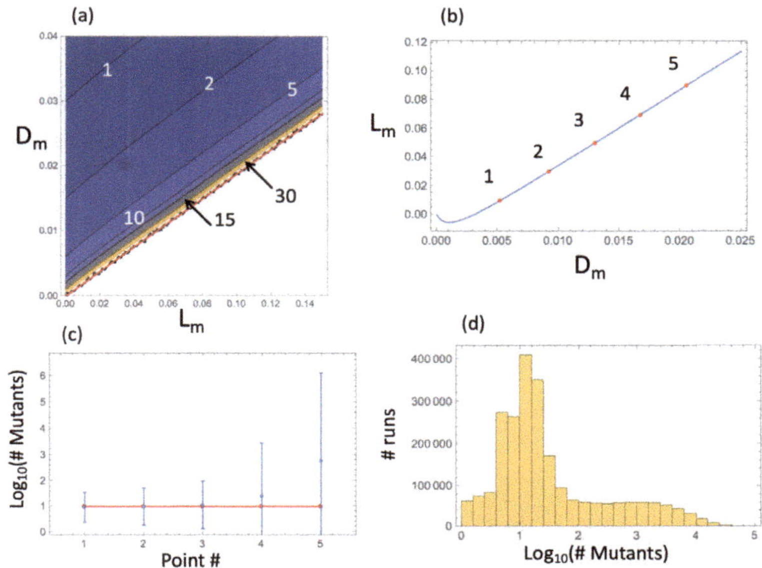

Figure 3. The level of mutants in the spatial (von Neumann) system: analytical approximation and numerical results. (**a**) The quantity $Nuv_{\mathrm{vN}}^{\mathrm{eq}}$ is presented as a contour plot as a function of L_m and D_m, for fixed values of L_w and D_w. Mutants are disadvantageous above the red line (inequality (12)). The contours' values are specified. (**b**) Solution L_m of equation $Nuv_{\mathrm{vN}}^{\mathrm{eq}} = 10$ as a function of D_m; the 5 points, used in panel (**c**), are marked in red and numbered. (**c**) The comparison of the predicted number of mutants, $Nuv_{\mathrm{vN}}^{\mathrm{eq}} = 10$, (horizontal red line) and simulated number of mutants in the 5 parameter pairs from panel (**b**). Simulated means and standard deviations are shown (out of 2.5×10^6 runs). (**d**) For the 5th parameter combination, the numerically obtained histogram of the number of mutants is shown. The rest of the parameters are: $u = 2 \times 10^{-5}$, $N = 10^5$, $L_w = 0.08$, and $D_w = 0.015$. See text for details.

5. Laws of Neutral and Advantageous Mutant Spread in Different Geometries

In the previous Section, we showed that (under the assumption of small mutation rates) the number of disadvantageous mutants in a spatially growing population scales with the total number of cells, as $\propto Nu$. If cells are advantageous or disadvantageous, they obey different scaling laws [19]. In this Section we present simple derivations for those laws in different geometries.

5.1. Derivation of the Growth Laws

Below, we consider several scenarios that differ by the dimensionality of the grid (2D or 3D) and by the direction of growth (planar growth vs. range expansion; see Table 2.

5.1.1. Two-Dimensional Flat Front

Let us first assume that the death rate of cells is equal to zero. Consider cells growing along the surface of a cylinder of width W. This represents a one-directional growth process, where during each generation, we assume that W new cells appear, and the the total population is given by $N = \mathcal{L}W$, where $\mathcal{L}$ represents the number of layers. The value of $\mathcal{L}$ is proportional to the number of generations, and thus to the physical time, t:

$$\mathcal{L} \propto t.$$

The following calculation estimates the growth law of mutants. Every time a new layer (of width W) is added, the mean number of new mutations is given by Wu. Suppose that mutants are neutral. Then, each such mutation will give rise to an array of daughter mutant cells of width 1; see Figure 4. The length of this array is given by $\mathcal{L} - i$, where i is the layer at which the mutation occurred. Therefore, the total expected number of neutral mutants is a cylinder of length $\mathcal{L}$ given by

$$M^{neut}_{\text{2D flat}} = \sum_{i=1}^{\mathcal{L}} Wu \sum_{j=i+1}^{\mathcal{L}} 1 = \frac{uW\mathcal{L}(\mathcal{L}-1)}{2} \approx \frac{uW\mathcal{L}^2}{2} = \frac{u}{2W}N^2, \tag{30}$$

where we assumed $\mathcal{L} \gg 1$. Note that in this derivation, we assumed that the number of mutants is small compared to the total population, and individual mutant clones do not interact. In a more precise calculation, the number of wild-type cells in each layer is smaller than W because of the existence of mutants, and thus, the rate of new mutant production is smaller than Wu. We, however, assume that $u\mathcal{L}W \ll 1$, such that the number of mutants is relatively small.

Note that the number of neutral mutants decreases with W; see Figure 5; the largest number of mutants is achieved in the case of $W = 1$, a one-dimensional expanding array of cells.

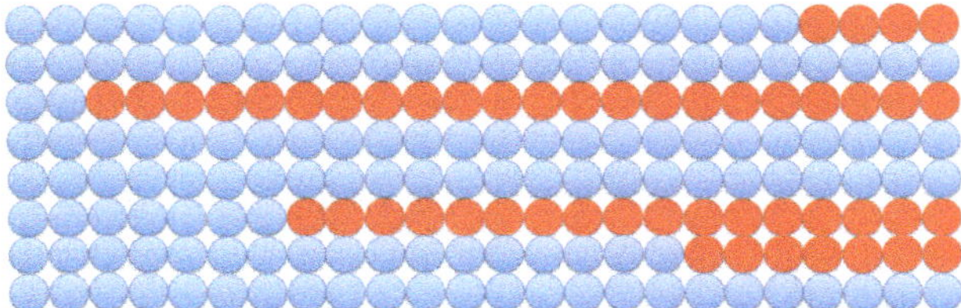

Figure 4. The conceptual model for mutant number calculations, the case of neutral mutants in a colony growing along the surface of a cylinder (2D flat front). Red circles denote mutants and blue circles wild-type cells.

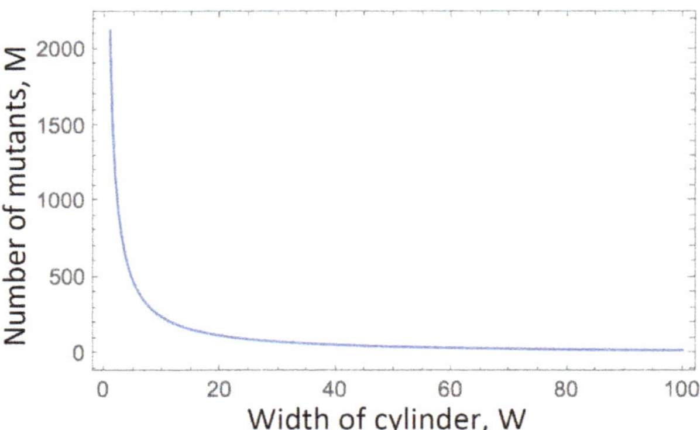

Figure 5. The number of mutants during a 2D flat front expansion decays with the front width. Equation (30) is presented with N = 10,000 and $u = 5 \times 10^{-5}$.

Next, let us consider advantageous mutants. In this case, each new mutant gives rise to a triangular clone. In the first layer, the width of the clone is 1; in the next layer it is $1 + s$; in the kth layer, it is $1 + (k-1)s$, where parameter $s \geq 0$ measures the advantage of the mutants (with $s = 0$ corresponding to neutral mutants). Therefore, we have

$$
\begin{aligned}
M^{adv}_{\text{2D flat}} &= \sum_{i=1}^{L} Wu \sum_{j=i+1}^{\mathcal{L}} (1 + (j - (i+1))s) = uW\mathcal{L}(\mathcal{L}-1)\left(\frac{1}{2} + \frac{s(\mathcal{L}-2)}{6}\right) \\
&\approx \frac{uWs\mathcal{L}^3}{6} = \frac{us}{6W^2}N^3,
\end{aligned}
\tag{31}
$$

where for the approximation, we assumed that $\mathcal{L}s \gg 1$. Furthermore, for this simple calculation to be valid, we need to assume that the wedges created by mutants do not come close to the cylinder's width, W, that is $\mathcal{L}s \ll W$. In particular, Equation (31) can be valid for small values of $W > 1$, but only for mutants that are neutral for practical purposes ($s \ll 1$).

Note that when N is fixed, the total number of cell divisions that the system has undergone is also fixed. The number of mutants however is vastly different depending on the spatial configuration. It is the highest for $W = 1$ (one row of cells) and decreases drastically with the width of the cylinder. This is consistent with the notion that spatial restrictions result in a heightened number of mutants, the 1D space ($W = 1$) being the most spatially restrictive system. The reason for this is that in 1D, a mutant, once created, blocks the whole range of expansion and prevents wild-type cells from reproducing. The wider the front, the weaker this effect. Further, we note that in the special case where $W = 1$, mutant advantage does not play a role, and the number of advantageous, neutral, and even disadvantageous mutants is given by the same formula, Equation (30).

In the derivations above, a zero death rate of cells was assumed. This means that the colony spreads as a solid mass, where all of the spots in the core are occupied. Including a nonzero death rate does not change the geometric argument presented here, because the only difference now is that the expanding population is "porous", such that the same number of cells occupies a larger number of spots on the grid. Therefore, adding a nonzero death rate does not alter the scaling laws derived here and in the other cases. For this reason, we present calculations assuming a zero death rate. Numerical calculations in Section 5.2 confirm that the scaling laws remain the same in the presence of nonzero death rates.

5.1.2. Two-Dimensional: Circular Range Expansion

Next, we turn to the dynamics of neutral mutants on a circle. Let us suppose that the radius of the circle is R and $N = \pi R^2$. The size of the colony increases via surface growth with $N \propto t^2$ and

$$R \propto t.$$

As the range expansion proceeds, the circular layer of radius r will on average give rise to $2\pi r u$ new mutations. Each mutation will result in a wedge expanding outwards. If the new mutation occurred in the layer with radius r, the number of mutating cells in layer r is 1. The number of mutants in the next layer is given by $\frac{r+1}{r}$, because under the assumption of mutant neutrality, the fraction of mutants in each new layer of radius $j > r$ (with surface $2\pi j$) should stay constant and equal to $\frac{1}{2\pi r}$. For layer j, the number of mutants is then given by j/r. This gives rise to the following calculation:

$$M^{neut}_{\text{2D range}} = \sum_{r=1}^{R} 2\pi r u \sum_{j=r+1}^{R} \frac{j}{r} = \frac{2}{3}\pi R(R^2 - 1)u \approx \frac{2\pi R^3 u}{3} = \frac{2u}{3\pi^{1/2}}N^{3/2}$$

(the approximation is valid for $R \gg 1$).

For advantageous mutants in a growing 2D circle, the fraction of mutants will grow with each layer:

$$M^{adv}_{\text{2D range}} = \sum_{r=1}^{R} 2\pi r u \sum_{j=r+1}^{R} (1+(j-(r+1))s)\frac{j}{r} = \pi R(R^2-1)u\left(\frac{2}{3}+\frac{1}{4}s(R-2)\right)$$
$$\approx \frac{\pi R^4 s u}{4} = \frac{su}{4\pi}N^2, \tag{32}$$

where we assumed $Rs \gg 1$. For this approximation to be valid, the mutant wedges should not exceed the circumference of the colony. Strictly speaking, this results in the condition $Rs << 2\pi R$, that is $s \ll 1$. For larger values of s, the events where the mutant covers the whole surface of the colony are no longer negligible.

5.1.3. Three-Dimensional Flat Front

In a 3D space, let us first consider a solid cylinder of constant radius R_0, where initially the cells are situated as a layer at the bottom of the cylinder and proceed to grow by adding layers of size πR_0^2. Each generation contributes $\pi R_0^2 u$ new mutants, and as the colony grows to length $\mathcal{L}$ (and volume $2\pi R_0^2 \mathcal{L}$), we have in the neutral case:

$$M^{neut}_{\text{3D flat}} = \sum_{i=1}^{\mathcal{L}} 2\pi R_0^2 u \sum_{j=i+1}^{\mathcal{L}} 1 = \pi R_0^2 u \mathcal{L}(\mathcal{L}-1) \approx \pi R_0^2 u \mathcal{L}^2 = \frac{u}{\pi R_0^2}N^2,$$

which is similar to the 2D flat front expansion. If the mutants are advantageous, then their number will increase from layer to layer, giving rise to conical wedges. This gives rise to the following calculation:

$$M^{adv}_{\text{3D flat}} = \sum_{i=1}^{\mathcal{L}} 2\pi R_0^2 u \sum_{j=i+1}^{\mathcal{L}} (1+(j-(i+1))s)^2$$
$$= \frac{\mathcal{L}(\mathcal{L}-1)\pi R_0^2 u}{6}\left[(\mathcal{L}^2-3\mathcal{L}+2)s^2+4(\mathcal{L}-2)s+6\right]$$
$$\approx \frac{\pi R_0^2 s^2 u \mathcal{L}^4}{6} = \frac{s^2 u}{6\pi^3 R_0^6}N^4,$$

where $\mathcal{L}s \gg 1$ for the approximation, and the approach is valid as long as the wedge radius is smaller than that of the cylinder, $s\mathcal{L} \ll R_0$.

5.1.4. Three-Dimensional Range Expansion

Next, we consider a 3D expanding sphere. For a sphere of radius R, we have $N = 4/3\pi R^3$ and the surface is given by $4\pi R^2$. The size of the colony increases via 3D surface growth with $N \propto t^3$. Each spherical layer of radius r will on average give rise to $4\pi r^2 u$ new mutations. Each mutation will result in a conical wedge expanding outwards. If the new mutation occurred in layer with radius r, the number of mutating cells in layer r is 1. The number of mutants in a layer of radius $j > r$ is given by $(j/r)^2$, because under the assumption of mutant neutrality, the fraction of mutants in each new layer should stay constant (and equal to $\frac{1}{4\pi r^2}$). Therefore, we write:

$$M^{neut}_{\text{3D range}} = \sum_{r=1}^{R} 4\pi r^2 u \sum_{j=r+1}^{R}\left(\frac{j}{r}\right)^2 = \pi R(R^2-1)(R+2/3)u \approx \pi R^4 u = \frac{3^{4/3}u}{\pi^{1/3}4^{4/3}}N^{4/3}$$

(the approximation is again valid for $R \gg 1$).

If the mutant in a growing 3D sphere is advantageous, the fraction in each layer will increase according to the fitness advantage s and stretch from layer to layer in the same way as for the neutral mutants. We therefore have

$$
\begin{aligned}
M^{adv}_{\text{3D range}} &= \sum_{r=1}^{R} 4\pi r^2 u \sum_{j=r+1}^{R} (1+(j-(r+1))s)^2 \left(\frac{j}{r}\right)^2 \\
&= \frac{\pi u R(R^2-1)}{90}\Big[(20R^3-48R^2-5R+42)s^2+(72R^2-90R-108)s+90R+60\Big] \\
&\approx \frac{2}{9}\pi s^2 u R^6 = \frac{s^2 u}{8\pi} N^2.
\end{aligned}
$$

As before, the approximation holds if $Rs \gg 1$. The method assumes that the mutant colony's size in each layer does not come close to the surface area, which amounts to the inequality $s \ll 1$.

5.1.5. Exponential (Non-Spatial, Mass-Action) Growth

Finally, for exponentially growing populations, similar formulas could be derived. In particular, for neutral mutants, we have

$$M^{neut}_{\text{exp}} = Nu \ln N,$$

and for advantageous mutants with advantage α (which is the ratio of the net growth rate of mutants and the net growth rate of wild-type cells), we have

$$M^{adv}_{\text{exp}} = \frac{\alpha}{(\alpha-1)2^{\frac{\alpha-1}{\alpha}}} N^{\frac{2\alpha-1}{\alpha}};$$

see [25], Equation (14c); see also Equation (13) there for a more general formula.

5.2. Comparison with Numerical Simulations

We ran numerical simulations to check the results derived here. Figures 6 and 7, which are described in detail below, illustrate the accuracy and applicability of the approximations derived above. They contain colored lines, which are the results of the numerical simulations, and black "guides for the eye", which represent power-law functions characterized by the powers predicted by the theory. The figures show that these scaling laws hold over large intervals of N, the population size.

Table 3. Simulation parameters for Figure 6.

Curve	Description	L_w	L_m	D_w	D_m
A	Neutral, no death	0.7	0.7	0	0
B	Neutral, with death	0.7	0.7	0.2	0.2
C	Adv, no death	0.7	0.9	0	0
D	Adv, no death, larger advantage	0.7	1.0	0	0
E	Adv by division, with death	0.7	0.8	0.2	0.2
F	Adv by death	0.7	0.7	0.2	0.1
G	Adv by death, wider front	0.7	0.7	0.2	0.1

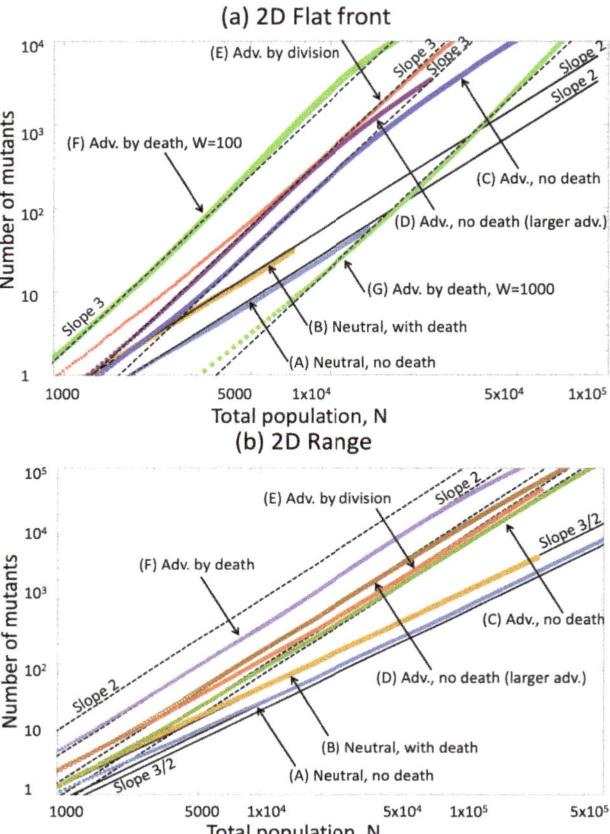

Figure 6. Neutral and advantageous mutants in a 2D colony under (**a**) a flat front expansion and (**b**) a range expansion. The number of mutants is plotted as a function of the total population, averaged over 1000 stochastic runs (standard error is too small to see). Cases (A,B) are neutral, and the corresponding solid black lines are guides to the eye, with (**a**) slope 2 and (**b**) slope 3/2 in the log–log plot. Cases (C–G) are advantageous, and the dashed lines are guides to the eye with (**a**) slope 3 and (**b**) slope 2. Curves (A–G) are described in Table 3. The rest of the parameters are $u = 5 \times 10^{-5}$, $W = 100$ (except G, where $W = 1000$).

Figure 6 shows results for the case of 2D systems, with flat front expansion presented in Figure 6a and range expansion in Figure 6b. Plotted are the average numbers of mutants (the vertical axis) measured at different population sizes, N (the horizontal axis). Different curves correspond to different parameter values, summarized in Table 3.

To simulate flat front expansion (Figure 6a), ABM simulations were performed in cylindrical geometry, with an $n \times W$ rectangular domain of width W. We started with an array of W wild-type cells at the left boundary of the domain and imposed periodic boundary conditions in the transversal direction. In each simulation, the cell population was allowed to grow to a size N, and the number of mutant cells at this size was recorded. Such simulations were performed repeatedly, and the average number of mutants was calculated. Simulation runs, in which the total cell population went extinct due to stochastic effects, were ignored.

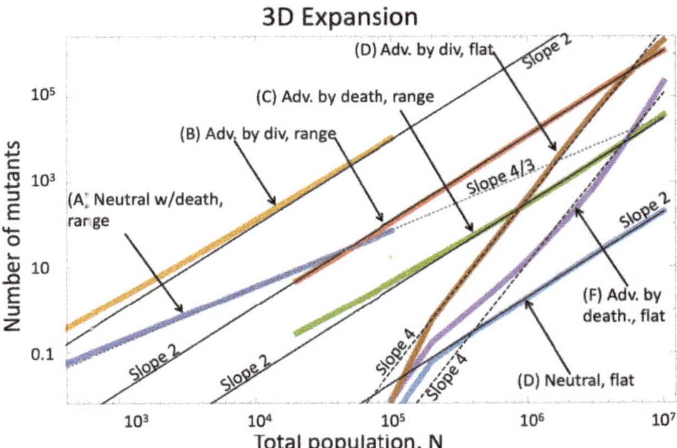

Figure 7. Mutants in the 3D expansion: the average number of mutants is plotted as a function of the total population. (A) Neutral mutants in a range expansion, with the corresponding dotted gray guide to the eye with slope 4/3 in the log–log plot. (B,C) Advantageous mutants in a range expansion, and the solid lines are guides to the eye with slope 2. (D) Neutral mutants in a 3D flat front expansion, and the solid guide to the eye has slope 2. (E,F) Advantageous mutants in a colony with a 3D flat expansion: the dashed guides to the eye have slope 4. Curve parameters are given in Table 4.

Table 4. Simulation parameters for Figure 7.

Curve	Description	L_w	L_m	D_w	D_m	u
A	Neutral, range	0.7	0.7	0.1	0.1	2×10^{-5}
B, yellow	Adv by division, range	0.4	0.8	0.1	0.1	2×10^{-5}
B, red	Adv by division, range	0.7	0.7	0.2	0.2	2×10^{-7}
C	Adv by death, range	0.7	0.7	0.2	0.1	2×10^{-7}
D	Neutral, flat	0.8	0.8	0.1	0.1	2×10^{-7}
E	Adv by division, flat	0.4	0.8	0.1	0.1	2×10^{-7}
F	Adv by death, flat	0.7	0.7	0.2	0.1	2×10^{-7}

Curves (A) and (B) in Figure 6a represent neutral mutants in the absence (A) and in the presence (B) of cell death. The black solid lines are guides for the eye with slope 2 in the log–log plot, representing the quadratic scaling law (see Table 2). Curves (C–G) represent advantageous mutants, and the dashed lines are guides for the eye with slope 3 in the log–log plot, representing the cubic scaling law (see Table 2). The different cases of advantageous mutants include systems with and without cell death, cases where mutants are advantageous by divisions (that is, have a larger division rate and the same death rate, compared with wild-type cells), and cases where mutants are advantageous by deaths (that is, have a smaller death rate and the same division rate, compared with wild-type cells). The cubic scaling law holds at least for part of the N values for all these cases; see [19] for more details.

To simulate range expansion (Figure 6b), we performed simulation on a square grid, starting with a single cell in the middle and letting the population expand outwards. Simulations resulting in population extinction were discarded, and all simulations were stopped before the boundary of the grid was reached. In Figure 6b, curves (A) and (B) again represent neutral mutants without and with cell death (with solid guides to the eye having slope 3/2). The rest of the curves again explore advantageous mutants under different assumptions, with dashed guides for the eye having slope 2. In all cases, the scaling laws in Table 2 are confirmed.

Finally, Figure 7 demonstrates numerical results for 3D expanding colonies (with parameters listed in Table 4). The different cases considered include neutral and advantageous mutants experiencing 3D flat and range expansion, with the advantage realized through differences in division and death rates.

6. Conclusions

In this paper, we reviewed some recent results on the behavior of populations in lattice models. Both homogeneous populations (that is, populations consisting of a single type of individuals) and evolving populations (wild-type and mutants) were considered, and some simple laws derived approximate population densities in different dimensionalities, geometries of growth, and lattice types. The methodologies developed here can be extended to other cases, for example Equation (1) of mutant density was derived for a 2D von Neumann grid, but the methodology can be generalized, e.g., to the Moore lattice and also to 3D systems.

Our results have further practical applications, for example for understanding the dynamics of drug-resistant mutants in solid tumors [26] or the laws of evolution in bacterial biofilms [27].

The laws described in this paper can be viewed as a way to reduce a complex (spatial, stochastic) process to a small number of important observables, such as the mean equilibrium population density or the expected number of mutants. These quantities are shown to obey some simple rules, which relate these observables to the microscopic (kinetic) parameters of the cellular turnover and to system geometry. Having these simple rules can be very useful, if one, for example, fits a stochastic agent-based modeling (ABM) to a set of data: some parameters can be extracted by applying these laws and solving for the unknown quantities.

Aspects of the theory presented here can be tested in biological experiments. Even though lattice models present a certain simplification of reality, the scaling laws of mutant growth (Table 2) are quite versatile, as they have been shown (numerically) to hold in models with different neighborhood structures and over a large variety of assumptions on cells' kinetic parameters. While some of these laws were tested previously (see, e.g., [18]), others, such as 3D flat and 3D range expansion laws, have not yet been confirmed experimentally.

Having simple laws that connect the macroscopic state with microscopic variables can be useful in interpreting experimental results. The laws in Table 1 describe the equilibrium density of populations of cells undergoing a turnover. By comparing the observed population density in experiments with the theoretical predictions, one could simply deduce the division-to-death ratio of cells growing in a given system. This can provide a useful quantitative measure of cells' state. In particular, the differences in cellular density can inform one of the changes in kinetic parameters that arise as a consequence of changes in the microenvironment. Examples of relevant experiments include investigations of the effect of various drug treatments on the cellular turnover.

Finally, our techniques can be extended to describing more complex systems. One potential application is deriving spatial laws of the microbial colony dynamics, such as bacteria growing in a microfluidic trap [28–30]. Deriving rules of global dynamics of microbial communities from local interaction rules is a task that is similar in spirit to the efforts that were described in the present paper.

Author Contributions: Conceptualization, N.L.K., I.A.R.-B., D.W.; methodology, N.L.K., I.A.R.-B., D.W.; software, N.L.K., I.A.R.-B., D.W.; formal analysis, N.L.K., I.A.R.-B., D.W.; investigation, N.L.K., I.A.R.-B., D.W.; writing—original draft preparation, N.L.K.; writing—review and editing, N.L.K., I.A.R.-B., D.W.; visualization, N.L.K., I.A.R.-B., D.W.; supervision, N.L.K., D.W.; funding acquisition, N.L.K., D.W. All authors have read and agreed to the published version of the manuscript.

Funding: This research was funded by NSF DMS grant number 1812601, by NSF DMS grant number 2152155, and by NSF-Simons Center for Multiscale Cell Fate Research.

Data Availability Statement: Not applicable.

Conflicts of Interest: The funders had no role in the design of the study; in the collection, analyses, or interpretation of data; in the writing of the manuscript, or in the decision to publish the results.

References

1. Kimura, M. On the probability of fixation of mutant genes in a population. *Genetics* **1962**, *47*, 713. [CrossRef] [PubMed]
2. Patwa, Z.; Wahl, L.M. The fixation probability of beneficial mutations. *J. R. Soc. Interface* **2008**, *5*, 1279–1289. [CrossRef] [PubMed]
3. Kepler, T.B.; Oprea, M. Improved inference of mutation rates: I. An integral representation for the Luria–Delbrück distribution. *Theor. Popul. Biol.* **2001**, *59*, 41–48. [CrossRef] [PubMed]
4. Dewanji, A.; Luebeck, E.; Moolgavkar, S.H. A generalized Luria–Delbrück model. *Math. Biosci.* **2005**, *197*, 140–152. [CrossRef] [PubMed]
5. Komarova, N.L.; Wu, L.; Baldi, P. The fixed-size Luria–Delbruck model with a nonzero death rate. *Math. Biosci.* **2007**, *210*, 253–290. [CrossRef]
6. Zheng, Q. Progress of a half century in the study of the Luria–Delbrück distribution. *Math. Biosci.* **1999**, *162*, 1–32. [CrossRef]
7. Bryder, D.; Rossi, D.J.; Weissman, I.L. Hematopoietic stem cells: The paradigmatic tissue-specific stem cell. *Am. J. Pathol.* **2006**, *169*, 338–346. [CrossRef]
8. Gestel, J.V.; Vlamakis, H.; Kolter, R. Division of labor in biofilms: The ecology of cell differentiation. In *Microbial Biofilms*; Ghannoum, M., Parsek, M., Whiteley, M., Mukherjee, P.K., Eds.; AMS Press: Washington, DC, USA, 2015; pp. 67–97. [CrossRef]
9. Webb, J.S.; Givskov, M.; Kjelleberg, S. Bacterial biofilms: Prokaryotic adventures in multicellularity. *Curr. Opin. Microbiol.* **2003**, *6*, 578–585. [CrossRef]
10. Komarova, N.L. Spatial stochastic models for cancer initiation and progression. *Bull. Math. Biol.* **2006**, *68*, 1573–1599. [CrossRef]
11. Durrett, R.; Moseley, S. Spatial Moran models I. Stochastic tunneling in the neutral case. *Ann. Appl. Probab.* **2015**, *25*, 104. [CrossRef]
12. Durrett, R.; Foo, J.; Leder, K. Spatial Moran models, II: Cancer initiation in spatially structured tissue. *J. Math. Biol.* **2016**, *72*, 1369–1400. [CrossRef] [PubMed]
13. Gralka, M.; Hallatschek, O. Environmental heterogeneity can tip the population genetics of range expansions. *Elife* **2019**, *8*, e44359. [CrossRef] [PubMed]
14. Waclaw, B.; Bozic, I.; Pittman, M.E.; Hruban, R.H.; Vogelstein, B.; Nowak, M.A. A spatial model predicts that dispersal and cell turnover limit intratumour heterogeneity. *Nature* **2015**, *525*, 261–264. [CrossRef] [PubMed]
15. Gralka, M.; Stiewe, F.; Farrell, F.; Möbius, W.; Waclaw, B.; Hallatschek, O. Allele surfing promotes microbial adaptation from standing variation. *Ecol. Lett.* **2016**, *19*, 889–898. [CrossRef] [PubMed]
16. Otwinowski, J.; Krug, J. Clonal interference and Muller's ratchet in spatial habitats. *Phys. Biol.* **2014**, *11*, 056003. [CrossRef]
17. Lavrentovich, M.O.; Wahl, M.E.; Nelson, D.R.; Murray, A.W. Spatially constrained growth enhances conversional meltdown. *Biophys. J.* **2016**, *110*, 2800–2808. [CrossRef]
18. Fusco, D.; Gralka, M.; Kayser, J.; Anderson, A.; Hallatschek, O. Excess of mutational jackpot events in expanding populations revealed by spatial Luria–Delbrück experiments. *Nat. Commun.* **2016**, *7*, 12760. [CrossRef]
19. Wodarz, D.; Komarova, N.L. Mutant Evolution in Spatially Structured and Fragmented Expanding Populations. *Genetics* **2020**, *216*, 191–203. [CrossRef]
20. Rodriguez-Brenes, I.A.; Wodarz, D.; Komarova, N.L. Beyond the pair approximation: Modeling colonization population dynamics. *Phys. Rev. E* **2020**, *101*, 032404. [CrossRef]
21. Gerlee, P.; Anderson, A.R. An evolutionary hybrid cellular automaton model of solid tumour growth. *J. Theor. Biol.* **2007**, *246*, 583–603. [CrossRef]
22. Wodarz, D.; Sun, Z.; Lau, J.W.; Komarova, N.L. Nearest-neighbor interactions, habitat fragmentation, and the persistence of host-pathogen systems. *Am. Nat.* **2013**, *182*, E94–E111. [CrossRef] [PubMed]
23. Komarova, N.L.; Shahriyari, L.; Wodarz, D. Complex role of space in the crossing of fitness valleys by asexual populations. *J. R. Soc. Interface* **2014**, *11*, 20140014. [CrossRef] [PubMed]
24. Kreger, J.; Brown, D.; Komarova, N.L.; Wodarz, D.; Pritchard, J. The role of migration in mutant evolution in fragmented populations. *bioRxiv* **2021**. [CrossRef]
25. Iwasa, Y.; Nowak, M.A.; Michor, F. Evolution of resistance during clonal expansion. *Genetics* **2006**, *172*, 2557–2566. [CrossRef]
26. Horswell, S.; Matthews, N.; Swanton, C. Cancer heterogeneity and "the struggle for existence": Diagnostic and analytical challenges. *Cancer Lett.* **2013**, *340*, 220–226. [CrossRef]
27. Banin, E.; Hughes, D.; Kuipers, O.P. Bacterial pathogens, antibiotics and antibiotic resistance. *FEMS Microbiol. Rev.* **2017**, *41*, 450–452. [CrossRef]
28. Karamched, B.; Ott, W.; Timofeyev, I.; Alnahhas, R.; Bennett, M.; Josić, K. Moran model of spatial alignment in microbial colonies. *Phys. D Nonlinear Phenom.* **2019**, *395*, 1–6. [CrossRef]
29. van Vliet, S.; Hauert, C.; Fridberg, K.; Ackermann, M.; Dal Co, A. Global dynamics of microbial communities emerge from local interaction rules. *PLoS Comput. Biol.* **2022**, *18*, e1009877. [CrossRef]
30. Koldaeva, A.; Tsai, H.F.; Shen, A.Q.; Pigolotti, S. Population genetics in microchannels. *Proc. Natl. Acad. Sci. USA* **2022**, *119*, e2120821119. [CrossRef]

MDPI

Review

Past and Present Trends in the Development of the Pattern-Formation Theory: Domain Walls and Quasicrystals

Boris A. Malomed [1,2]

[1] Department of Physical Electronics, School of Electrical Engineering, Faculty of Engineering, and Center for Light-Matter Interaction, Tel Aviv University, Ramat Aviv, Tel Aviv P.O. Box 39040, Israel; malomed@tauex.tau.ac.il

[2] Instituto de Alta Investigación, Universidad de Tarapacá, Casilla 7D, Arica, Chile

Abstract: A condensed review is presented for two basic topics in the theory of pattern formation in nonlinear dissipative media: (i) domain walls (DWs, alias grain boundaries), which appear as transient layers between different states occupying semi-infinite regions, and (ii) two- and three-dimensional (2D and 3D) quasiperiodic (QP) patterns, which are built as a superposition of plane–wave modes with incommensurate spatial periodicities. These topics are selected for the present review, dedicated to the 70th birthday of Professor Michael I. Tribelsky, due to the impact made on them by papers of Prof. Tribelsky and his coauthors. Although some findings revealed in those works may now seem "old", they keep their significance as fundamentally important results in the theory of nonlinear DW and QP patterns. Adding to the findings revealed in the original papers by M.I. Tribelsky et al., the present review also reports several new analytical results, obtained as exact solutions to systems of coupled real Ginzburg–Landau (GL) equations. These are a new solution for symmetric DWs in the bimodal system including linear mixing between its components; a solution for a strongly asymmetric DWs in the case when the diffusion (second-derivative) term is present only in one GL equation; a solution for a system of three real GL equations, for the symmetric DW with a trapped bright soliton in the third component; and an exact solution for DWs between counter-propagating waves governed by the GL equations with group-velocity terms. The significance of the "old" and new results, collected in this review, is enhanced by the fact that the systems of coupled equations for two- and multicomponent order parameters, addressed in this review, apply equally well to modeling thermal convection, multimode light propagation in nonlinear optics, and binary Bose–Einstein condensates.

Keywords: Ginzburg-Landau equations; thermal convection; quasiperiodic patterns

Citation: Malomed, B.A. Past and Present Trends in the Development of the Pattern-Formation Theory: Domain Walls and Quasicrystals. *Physics* **2021**, *3*, 1015–1045. https://doi.org/10.3390/physics3040064

Received: 16 September 2021
Accepted: 28 October 2021
Published: 10 November 2021

Publisher's Note: MDPI stays neutral with regard to jurisdictional claims in published maps and institutional affiliations.

1. Introduction

1.1. The Objective of This Paper

This text was written as a contribution for a festschrift devoted to the celebration of fifty years of the work of Prof. Mikhail Tribelsky in theoretical physics (the name known as Mikhail/Michael/Michel/Miguel/Michele/Michal/Mikael/Mikkel/Mitxel/... is considered as the oldest masculine name used in modern languages; the meaning of its original form in Hebrew is "Who (is) like El (God)?", which implies a response "no one can be likened to God"). Apart from his fundamental contributions to optics, especially to the theory of the nonresonant light–matter interaction [1,2] and light scattering by small particles [2–7], an essential topic in the works of Prof. Tribelsky is the theory of pattern formation in nonlinear dissipative media. In particular, two important subjects considered in his publications are domain walls (DWs, alias grain boundaries), i.e., stationary stripes separating two domains which are filled by different stable patterns, and quasiperiodic (QP) patterns, alias dissipative two-dimensional (2D) quasicrystals. It is relevant to mention that the fundamental papers of Prof. Tribelsky on the former and latter topics, viz., Refs. [8,9], are, respectively, his second and sixth best-cited publications, according to the data provided by Web of Science. The objective of this paper is to produce a condensed review of basic results reported in those older but still significant works, and outline directions of subsequent work initiated by the results reported in them. The review also includes some new exact analytical results for DWs, which offer a natural extension of the analysis initiated in Ref. [8] (in a detailed form, the new results will be reported elsewhere [10]). The presentation given in this review has a personal flavor, due to the fact that the present author was Mikhail's collaborator in projects which produced the above-mentioned original publications.

In addition (Ref. [8]) it is relevant to mention a still earlier paper [11], where we addressed a well-known model equation, which is usually called the real Ginzburg–Landau (GL) equation (the name originates from the phenomenological theory of superconductivity elaborated by Ginzburg and Landau 70 years ago [12]). The usual scaled form of the real GL equation is as follows:

$$\frac{\partial u}{\partial t} = u + \frac{\partial^2 u}{\partial x^2} - |u|^2 u, \tag{1}$$

where x and t are the spatial coordinate and time measured in scaled units.

In fact, the order parameter $u(x,t)$ governed by Equation (1) is a complex function; the equation is called "real" because its coefficients are real (therefore, by means of scaling, all coefficients in Equation (1) are set to be ± 1). The first, second, and third terms on the right-hand side of Equation (1) represent, respectively, the linear gain, diffusion/viscosity (dispersive linear loss), and nonlinear loss. The real GL equation is a universal model for many nonlinear dissipative media, such as the Rayleigh–Bénard convective instability in a shallow layer of a liquid heated from below [13] and instability of a plane laser evaporation front [1].

Note that Equation (1) may be represented in the gradient form as follows:

$$\frac{\partial u}{\partial t} = -\frac{\delta L}{\delta u^*}, \tag{2}$$

where $\delta/\delta u^*$ stands for the functional (Freché) derivative, and

$$L = \int_{-\infty}^{+\infty} \left(-|u|^2 + \left| \frac{\partial u}{\partial x} \right|^2 + \frac{1}{2}|u|^4 \right) dx \tag{3}$$

is a real Lyapunov functional. A consequence of the gradient representation is that L may only decrease or stay constant in the course of the evolution, $dL/dt \leq 0$. This constraint simplifies the dynamics of the real GL equation.

Equation (1) gives rise to a family of simple stationary plane–wave (PW) solutions:

$$u(x) = \sqrt{1-k^2}\exp(ikx), \tag{4}$$

where real wavenumber k takes values in the existence band,

$$-1 < k < +1. \tag{5}$$

A nontrivial issue is stability of the PW solutions against small perturbations. It can be naturally addressed by rewriting Equation (1) in the Madelung form (sometimes called the hydrodynamic representation), substituting

$$u(x,t) = A(x,t)\exp(i\phi(x,t)), \tag{6}$$

where A and ϕ are the real amplitude and phase. The substitution splits Equation (1) into a pair of real equations:

$$\frac{\partial A}{\partial t} = A + \frac{\partial^2 A}{\partial x^2} - A\left(\frac{\partial \phi}{\partial x}\right)^2 - A^3, \tag{7}$$

$$A\frac{\partial \phi}{\partial t} = A\frac{\partial^2 \phi}{\partial x} + 2\frac{\partial A}{\partial x}\frac{\partial \phi}{\partial x}. \tag{8}$$

In terms of the latter system, the PW solution (4) is written as follows:

$$A = \sqrt{1-k^2}, \phi(x) = kx. \tag{9}$$

In paper [11] (incidentally, it is the ninth best-cited publication of M.I. Tribelsky), the stability of solution (9) was explored by means of linearization of Equations (7) and (8) against small perturbations of the amplitude and phase. We thus found that the stability region in the existence band (5) is

$$-1/\sqrt{3} \le k \le +1/\sqrt{3}. \tag{10}$$

In this region, the squared amplitude of the PW solution, $A^2(k)$, exceeds $2/3$ of its maximum value, $A^2_{\mathrm{max}} \equiv 1$, which corresponds to $k = 0$:

$$A^2 \equiv 1 - k^2 \ge 2/3. \tag{11}$$

At that time, we were not aware of the fact that this result, in the form of Equation (10), was established much earlier [14] by W. Eckhaus (Wiktor Eckhaus (1930–2000) was born in Poland, where he had survived Holocaust; after WWII, he moved to the Netherlands, where he had eventually become a professor at the Utrecht University). It is now commonly known as the Eckhaus stability criterion (ESC). Later, we learned that some other people entering this research area had also independently rediscovered the ESC (this fact suggested our coauthor in Refs. [8,9], Prof. Alexander Nepomnyashchy, to formulate a *Nepomnyashchy criterion*: a necessary condition for successful work on the pattern-formation theory is the ability of the researcher to re-derive the ESC from the scratch).

1.2. Complex Ginzburg–Landau Equations: The Formulation, Plane Waves, and Dissipative Solitons

Before proceeding to the discussion of particular topics included in this review, it is relevant to briefly recapitulate the main principles concerning complex GL equations as a class of fundamental models underlying the theory of pattern formation under the combined action of linear gain and loss (including the diffusion/viscosity), linear wave dispersion, nonlinear loss, and nonlinear dispersion. In the case of cubic nonlinearity, the generic form of this equation is as [15,16]

$$\frac{\partial u}{\partial t} = gu + (a + ib)\frac{\partial^2 u}{\partial x^2} - (d + ic)|u|^2 u, \tag{12}$$

cf. its counterpart (1) with real coefficients. Here, constants $g > 0$, $a \ge 0$, and $d > 0$ represent, respectively, the linear gain, diffusion coefficient, and nonlinear loss. Further, coefficients b and c, which may have any sign, control the linear and nonlinear dispersion, respectively. Coefficient g in Equation (CCGL) may include an imaginary part too, but such a frequency term can be trivially removed by a transformation, $u(x,t) \equiv u(x,t)\exp[i\mathrm{Im}(g)t]$.

By means of obvious rescaling of t, x, and u, one can fix three coefficients in Equation (12):

$$g = a = d = 1, \tag{13}$$

unless the equation does not include the diffusion term, in which case $a = 0$ is set. Equation (12) is written in the 1D form, while its multidimensional version is obtained replacing $\partial^2 u/\partial x^2$ by the Laplacian, $\nabla^2 u$.

Unlike Equation (1), the complex GL Equation (12) does not admit a gradient representation (see Equation (2)). In the case of relatively small real parts of the coefficients, i.e., $a \ll |b|, d \ll |c|$, Equation (12) may be treated as a perturbed version of the nonlinear Schrödinger (NLS) equation. Methods of the perturbation theory for NLS equations were developed in detail long ago [17].

The ubiquity of the complex GL equations is stressed by the title of the major review by Aranson and Kramer [15], The world of the complex Ginzburg-Landau equation—indeed, the great number of particular forms of such equations, their various realizations and applications, and the great number of solutions, obtained by means of numerical and approximate analytical methods, form a "world" in itself. As concerns applications, complex GL equations emerge not only in such areas as optics of laser cavities [18–21], where they can be directly derived as basic physical models, with $u(x,t)$ being a slowly varying amplitude of the optical field, but also in many other areas of physics (hydrodynamics, electron-hole plasmas in semiconductors, gas-discharge plasmas, chemical waves, etc.). In many cases, the underlying systems of basic equations are complicated, but complex GL equations may be derived from them as asymptotic equations for long-scale small-amplitude (but, nevertheless, essentially nonlinear) excitations [22–24]. In some cases, equations of the complex-GL type may also be quite useful as phenomenological models [15,16].

While DW states are supported by a finite-amplitude PW background, it is relevant to mention that complex GL equations may give rise to localized states (dissipative solitons [25–31]). In particular, Equation (12) admits an exact solution,

$$u = A\left[\cosh(\kappa x)\right]^{-(1+i\mu)}\exp(-i\omega t), \tag{14}$$

with a single set of parameter values, A, κ, μ, and ω, given by cumbersome expressions [32,33]. If the complex GL equation reduces to a perturbed NLS equation, the dissipative soliton (14) can be obtained from the NLS soliton by means of the perturbation theory, under condition $bc < 0$ (otherwise, the underlying NLS equation does not have bright-soliton solutions). However, solution (14) is always unstable, as the linear gain in Equation (12), represented by $g > 0$, makes the zero background around the soliton unstable.

Dissipative solitons of this type may be effectively stabilized, in a nonstationary form, in a model including time-periodic alternation of linear gain and loss, which implies replacing the constant coefficient g in Equation (12) by function $g(t)$ periodically changing between positive and negative values; in particular, it may be taken as a periodic array of amplification pulses on top of a constant lossy background:

$$g(t) = G\sum_{n=-\infty}^{+\infty}\delta(t-\tau n) - g_0, \tag{15}$$

with $G > 0$ and $g_0 > 0$, τ being the amplification period [34]. Another option for the stabilization is the use of the dispersion management, i.e., replacing constant dispersion coefficient b in Equation (12) by function $b(t)$, which periodically jumps between positive and negative values, cf. Equation (15) [35].

The fact that the dissipative soliton (14) may be considered an extension of bright NLS solitons suggests that Equation (12) may also support a solution resembling the dark soliton of the NLS equation with the self-defocusing nonlinearity. Indeed, such solutions were found by Nozaki and Bekki in the form of [36]. Although the holes, as well as DWs, are supported by a stable PW background, they are completely different states, as DWs separate different PWs (see below), while the hole is built into a single PW.

A more sophisticated version of the complex GL equation admits the existence of stable stationary dissipative solitons if the zero solution is stable, i.e., the linear term must be lossy, corresponding to $g < 0$ in Equation (12). In this case, it is necessary to include the cubic gain and quintic loss (the latter term prevents the blowup). Thus, one arrives at the complex GL equation with the cubic-quintic nonlinearity, which was first introduced by Petviashvili and Sergeev [37] (actually, as a 2D equation) in the following form:

$$\frac{\partial u}{\partial t} = gu + (a + ib)\frac{\partial^2 u}{\partial x^2} - (d + ic)|u|^2 u - (f + ih)|u|^4 u, \tag{16}$$

with $g < 0$, $a \geq 0$, $d < 0$, and $f > 0$, cf. Equation (12).

It follows from Equation (16) that the interplay of the gain and loss terms in Equation (16) allows the generation of nonzero states under the condition that the cubic-gain strength exceeds a minimum value necessary to compensate the effect of the losses:

$$|d| > (|d|)_{\min} = 2\sqrt{|g|f}. \tag{17}$$

Further, using the rescaling freedom, one can normalize Equation (16) by setting

$$-g = a = -d \equiv 1, \tag{18}$$

cf. Equation (13) in the case of Equation (12). Then, condition (17) amounts to $f < 1/4$.

Stable dissipative solitons as solutions of Equation (16), in the case when they may be considered to be a perturbation of NLS solitons, were first predicted in Ref. [25] and later rediscovered in Ref. [38]. Then, stable dissipative soliton solutions of Equation (16) were found in the opposite limit, when the dispersive terms in this equation may be treated as small perturbations. In this case, the dissipative solitons are broad (nearly flat) states, bounded by sharp edges in the form of a kink and antikink [39–41].

The complex GL Equation (16), subject to normalization (18), generates a family of PW solutions, where the wavenumber takes values in the same interval (5) as above:

$$\psi = \sqrt{1 - k^2}\exp(ikx - i\omega t),\ \omega = c + (b - c)k^2, \tag{19}$$

cf. stationary solutions (4) of the real GL equation. The stability of these flat states against long-wave perturbations can be investigated analytically, leading to a generalization of the ESC (cf. Equation (10)) [42]:

$$k^2 \leq (1 + bc)/\left(3 + 2c^2 + bc\right) \tag{20}$$

The full stability of solutions (19) was investigated in a numerical form [15,16]. Note that condition (20) cannot hold unless the dispersion coefficients in Equation (16), normalized as per Equation (18), satisfy the Benjamin–Feir–Newell (BFN) condition,

$$1 + bc > 0. \tag{21}$$

If this condition does not hold, unstable PWs develop phase turbulence, with $|\psi|$ staying roughly constant, while the phase of the complex order parameter, $\phi(x,t) \equiv \arg\{u(x,t)\}$, demonstrates spatiotemporal chaos. Just below the BFN instability threshold, i.e., at $0 < -(1 + bc) \ll 1$ (see Equation (21)), the chaotic evolution of the phase gradient $p \equiv \phi_x$ obeys the Kuramoto–Sivashinsky equation [43,44], whose scaled form is

$$p_t + p_{xx} + p_{xxxx} + pp_x = 0. \tag{22}$$

Deeper into the region of $1 + bc < 0$, the instability creates defects of the wave field, at which $|u(x,t)| = 0$, and eventually leads to the onset of defect turbulence [15,16]. Further evolution may lead to emergence of regularly arranged train-shaped patterns in the turbulent states [45].

1.3. The Structure of This Paper

The rest of the paper is divided into Sections 2 and 3, which are followed by concluding Section 4. Section 2 addresses the concept of DWs and its further developments, starting from Ref. [8]. The DWs considered in that work were constructed as solutions of a system of two nonlinearly coupled real GL equations, which model the interaction of two families of simplest roll patterns (quasi-1D spatially periodic structures) in the Rayleigh–Bénard convection. This setup is controlled by the overcriticality, which is defined as follows:

$$\varepsilon = (\mathrm{Ra} - \mathrm{Ra}_{\mathrm{crit}})/\mathrm{Ra}_{\mathrm{crit}}\,, \tag{23}$$

where Ra is the Rayleigh number, and $\mathrm{Ra}_{\mathrm{crit}}$ is its critical value at the threshold of the convective instability of the fluid layer heated from below. DWs in convection patterns were predicted as linear defects (grain boundaries) [13,46,47], and were directly observed in experiments, both as DWs proper and more complex structures, formed by junctions of DWs. Typical examples of the experimentally observed patterns, borrowed from Ref. [48], are presented in Figure 1.

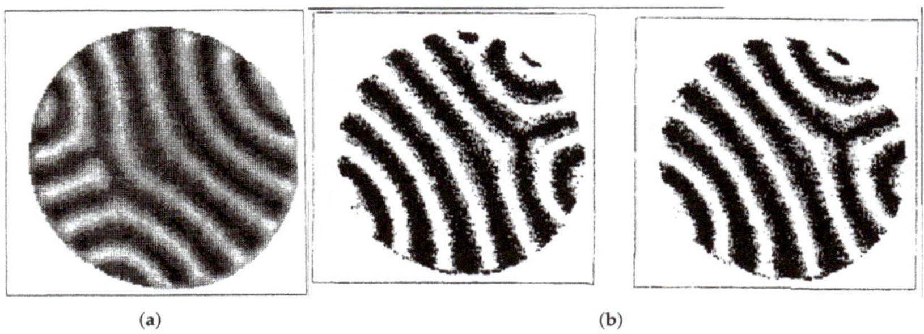

(a) (b)

Figure 1. (**a**) An experimentally observed pattern of rolls in the Rayleigh–Bénard convection, which demonstrates a junction of domain walls (grain boundaries). The pattern corresponds to overcriticality $\varepsilon = 1$; see Equation (23). (**b**) Similar patterns observed at $\varepsilon = 1.8$ (**left**) and 2 (**right**). Reprinted with permissions from Ref. [48].

It is relevant to stress that the concept of grain boundaries is known, in a great variety of different realizations, as a very general one in condensed-matter physics [49–55]. In most cases, the nature of such objects is different from that in thermal convection and other nonlinear dissipative media. Nevertheless, the phenomenology of the grain boundaries in completely different physical systems has many common features.

The DW states were constructed in Ref. [8] as solutions of two coupled real GL equations for amplitudes of PWs connected by the DW. In a particular case, such a solution for a symmetric DW is available in an exact analytical form, see Equations (50) and (51) below. It is also demonstrated that the symmetric DW may play the role of a potential well, which traps an additional small-amplitude component, in the form of a bright soliton, thus making the structure of the DW more complex, as shown below by Equations (72)–(75) and Figure 5. Further, a newly derived extension of the exact solution is included, for the case when the symmetrically coupled real GL equations include linear-mixing terms (see Equation (54) below), and a new exact solution for a strongly asymmetric DW, in the case when only one real GL equation includes the diffusion term (second derivative). This solution is given below by Equations (64)–(68) and Figure 4.

At the level of stationary solutions, the same coupled equations which model the grain boundaries in thermal convection predict DWs in optics, as boundaries between spatial or temporal domains occupied by PWs representing different polarizations or different carrier frequencies of light [56]. These equations also produce DW states in binary Bose–Einstein condensates (BECs) composed of immiscible components [57].

Still earlier, approximate solutions similar to DWs were constructed in the framework of a single complex GL equation [58]. Such solutions represent stationary sources of stable PWs with wavenumbers $\pm k$ (see Equation (19)) emitted in opposite directions (while the above-mentioned holes are sinks absorbing colliding PWs with opposite wavenumbers). A special case corresponds to complex GL Equation (12) without the diffusion term, i.e., with $a = 0$. In that case, DWs may be approximately reduced to shock waves governed by an effective Burgers equation for a local wavenumber [59]. These results are also included, in a brief form, in Section 2. As an extension of the topic, this section also addresses DWs between semi-infinite domains filled by counterpropagating traveling waves (this is possible, in particular, in thermal convection in a layer of a binary fluid heated from below [60–63]). Furthermore, Section 2 includes a newly found exact solution for the DW between traveling waves produced by a system of coupled real GL equations that include group-velocity terms (see Equations (85)–(90) below).

Section 3 summarizes some theoretical results for QP patterns in 2D and 3D nonlinear dissipative media, the study of which was initiated in Ref. [9]. In particular, included are findings for stable QP states produced by combinations of four spatial modes in a laser cavity with different 3D wave vectors [64]. Another possibility to produce a spatially confined four-mode (eight-fold) QP structure, briefly considered in Section 3, is offered by the overlap of two square-shaped (two-mode) patterns under the angle of 45° in a transient layer between the patterns [65]. This possibility is a combination of the two main topics considered in this review, viz., DWs and QP patterns.

The review is completed by Section 4, which summarizes basic results and briefly outlines new possibilities in this area.

2. DW (Domain-Wall) Patterns

2.1. The Source Pattern Generated by the Single Complex GL Equation

2.1.1. The Generic Case

To produce approximate solutions to Equation (12), it is convenient to rewrite it in the Madelung form (6), which yields the following system of equations for real amplitude A and phase ϕ:

$$\frac{\partial A}{\partial t} = A - A^3 + \frac{\partial^2 A}{\partial x^2} - A\left(\frac{\partial \phi}{\partial x}\right)^2 - 2b\frac{\partial A}{\partial x}\frac{\partial \phi}{\partial x} - bA\frac{\partial^2 \phi}{\partial x^2}, \tag{24}$$

$$A\frac{\partial \phi}{\partial t} = 2\frac{\partial A}{\partial x}\frac{\partial \phi}{\partial x} + A\frac{\partial^2 \phi}{\partial x^2} - cA^3 + b\frac{\partial^2 A}{\partial x^2} - bA\left(\frac{\partial \phi}{\partial x}\right)^2 \tag{25}$$

(recall that the coefficients of Equation (12) are subject to normalization conditions (13)). As shown in Ref. [58], a stationary solution of the DW type, which represents a source of PWs emitted in the directions of $x \to \pm\infty$, can be looked for assuming that the dispersion coefficients b and a are small, and the local amplitude, $A(x)$, and wavenumber, $p(x) \equiv \partial\phi/\partial x$, are slowly varying functions of x (the "nonlinear geometric-optics approximation", alias the "eikonal approximation"). In the lowest order, all derivatives and dispersion terms may be neglected in Equation (24), reducing it merely to $A^2 \approx 1 - p^2$, cf. Equation (4). Next, this approximation is substituted in Equation (25), with the phase taken as follows:

$$\phi(x,t) = -\left(c + (b-c)k^2\right)t + \int p(x)dx, \tag{26}$$

where it is assumed that the asymptotic values of the wavenumber are

$$p(x \to \pm\infty) = \pm k \tag{27}$$

(hence, the frequency in expression (26) is the same as in Equation (19)). Keeping the lowest-order small terms with respect to the small dispersive coefficients and small derivative dp/dx of the slowly varying local wavenumber leads to the following approximate equation:

$$\frac{1-3p^2}{1-p^2}\frac{dp}{dx} = (c-b)\left(k^2-p^2\right). \tag{28}$$

The DW solution to Equation (28) can be obtained in an implicit form, which yields x as a function of p, satisfying the boundary conditions (27):

$$2k\ln\frac{1-p}{1+p} + \left(1-3k^2\right)\ln\frac{k-p}{k+p} = 2k(b-c)\left(1-k^2\right)x. \tag{29}$$

The solution can be easily cast in an explicit form under condition $k^2 \ll 1$:

$$p(x) \approx k\tanh[(c-b)kx]. \tag{30}$$

This form clearly demonstrates that the DW may be indeed construed as an emitter of waves from the center, where $p(x=0)=0$, to $x \to \pm\infty$, in agreement with Equation (28). The explicit solutions, as well as the implicit ones (29), constitute a family parameterized by free constant k.

In the real GL equation, with $b=c=0$, as well as in the case when the linear and nonlinear dispersions exactly cancel each other, $b=c$, Equation (28) cannot produce a stationary DW solution. As shown in Ref. [58], in that case, initial configurations in the form resembling expression (30), i.e., a step-shaped profile of the local wavenumber, give rise to nonstationary solutions, which may be approximated by means of characteristics and caustics of a quasi-linear evolution equation for $p(x,t)$.

2.1.2. Domain Walls as Shock Waves in the Diffusion-Free Complex GL Equation

The consideration of DWs should be performed differently in the special case of the complex GL Equation (12) with $a=0$, which does not include the diffusion term. Taking into regard normalization (13), the respective equation takes the following form:

$$\frac{\partial u}{\partial t} = u + ib\frac{\partial^2 u}{\partial x^2} - (1+ic)|u|^2 u \tag{31}$$

(in fact, one can additionally rescale coordinate x here, to set $b=\pm 1$). This form of the equation admits free motion of various modes [26,59]. In this case, the Madelung substitution (6) leads, instead of the amplitude-phase Equations (24) and (25), to the following system:

$$\frac{\partial A}{\partial t} = A - A^3 - 2b\frac{\partial A}{\partial x}\frac{\partial \phi}{\partial x} - bA\frac{\partial^2 \phi}{\partial x^2}, \tag{32}$$

$$A\frac{\partial \phi}{\partial t} = -cA^3 + b\frac{\partial^2 A}{\partial x^2} - bA\left(\frac{\partial \phi}{\partial x}\right)^2. \tag{33}$$

Further, the lowest approximation of the nonlinear geometric optics, applied to Equation (32), yields

$$A^2 \approx 1 - b\frac{\partial p}{\partial x}. \tag{34}$$

The substitution of this in Equation (33) leads, after simple manipulations (including the division by A and differentiation with respect to x, in order to replace $\partial\phi/\partial t$ by $\partial p/\partial t$), to the Burgers equation [59]:

$$\frac{\partial p}{\partial t} = bc\frac{\partial^2 p}{\partial x^2} - 2bp\frac{\partial p}{\partial x}. \tag{35}$$

The usual shock-wave solutions of Equation (35) give rise to a family of DWs with two independent parameters, viz., wall thickness $\xi > 0$ and speed s, which may be positive or negative:

$$p(x,t) = \frac{s}{2b} - \frac{c}{\xi}\tanh\left(\frac{x-st}{\xi}\right). \tag{36}$$

The appearance of the second free parameter, s, in this solution corresponds to the above-mentioned fact that Equation (31) admits free motion of patterns produced by this variant of the complex GL equation.

2.2. DWs in Systems of Real Coupled GL Equations: Old and New Solutions

2.2.1. The Setting

The starting point of the analysis developed in Ref. [8] was a general expression for the distribution of the complex order parameter in the 2D system (e.g., the amplitude of the convective flow):

$$U(x,y;t) = \sum_{l=1}^{N} u_l(x,y;t)\exp(i\mathbf{n}_l\cdot\mathbf{R}), \tag{37}$$

where $\mathbf{R} = (x,y)$. Equation (37) implies that the order-parameter field is a superposition of N plane-wave modes (often called rolls, in the context of the convection theory) with wave vectors $\mathbf{n}_l$, and $u_l(x,y)$ are slowly varying amplitudes of these modes. Stationary states produced by the real GL equations may be looked for in the real form too as follows:

$$u_l(x,y) \equiv r_l(x,y), \arg(u_l) = 0. \tag{38}$$

as the evolution of phases $\arg(u_l)$ is trivial in this case.

It is relevant to mention that patterns similar to the rolls (known under the same name) are produced by the Lugiato–Lefever (LL) equation and its varieties. The basic LL equation may be considered the NLS equation for amplitude $u(x,t)$ of the optical field in a laser cavity, which includes the linear-loss coefficient, $\gamma > 0$, a real cavity-mismatch parameter, $\theta \gtrless 0$, and a constant pump field, u_0 [66]:

$$\frac{\partial u}{\partial t} = -\gamma u + u_0 + i\left(|u|^2 - \theta\right)u + i\frac{\partial^2 u}{\partial x^2}. \tag{39}$$

Roll patterns were studied in detail in various forms of LL models [67–69]. DWs also occur in these systems [70,71].

As is illustrated by Figure 2a, the simplest possibility of the realization of patterns represented by Equation (37) is the superposition of $N = 2$ modes, each one filling, essentially, a half-plane bounded by the DW. In this case, real amplitudes $r_{1,2}$ are slowly varying functions of only one coordinate, x, directed perpendicular to the DW. The respective boundary conditions (b.c.) are as follows:

$$\begin{aligned} r_1(x\to-\infty) &= r_2(x\to+\infty) = \text{const} \neq 0, \\ r_1(x\to+\infty) &= r_2(x\to-\infty) = 0. \end{aligned} \tag{40}$$

The scaled form of stationary (time-independent) coupled real GL equations for slowly varying amplitudes $r_1(x)$ and $r_2(x)$, corresponding to the bimodal DW configuration defined as per Figure 2a and Equation (40), is [8] (see also Refs. [13,46]):

$$D_1\frac{d^2r_1}{dx^2} + r_1\left(1 - r_1^2 - Gr_2^2\right) = 0, \tag{41}$$

$$D_2\frac{d^2r_2}{dx^2} + r_2\left(1 - r_2^2 - Gr_1^2\right) = 0. \tag{42}$$

Here, the effective diffusion coefficients are

$$D_{1.2} = \cos^2\theta_{1,2} \tag{43}$$

(see Figure 2a), and $G > 0$ is an effective coefficient of the cross-interaction between different plane waves, while the self-interaction coefficient is scaled to be 1.

The symmetric configuration corresponds to Figure 2a with the following:

$$\theta_1 = -\theta_2, \tag{44}$$

which implies $D_1 = D_2 \equiv D$, according to Equation (43). Naturally, the symmetric case plays an important role in the analysis, as shown below.

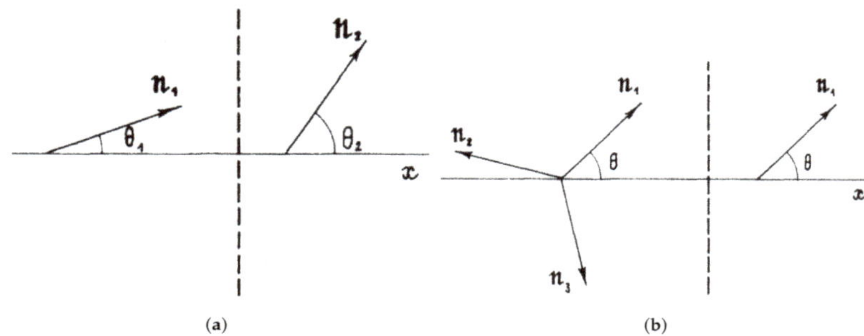

Figure 2. The scheme of the formation of the DW (domain wall) between two-dimensional patterns in the Rayleigh–Bénard convection and similar settings. (**a**) The DW between plane waves (rolls) with wave vectors oriented under angles θ_1 and θ_2 with respect to the x axis, see Equation (37). The respective amplitudes $r_{1,2}(x)$ satisfy Equations (41) and (42) and are subject to b.c. (40). The position of the DW is shown by the vertical dashed line. An example of the DW profile is displayed below in Figure 3a. (**b**) The same as in (**a**) for the DW between hexagons (the triple-mode pattern) and single-mode rolls. Figure is reprinted with permissions from Ref. [8].

It is relevant to mention that coupled Equations (41) and (42) may be considered as formal equations of motion for a mechanical system with two degrees of freedom, while x plays the role of formal time. This system keeps a constant value of its (formal) Hamiltonian,

$$h = \frac{1}{2}\sum_{j=1,2}\left[D_j\left(\frac{dr_j}{dx}\right)^2 + r_j^2 - \frac{1}{2}r_j^4\right] - \frac{G}{2}r_1^2r_2^2. \tag{45}$$

DW solutions can be readily found as numerical solutions of coupled Equations (41) and (42), subject to b.c. (40). A characteristic example of the solution is displayed in Figure 3a. In fact, the existence of the DWs in the framework of Equations (41) and (42) may be understood as the immiscibility of the modes whose amplitudes are produced by these equations. The general condition for the immiscibility, written in the present notation, is well known:

$$G > 1, \tag{46}$$

i.e., the strength of the mutual repulsion of the two components must exceed the strength of their self-repulsion [72].

Hexagonal states in the Rayleigh–Bénard convection are produced by a superposition of three plane waves, with angles 120° between their wave vectors. Such patterns are stable if, in addition to the cubic inter-mode interaction in Equations (41) and (42), the respective system of three GL equations for local amplitudes $r_{1,2,3}(x)$ includes resonant quadratic terms:

$$D_1\frac{d^2r_1}{dx^2} + r_1\left[1 - r_1^2 - G\left(r_2^2 + r_3^2\right)\right] + \nu r_2 r_3 = 0 \tag{47}$$

plus two complementing equations obtained from Equation (47) by cyclic permutations of subscripts $(1,2,3)$, where ν is a coefficient of the resonant interaction. In the theory of the thermal convection, the quadratic terms represent the effects beyond the framework of the basic Boussinesq approximation [73,74]. The numerical solution of Equation (47) produces DWs connecting single-mode rolls and the hexagonal pattern [8], see an illustration in Figure 2b and the corresponding pattern displayed in Figure 3b. It was also demonstrated that DWs are possible between two bimodal (square-shaped) patterns, each one composed of two plane waves with perpendicular orientations. In that case, the DW appears as a boundary between two half planes filled by square patterns with different orientations [8,65]; see further details below in Equations (125)–(128) and Figure 9. Further, in Ref. [75], a spatially inhomogeneous model in which the cross-interaction coefficient is a function of the coordinate $G = G(x)$ was introduced, making it possible to construct stable DWs between the single- and bimodal patterns.

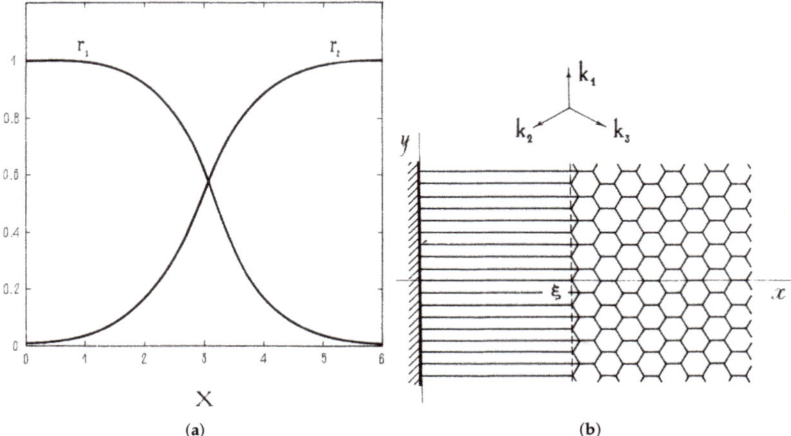

(a) (b)

Figure 3. (**a**) A typical profile of the DW between different plane-wave (roll) families. (**b**) The structure of the DW between the plane-wave and hexagonal patterns (in this panel, $\mathbf{k}_{1,2,3}$ are identical to $\mathbf{n}_l$ in Equation (37)). Figure is reprinted with permissions from Ref. [8].

2.2.2. Original Analytical Results

An analytically tractable case is the symmetric one, with $D_1 = D_2 \equiv D$, and

$$0 < G - 1 \ll 1 \tag{48}$$

(recall that $G > 1$ is a necessary condition for the existence of DWs). The analysis makes it possible to reduce the coupled GL equations to an effective sine-Gordon equation for a slowly varying inter-component phase $\chi(x)$, and thus produce an approximate analytical DW solution with a large width of the transient layer, $L \sim (G-1)^{-1/2}$ [8]:

$$\left\{ \begin{array}{c} r_1(x) \\ r_2(x) \end{array} \right\} \approx \left\{ \begin{array}{c} \cos\chi(x) \\ \sin\chi(x) \end{array} \right\}, \chi = \arctan\left(\exp\left(\sqrt{\frac{G-1}{D}}x\right)\right). \tag{49}$$

In the particular case of

$$G = 3, \tag{50}$$

the symmetric version of Equations (41) and (42) admits an exact DW solution [8]:

$$\left\{ \begin{array}{c} r_1(x) \\ r_2(x) \end{array} \right\} = \frac{1}{2} \left\{ \begin{array}{c} 1 - \tanh\left(x/\sqrt{2D}\right) \\ 1 + \tanh\left(x/\sqrt{2D}\right) \end{array} \right\}, \tag{51}$$

which is obviously compatible with b.c. (40).

2.2.3. New Analytical Results

Symmetric DWs

Precisely the same real time-independent equations as Equations (41) and (42) appear in nonlinear optics as a stationary version of coupled NLS equations in bimodal waveguides, with r_1 and r_2 being the local amplitudes of electromagnetic waves carrying different wavelengths or different polarizations of light [76]. In the latter case, typical values of G are $2/3$ or 2 for the linear or circular polarizations of the light, respectively. Other values are possible too, in photonic-crystal waveguides [77]. Similarly, the stationary real equations naturally appear as the time-independent version of coupled Gross–Pitaevskii equations for mean-field wave functions of binary BECs in ultracold atomic gases [78].

Thus, the same solutions considered here may represent optical DWs [56], as well as DWs separating two immiscible species in the BEC [57]. Further, the coupled equations in optics and BEC models may also include linear mixing between the interacting modes. In particular, this effect is produced by a twist applied to the bulk optical waveguide. A similar effect in binary BEC, viz., mutual inter-conversion of two atomic states, which form the binary BEC, may be induced by the resonant radio-frequency field [79]. The respectively modified symmetric system of Equations (41) and (42) is

$$D\frac{d^2r_1}{dx^2} + r_1\left(1 - r_1^2 - Gr_2^2\right) + \lambda r_2 = 0, \tag{52}$$

$$D\frac{d^2r_2}{dx^2} + r_2\left(1 - r_2^2 - Gr_1^2\right) + \lambda r_1 = 0, \tag{53}$$

where real λ is the linear-coupling coefficient. In fact, Equations (52) and (53) apply to the Rayleigh–Bénard convection too, in the case when periodic corrugation of the bottom of the convection cell, with amplitude $\sim \lambda$ and wave vector $\mathbf{n}_1 + \mathbf{n}_2$ (see Equation (37)), gives rise to the effect of the linear cross-gain. It is used in many laser setups that are similar to thermal convection [80,81].

The system of Equations (52) and (53) with $G = 3$ admits an exact DW solution, which is an extension of its counterpart (51):

$$\left\{ \begin{array}{c} r_1(x) \\ r_2(x) \end{array} \right\} = \frac{1}{2} \left\{ \begin{array}{c} \sqrt{1+\lambda} - \sqrt{1-\lambda}\tanh\left(\sqrt{\frac{1-\lambda}{2D}}x\right) \\ \sqrt{1+\lambda} + \sqrt{1-\lambda}\tanh\left(\sqrt{\frac{1-\lambda}{2D}}x\right) \end{array} \right\}. \tag{54}$$

Due to the action of the linear mixing, b.c. (40) is replaced by the following one:

$$\begin{aligned} r_1(x \to -\infty) = r_2(x \to +\infty) &= \frac{1}{2}\left(\sqrt{1+\lambda} + \sqrt{1-\lambda}\right), \\ r_1(x \to +\infty) = r_2(x \to -\infty) &= \frac{1}{2}\left(\sqrt{1+\lambda} - \sqrt{1-\lambda}\right). \end{aligned} \tag{55}$$

The exact solution given by Equations (52)–(55) was not reported in previous publications.

The Effect of the Confining Potential

The above-mentioned realization of the coupled real GL equations in terms of the binary BEC should include, in the general case, a trapping harmonic-oscillator (HO) potential, which is normally used in the experiment [78]. The accordingly modified system of Equations (52) and (53) is

$$D\frac{d^2r_1}{dx^2} + r_1\left(1 - r_1^2 - Gr_2^2\right) + \lambda r_2 = \frac{\aleph^2}{2}x^2 r_1, \tag{56}$$

$$D\frac{d^2r_2}{dx^2} + r_2\left(1 - r_2^2 - Gr_1^2\right) + \lambda r_1 = \frac{\aleph^2}{2}x^2 r_2, \tag{57}$$

where $\aleph^2$ is the strength of the OH potential. DW solutions of the system of Equations (56) and (57) were addressed in Ref. [82]. A rigorous mathematical framework for the analysis of such solutions in the absence of the linear coupling ($\lambda = 0$) was elaborated in Ref. [83].

If the HO trap is weak enough, viz., $\aleph^2 \ll 4/(1-\lambda)$, the DW trapped in the OH potential takes nearly constant values that are close to those in Equation (55) in the region of

$$2D/(1-\lambda) \ll x^2 \ll 8D/\aleph^2. \tag{58}$$

At $x^2 \to \infty$, solutions generated by Equations (56) and (57) decay similar to eigenfunctions of the HO potential in quantum mechanics, viz.,

$$r_{1,2} \approx \varrho_{1,2}|x|^\beta \exp\left(-\frac{\aleph}{2\sqrt{2D}}x^2\right), \tag{59}$$

$$\beta = \frac{1+\lambda}{\sqrt{2D}\aleph} - \frac{1}{2}, \tag{60}$$

where $\varrho_{1,2}$ are constants. In the case of $\lambda = 0$, the asymptotic tails (59) follow the structure of solution (51), i.e., $\varrho_1(x \to +\infty) = \varrho_2(x \to -\infty) = 0$ and $\varrho_1(x \to -\infty) = \varrho_2(x \to +\infty) \neq 0$. On the other hand, the presence of the linear mixing, $\lambda \neq 0$, makes the tail symmetric with respect to the two components, $\varrho_1(|x| \to \infty) = \varrho_2(|x| \to \infty) \neq 0$. Note that $\beta = 0$ in Equation (60) with $\lambda = 0$ is tantamount to the case when values of $\aleph$ and D in Equations (56) and (57) correspond to the ground state of the HO potential.

Exact Asymmetric DWs

Another possibility to add a new analytical solution for DWs appears in the limit case of the extreme asymmetry in the system of Equations (41) and (42), which corresponds to $D_2 = 0$ and $D_1 \equiv D > 0$, i.e., the DW between two roll families one of which has the wave vector perpendicular to the x axis (see Equation (43)):

$$D\frac{d^2r_1}{dx^2} + r_1\left(1 - r_1^2 - Gr_2^2\right) = 0, \tag{61}$$

$$r_2\left(1 - r_2^2 - Gr_1^2\right) = 0. \tag{62}$$

In fact, the form of Equation (62), in which the second derivative drops out, corresponds to the well-known Thomas–Fermi approximation (TF) in the BEC theory. In the framework of the TF approximation, the kinetic-energy term in the Gross–Pitaevskii equation is neglected, in comparison with larger ones, representing a trapping potential and the (self-repulsive) nonlinearity [78]. In the present case, corresponding to $\theta_2 = 90^\circ$, i.e., $D_2 = 0$ in Equation (43), is not an approximation but the exact special case. As concerns the application of Equations (41) and (42) to BEC, with the kinetic energy coefficients $D_{1,2} = \hbar^2/(2m_{1,2})$ in physical units, where $m_{1,2}$ are atomic masses of the two components of the heteronuclear binary condensate, Equations (61) and (62) correspond to a semi-TF approximation, representing a mixture of light (small m_1) and heavy (large m_2) atoms, e.g., a ^{7}Li–^{87}Rb diatomic gas [84].

Obviously, Equation (62) yields two solutions, viz., either

$$r_2 = 0, \tag{63}$$

or the one featuring the quasi-TF relation,

$$r_2^2(x) = 1 - Gr_1^2(x). \tag{64}$$

In the former case, Equation (61) with $r_2 = 0$ yields a solution in the form of the usual dark soliton, while in the latter case, the substitution of expression (64) in Equation (61) produces a bright soliton solution. These solutions may be "dovetailed" at a stitch point,

$$x = x_0 \equiv -\sqrt{2D}\ln\left(\frac{\sqrt{G}+1}{\sqrt{G}-1}\right), \tag{65}$$

which is defined by condition $r_1^2(x) = 1/G$ (see Equation (64). The global form of the solution, which complies with b.c. (40), is

$$r_1(x) = \begin{cases} -\tanh\left(x/\sqrt{2D}\right), \text{ at } -\infty < x < x_0, \\ \sqrt{\frac{2}{G+1}}\operatorname{sech}\left[\sqrt{\frac{G-1}{D}}(x-\xi)\right], \text{at } x_0 < x < +\infty, \end{cases} \tag{66}$$

$$r_2(x) = \begin{cases} 0, \text{ at } -\infty < x < x_0, \\ \sqrt{1 - Gr_1^2(x)}, \text{at } x_0 < x < +\infty \end{cases} \tag{67}$$

(in the context of BEC, a similar solution was reported in Ref. [85]).

Finally, the virtual center of bright-soliton segment of $r_1(x)$ is located at

$$x = \xi \equiv x_0 - \sqrt{\frac{D}{G-1}}\ln\left(\sqrt{\frac{2G}{G+1}} + \sqrt{\frac{G-1}{G+1}}\right) \tag{68}$$

(the exact solution (66) makes use of the "tail" of the bright soliton in the region of $x \geq x_0$, which does not include the central point, $x = \xi$). The distance $x_0 - \xi$, determined by Equation (68), defines the effective width of the strongly asymmetric DW. Note that, as seen in Equations (65) and (66), this exact solution exists under the condition of $G > 1$, which is the above-mentioned immiscibility condition.

It is easy to check that expression (66) satisfies the continuity conditions for $r_1(x)$ and dr_1/dx at $x = x_0$, and expression (67) provides the continuity of $r_2(x)$ at the same point. The continuity of dr_2/dt at $x = x_0$ is not required, as Equation (62) does not include derivatives. A typical example of the exact solution is displayed for $D = 1$ and $G = 2$ in Figure 4.

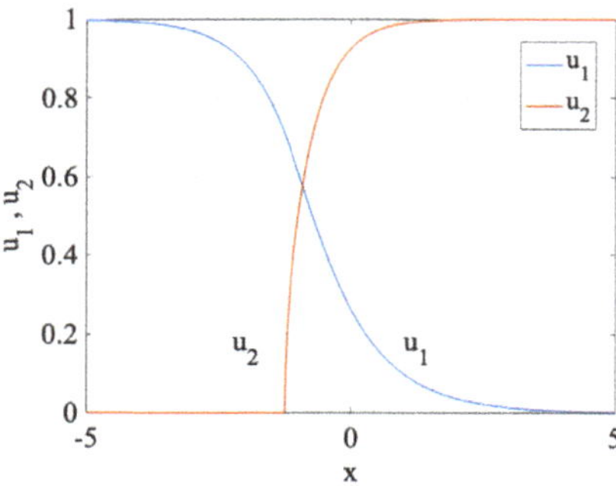

Figure 4. An example of the asymmetric DW, as given by Equations (65)–(68), for $D = 1$ and $G = 2$ (here, u_1 and u_2 stand for r_1 and r_2 in the analytical solution). Note that the coordinate of the stitch point is, in this case, $x_0 \approx -1.25$ as per Equation (65), and the "virtual center" of the bright-soliton segment of $u_1(x)$ is located at $\xi \approx -1.80$ as per Equation (68).

A remarkable fact is that, unlike the above-mentioned exact solutions (51) and (54), which exist solely at $G = 3$ (see Equation (50)), the one given by Equations (65)–(68) exists as a generic one for all values of $G > 1$.

2.2.4. DW–Bright-Soliton Complexes

An Exact Solution for the Composite State

The DW formed by two immiscible PWs may serve as an effective potential for trapping an additional PW mode. To address this possibility, it is relevant to consider the symmetric configuration, with $D_1 = D_2 \equiv D$ (see Equation (43)), and wave vector $\mathbf{k}_v$ of the additional PW mode, $v(x)$, directed along the bisectrix of the angle between the DW-forming wave vectors $\mathbf{k}_1$ and $\mathbf{k}_2$, i.e., along axis x (hence, Equation (43) yields $D_v = 1$). The corresponding system of three coupled stationary real GL equations is

$$D\frac{d^2u_1}{dx^2} + u_1\left(1 - u_1^2 - Gu_2^2 - gv^2\right) = 0, \tag{69}$$

$$D\frac{d^2u_2}{dx^2} + u_2\left(1 - u_2^2 - Gu_1^2 - gv^2\right) = 0, \tag{70}$$

$$\frac{d^2v}{dx^2} + \left(1 - v^3 - g\left(u_1^2 + u_2^2\right)\right)v = 0, \tag{71}$$

where $g > 0$ is the constant of the nonlinear interaction between components $u_{1,2}$ and v.

The system of Equations (69)–(71) admits the following exact solution in the form of the DW of components $u_{1,2}(x)$ coupled to a bright-soliton profile of $v(x)$:

$$\left\{ \begin{array}{c} u_1(x) \\ u_2(x) \end{array} \right\} = \frac{1}{2}\left\{ \begin{array}{c} 1 - \tanh\left(\sqrt{g-1}x\right) \\ 1 + \tanh\left(\sqrt{g-1}x\right) \end{array} \right\}, \tag{72}$$

$$v(x) = \sqrt{2 - \frac{3}{2}g}\,\mathrm{sech}\left(\sqrt{g-1}x\right). \tag{73}$$

This solution is valid under the condition that coefficients G and D in Equations (69) and (70) take the following particular values:

$$G = 3 - 8g + 6g^2, \tag{74}$$

$$D = \frac{1}{2}(3g - 1). \tag{75}$$

As it follows from Equation (73), this solution contains free parameter g, which may vary in a narrow interval as follows:

$$1 < g < 4/3 \tag{76}$$

(see also Equation (82) below). According to Equations (74) and (75), the interval (76) corresponds to coefficients G and D varying in intervals

$$1 < G < 3;\ 1 < D < 3/2. \tag{77}$$

Thus, adding the v component lifts the degeneracy of the exact DW solution (51), which exists solely at $G = 3$.

Recall that in the model of convection patterns, D cannot take values $D > 1$, which disagrees with Equation (77). However, values $D > 1$ are relevant for systems of Gross–Pitaevskii equations for the heteronuclear three-component BEC. In the latter case, D is the ratio of atomic masses of the different species which form the triple immiscible BEC. Similarly, D is the ratio of values of the normal group-velocity dispersion of copropagating waves in the temporal-domain realization of the real GL equations in nonlinear fiber optics [56]. In the latter case, values $D > 1$ are relevant too.

An example of the DW–bright-soliton complex is displayed in Figure 5 for $g = 7/6$. In this case Equations (74) and (75) yield $G = 11/6$ and $D = 5/4$ (according to Equations (74) and (43)).

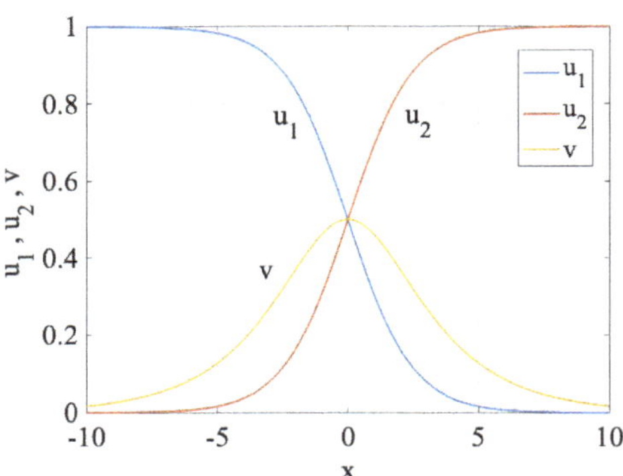

Figure 5. An example of the exact solution for the DW–bright-soliton complex, given by Equations (72) and (73), for $g = 7/6, G = 11/6$, and $D = 5/4$.

The Bifurcation of the Creation of the Composite State in the General Case

If relation (74) is not imposed on the interaction coefficients g and G, the solution for the composite state cannot be found in an exact form. Nevertheless, it is possible to identify bifurcation points at which component v with an infinitesimal amplitude appears. To this end, Equation (71) should be used in the form linearized with respect to v as follows:

$$\frac{d^2v}{dx^2} + \left\{1 - g\left[u_1^2(x) + u_2^2(x)\right]\right\}v = 0. \tag{78}$$

This linear equation can be exactly solved for $u_1(x) = u_2(x)$ given by expression (51) in the case of $G = 3$, while parameters D and g may take arbitrary values. Indeed, using the commonly known solution for the Pöschl–Teller potential in quantum mechanics, it is easy to find that Equation (78), with the effective potential corresponding to expression (51), gives rise to eigenmodes in the following form:

$$v(x) = \text{const} \cdot \left[\text{sech}\left(x/\sqrt{2D}\right)\right]^{\alpha}, \tag{79}$$

at a special value of the interaction coefficient, which identifies the bifurcation producing the composite state:

$$g_{\text{bif}} = D^{-1}\left(1 + 2D \mp \sqrt{1 + 2D}\right), \tag{80}$$

the respective value of power α in expression (79) being the following:

$$\alpha = \sqrt{2\left(1 + D \mp \sqrt{1 + 2D}\right)}. \tag{81}$$

The values given by Equations (80) and (81) with the top sign from $\mp$ correspond to the bifurcation, creating a fundamental composite state (the ground state, in terms of the quantum mechanical analog) at $g > g_{\text{bif}}$, while the bottom sign represents a higher-order bifurcation (alias, the second excited state, in the language of quantum mechanics; the first excited state, which is not considered here, is a spatially odd mode). While it is obvious that the fundamental bifurcation provides a stable composite state, it is plausible that the ones produced by higher-order bifurcations are unstable.

Lastly, varying coefficient D of the modes forming the underlying DW between $D = 0$ and $D = \infty$ (recall that the convection model corresponds to $D < 1$, while the realizations in BEC and optics admit $D > 1$). Equation (80) demonstrates monotonous variation of the bifurcation point in interval

$$g_{\text{bif}}(D = 0) \equiv 1 < g_{\text{bif}} < 2 \equiv g_{\text{bif}}(D \to \infty). \tag{82}$$

It extends interval (76) in which exact composite states with a finite amplitude were found above, see Equations (72)–(75).

2.3. Domain Walls between Traveling Waves

2.3.1. The Setting

An essential extension of the above results for DWs, produced by coupled Equations (41) and (42), was reported in Ref. [86], which addressed a system of coupled GL equations for counter-propagating traveling waves, such as those occurring in binary-fluid convection [60–63,87]. The system is composed of two equations of type (12) (subject to normalization (13)), coupled by complex cubic terms with coefficients G and H:

$$\frac{\partial u_1}{\partial t} + s\frac{\partial u_1}{\partial x} = u_1 + (1+ib)\frac{\partial^2 u_1}{\partial x^2} - (1+ic)|u_1|^2 u_1 - (G+iH)|u_2|^2 u_1, \tag{83}$$

$$\frac{\partial u_2}{\partial t} - s\frac{\partial u_1}{\partial x} = u_2 + (1+ib)\frac{\partial^2 u_2}{\partial x^2} - (1+ic)|u_2|^2 u_2 - (G+iH)|u_1|^2 u_2, \tag{84}$$

where $-s$ and $+s$ are group velocities of the counter-propagating waves, u_1 and u_2 (real coefficient H represents the cross-phase modulation (XPM), in terms of optics [76]). A natural approach to constructing DW solutions of the system of Equations (83) and (84) is to use the lowest approximation, which neglects imaginary parts of coefficients in the equations, but keeps the group-velocity terms. In this approximation, the order parameters are real, $u_{1,2}(x) \equiv r_{1,2}(x)$ obeying the time-independent version of Equations (83) and (84):

$$+s\frac{dr_1}{dx} = \frac{d^2 r_1}{dx^2} + r_1\left(1 - r_1^2 - Gr_2^2\right), \tag{85}$$

$$-s\frac{dr_2}{dx} = \frac{d^2 r_2}{dx^2} + r_2\left(1 - r_2^2 - Gr_1^2\right). \tag{86}$$

Note that, unlike similar Equations (41) and (42), Equations (85) and (86) cannot be derived from a formal Hamiltonian, cf. Equation (45).

2.3.2. A (New) Exact Analytical Solution

To illustrate the structure of the DW state in this approximation, it is relevant to produce a particular exact solution of the system of Equations (85) and (86), cf. the above-mentioned solution (51):

$$\left\{ \begin{array}{c} r_1(x) \\ r_2(x) \end{array} \right\} = \frac{1}{2}\left\{ \begin{array}{c} 1 - \tanh\left(\left(\sqrt{8+s^2}+s\right)(x/4)\right. \\ 1 + \tanh\left(\left(\sqrt{8+s^2}+s\right)(x/4)\right. \end{array} \right\}. \tag{87}$$

This solution (which was not reported in earlier studies) exists if the following relation holds between the cross-interaction coefficient G and group velocity v:

$$G - 3 = s\left(\sqrt{8+s^2}+s\right), \tag{88}$$

or, inversely,

$$s = \frac{G-3}{\sqrt{2(G+1)}}. \tag{89}$$

cf. Equation (50). It follows from the form of solution (87) and Equations (88) and (89) that

$$\mathrm{sgn}(s) = \mathrm{sgn}(G - 3), \tag{90}$$

i.e., the exact solution represents a sink of traveling waves ($s > 0$) for $G > 3$, and a source ($s < 0$) for $G < 3$. Note that the solution of the latter type exists even in the case of $G < 1$, when the two components are miscible, cf. Equation (46). In this case, the mixing is prevented by the opposite group velocities, which pull the components apart.

2.3.3. The Sink or Source Coupled to a Bright Soliton in an Additional Component

It is possible to consider a system including the counterpropagating traveling waves coupled to an additional standing one. This is a natural counterpart of the three-component system based on Equations (69)–(71). The traveling waves, which can trap the additional standing one, $v(x)$, are described by the following generalization of Equations (85) and (86):

$$+s\frac{du_1}{dx} = \frac{d^2u_1}{dx^2} + u_1\left(1 - u_1^2 - Gu_2^2 - gv^2\right), \tag{91}$$

$$-s\frac{du_2}{dx} = \frac{d^2u_2}{dx^2} + u_2\left(1 - u_2^2 - Gu_1^2 - gv^2\right), \tag{92}$$

while the equation for the standing component is

$$\frac{d^2v}{dx^2} + \left(1 - v^2 - g\left(u_1^2 + u_2^2\right)\right)v = 0, \tag{93}$$

cf. Equation (71). An exact solution of Equations (91)–(93) can be found for free parameters g and s:

$$u_{1,2}(x) = \frac{1}{2}\left(1 \mp \tanh\left(\sqrt{g-1}x\right)\right), \tag{94}$$

$$v(x) = \sqrt{2 - \frac{3}{2}g\mathrm{sech}\left(\sqrt{g-1}x\right)}, \tag{95}$$

$$G - 3 = 2g(3g - 4) + 4s\sqrt{g-1}, \tag{96}$$

$$D = \frac{s}{2\sqrt{g-1}} + \frac{1}{2}(3g - 1), \tag{97}$$

cf. Equations (72)–(75). As it is seen from Equation (96), the interaction with the soliton-shaped standing wave shifts the boundary between the sink and source of the traveling waves off the above-mentioned point, $G = 3$.

3. Two- and Three-Dimensional Quasiperiodic Patterns

Quasicrystals, as stable 3D ordered states of metallic alloys, whose atomic lattice is spatially quasiperiodic (QP), were discovered by D. Shechtman et al. [88]. For this discovery, Shechtman was awarded with the Nobel Prize in chemistry (2011). Then, a 2D quasicrystalline structure was also experimentally demonstrated in alloys [89]. The work on this topic remains very active in diverse branches of condensed-matter physics [90–94], as well as in other physical systems, which offer a natural realization of QP patterns, such as dissipative structures [95], photonics [96–99] and phononics [100].

The objective of this section is to summarize the results for stable 2D and 3D patterns with the quasicrystalline structure that were predicted as stable non-equilibrium dynamical structures (rather than equilibrium states of matter) in nonlinear dissipative media.

3.1. 2D Octagonal (Eight-Mode) and Decagonal (Ten-Mode) Quasicrystals

Following Ref. [9], generic 2D patterns of a real order parameter, such as one representing the convection flow, are defined by means of complex amplitudes r_l of PWs which build them, cf. Equation (37):

$$U(x,y,t) = \sum_{l=1}^{2N} u_l(t) \exp(i\mathbf{n}_l \cdot \mathbf{R}), \tag{98}$$

where the set of $2N$ vectors $\mathbf{n}_l$ is a star with angles π/N between adjacent ones. Note that the vectors satisfy the following relation:

$$\mathbf{n}_{l+N} = -\mathbf{n}_l, \text{ for } l = 1,2,\ldots,N. \tag{99}$$

Amplitudes $u_l(t)$ are, generally speaking, complex variables,

$$u_l(t) = A_l(t) \exp(i\varphi_l(t)) \tag{100}$$

(cf. Equation (6)), subject to constraint $u_{l+N} = u_l^*$, which, along with Equation (99), proves that the order-parameter distribution (98) is real.

For the lowest-order quasi-crystalline patterns, such as those corresponding to $N = 4$ (octagonal) and $N = 5$ (decagonal) ones, the phase evolution is trivial, making it possible to disregard φ_l in Equation (100). The resulting system of evolution equations for the real amplitudes, including the usual linear gain, $\gamma_0 > 0$, and cubic loss (cf. Equation (1)), is

$$\frac{dA_l}{dt} = \left(\gamma_0 - \sum_{m=1}^{N} T_{l-m} A_m^2\right) A_l \equiv -\frac{\partial L}{\partial A_l}, \tag{101}$$

where $T_{l-m} > 0$ are coefficients of the cubic lossy nonlinearity, subject to normalization $T_0 = 1$, and the Lyapunov function is

$$L = -\frac{\gamma_0}{2} \sum_{l=1}^{N} A_l^2 + \frac{1}{4} \sum_{l,m=1}^{N} T_{l-m} A_l^2 A_m^2, \tag{102}$$

cf. Equations (2) and (3). Detailed analysis of the results, produced with the help of Equation (101), is presented in Ref. [9]); see also some preliminary results in Refs. [101,102].

Spatially quasiperiodic patterns of the octagonal ($N = 4$) and decagonal ($N = 5$) types are displayed, respectively, in Figures 6a and 7a. It is seen that they are built as compositions of rhombuses of different shapes, and the presence of the overall octagonal or pentagonal structure is evident.

Solutions of Equation (101) for $N = 4$ depend on two independent nonlinearity coefficients, $T_1 = T_3$ and T_2. In this case, there are four distinct stationary solutions which have their stability areas: rolls, with

$$A_1 = \sqrt{\gamma_0}, A_{2,3,4} = 0; \tag{103}$$

a square lattice, with

$$A_{1,3} = \sqrt{\gamma_0/(1+T_2)}, A_{2,4} = 0; \tag{104}$$

an anisotropic rectangular lattice, with the aspect ratio $\tan(\pi/8) = \sqrt{2} - 1 \approx 0.414$, amplitudes

$$A_{1,2} = \sqrt{\gamma_0/(1+T_1)}, A_{3,4} = 0; \tag{105}$$

and the octagonal quasicrystal, with

$$A_{1,2,3,4} = \sqrt{\gamma_0/(1+2T_1+T_2)}. \tag{106}$$

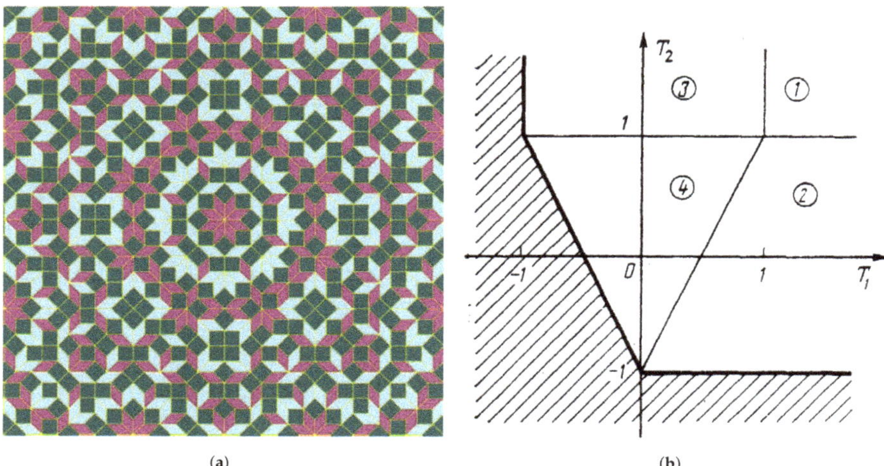

Figure 6. (**a**) The shape of the octagonal quasiperiodic pattern, reprinted with permissions from Ref. [103]. (**b**) The stability chart for patterns composed of four amplitudes, $A_{1,2,3,4}$, in the plane of nonlinearity coefficients, T_1 and T_2, of Equation (101), reprinted with permissions from Ref. [9]. Stability areas of the rolls (103), squares (104), rectangles (105), and octagonal quasicrystal (106) are denoted by encircled numbers 1, 2, 3, and 4, respectively.

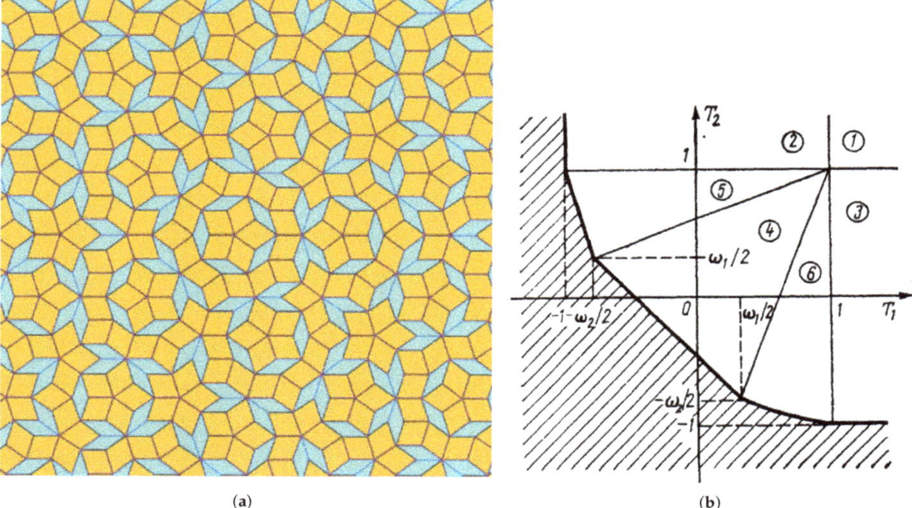

Figure 7. (**a**) The shape of the decagonal quasiperiodic pattern, reprinted with permissions from Ref. [103]. (**b**) The stability chart for patterns composed of five amplitudes, $A_{1,2,3,4,5}$, in the plane of nonlinearity coefficients, T_1 and T_2, of the respective system of equations (101), reprinted with permissions from Ref. [9]. Stability areas of the rolls (107), rectangles (108) and (109), decagonal quasicrystal (110), and semi-periodic states (111) and (112) are denoted by encircled numbers 1, 2, 3, 4, 5, and 6, respectively. Constants ω_1 and ω_2, marked in panel (b), are defined by Equation (113).

In addition to that, Equation (101) also has a stationary semi-periodic solution, which is quasiperiodic in direction $\mathbf{n}_2$ and periodic along $\mathbf{n}_4$, with $A_1 = A_2 \neq 0, A_3 \neq 0$, and $A_4 = 0$. However, the latter solution is completely unstable.

The full stability chart for stationary solutions (103)–(106) can be readily found in an analytical form. It is displayed in Figure 6b.

For $N = 5$, the solutions of Equation (101) also depend on two independent nonlinearity coefficients, $T_1 = T_4$ and $T_2 = T_3$. These equations produce six different species of stable stationary patterns. These are rolls, with

$$A_1 = \sqrt{\gamma_0}, A_{2,3,4,5} = 0; \tag{107}$$

two different species of rectangular lattices,

$$A_{1,2} = \sqrt{\gamma_0/(1+T_1)}, A_{3,4,5} = 0, \tag{108}$$

$$A_{1,3} = \sqrt{\gamma_0/(1+T_2)}, A_{2,4,5} = 0; \tag{109}$$

the decagonal quasicrystal,

$$A_{1,2,3,4,5} = \sqrt{\gamma_0/(1+2T_1+2T_2)}; \tag{110}$$

and two species of semi-periodic patterns, that are quasiperiodic in one direction and periodic in the other:

$$A_{1,3} = \sqrt{\frac{\gamma_0(1-T_1)}{1+T_2-2T_1^2}}, A_2 = \sqrt{\frac{\gamma_0(1+T_2-2T_1)}{1+T_2-2T_1^2}}, A_{4,5} = 0, \tag{111}$$

$$A_{1,5} = \sqrt{\frac{\gamma_0(1-T_2)}{1+T_1-2T_2^2}}, A_3 = \sqrt{\frac{\gamma_0(1+T_1-2T_2)}{1+T_1-2T_2^2}}, A_{2,4} = 0. \tag{112}$$

In addition to that, there is another semi-periodic solution, with $A_1 = A_2 \neq 0$, $A_3 = A_4 \neq 0$, and $A_5 = 0$, but it is completely unstable.

The full stability chart for this set of solutions was also found in an analytical form, as shown in Figure 7b. In this figure, the constants are

$$\omega_1 = \frac{\sqrt{5}-1}{2} \approx 0.618, \omega_2 = \omega_1 + 1. \tag{113}$$

Note that, unlike the situation for the octagonal setting ($N = 4$) displayed in Figure 6b, the stability areas for the decagonal ($N = 5$) quasicrystal (110) and periodic patterns (108), (109) are not adjacent to each other in Figure 7b, being separated by regions of stable semi-periodic states (111) and (112) (recall that all semi-periodic states are unstable in the case of $N = 4$).

It is relevant to mention that if higher-order nonlinear terms are added to the system of Equations (101), the sharp boundaries between stability areas of different patterns in Figures 6b and 7b may be modified. In particular, there may appear narrow strips of bistability (which is impossible in the framework of Equation (101)), as well as strips populated by more complex patterns, instead of the sharp lines [9].

3.2. Dodecagonal Quasicrystals ($N = 6$)

In the case of the twelve-mode patterns, corresponding to $N = 6$ in Equation (101), quadratic nonlinearity, with coefficient $\nu > 0$, should be included too, as the corresponding set of six wave vectors contains two resonant triads that may be naturally coupled by the quadratic terms (cf. Equation (47)):

$$\mathbf{n}_1 + \mathbf{n}_5 + \mathbf{n}_9 = \mathbf{n}_2 + \mathbf{n}_6 + \mathbf{n}_{10} = 0 \tag{114}$$

(recall that only six wave vectors are actually different, according to Equation (99)). In this case, the dynamics of phases of the complex amplitudes (6) cannot be disregarded, giving rise to phason modes [104–106]. Accordingly, Equation (101) is replaced by a coupled system of evolution equations for the real amplitudes and phases [9]:

$$\frac{dA_l}{dt} = \left(\gamma_0 - \sum_{m=1}^{N} T_{l-m} A_m^2\right) A_l + \nu A_{n+4} A_{n+8} \cos \Phi_n, \tag{115}$$

$$A_l \frac{d\varphi_l}{dt} = -\nu A_{l+4} A_{l+8} \sin \Phi_l, \tag{116}$$

$$\Phi_l \equiv \varphi_l + \varphi_{l+4} + \varphi_{l+8}, \tag{117}$$

with $\varphi_l \equiv \varphi_{l-12}$ for $l > 12$.

These equations give rise to the following stationary solutions for the dodecagonal quasicrystals, with equal values of real amplitudes A_l, and coinciding values of Φ_l for both resonantly coupled triads (114):

$$\cos \Phi_l = \pm 1, \tag{118}$$

$$A_l = \pm (2Q_0)^{-1} \left(\nu \pm \sqrt{\nu^2 + 4\gamma_0 Q_0}\right), \tag{119}$$

$$Q_0 \equiv 1 + 2T_1 + 2T_2 + T_3. \tag{120}$$

The analysis of the stability of these solutions in the framework of Equations (115)–(117) demonstrates that they may be stable only under conditions $Q_0 > 0$ (see Equation (120)) and the following one:

$$Q_3 \equiv 1 + T_3 - T_1 - T_2 > 0. \tag{121}$$

If these conditions hold, the amplitude of stable quasicrystals exceeds a minimum value,

$$A_l \geq A_{\min} \equiv (1/2) \max\{\nu/Q_0, \nu/Q_3\}. \tag{122}$$

Further, there is no stability constraint for the largest value of the amplitude, provided that the following combinations of the nonlinearity coefficients are positive:

$$Q_{1,2} \equiv 1 - T_2 \pm (T_1 - T_3) \geq 0. \tag{123}$$

Otherwise, the stability imposes the following limit on the amplitude:

$$A \leq A_{\max} \equiv \min\{-\nu/Q_1, -\nu/Q_2\} \tag{124}$$

(if only one combination Q_1 or Q_2 is negative, then only this one determines the upper limit for the stability, as per Equation (124)).

The existence and stability results for the amplitude of the dodecagonal quasicrystals is summarized in Figure 8b. Note that in the presence of the resonant interaction mediated by the quadratic term in Equation (115), the solution appears as a subcritical [107] one, with a finite value of the amplitude at $\gamma_0 < 0$, i.e., when this coefficient represents linear loss rather than gain.

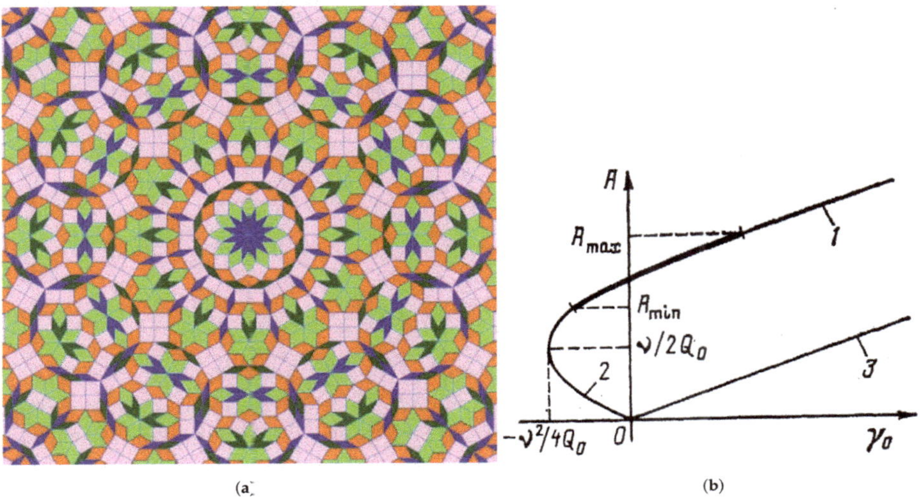

(a) (b)

Figure 8. (**a**) The shape of the dodecagonal quasiperiodic pattern, reprinted with permissions from Ref. [103]. (**b**) The pattern's amplitude, given by Equation (119) vs. the strength of the linear gain ($\gamma_0 > 0$) or loss ($\gamma_0 < 0$), reprinted with permissions from Ref. [9]. Stable and unstable solutions are represented, respectively, by bold and thin lines. Branches 1, 2 and 3 pertain, respectively, to the solutions with $\cos\Phi_l = +1$ and -1 in Equation (118). Parameters Q_0, $A_{\min}$, and $A_{\max}$ are defined as per Equations (120), (122) and (124), respectively.

3.3. A Quasicrystalline Layer between Orthogonally Oriented Square-Lattice Patterns

While, as shown in Figure 6b, square-lattice and octagonal quasiperiodic patterns cannot coexist as stable ones in the system with $N = 4$, it was demonstrated in Ref. [65] that a sufficiently broad stripe filled by an effectively stable nearly-octagonal quasiperiodic pattern may be realized as a transient layer between stable square-lattice patterns mutually oriented under the angle of 45°, as schematically shown in Figure 9a. For this configuration (which naturally combines the two main topics of the present review, viz., the DWs and QP patterns), one may naturally adopt equal amplitudes corresponding to wave vectors $\mathbf{k}_3$ and $\mathbf{k}_4$:

$$A_3(x) = A_4(x) \equiv A(x), \tag{125}$$

while amplitudes B_1 and B_2 related to $\mathbf{k}_1$ and $\mathbf{k}_2$ are different, the effective diffusion coefficient for the latter one being zero, as per Equation (43). The corresponding system of stationary real GL equations, naturally extending Equations (52), (53), (61), (62) and (101), takes the following form:

$$\frac{1}{2}\frac{d^2A}{dx^2} + A - A^3 - \left(T_1B_1^2 + T_1B_2^2 + T_2A^2\right)A = 0, \tag{126}$$

$$\frac{d^2B_1}{dx^2} + B_1 - B_1^3 - \left(2T_1A^2 + T_2B_2^2\right)B_1 = 0, \tag{127}$$

$$B_2 - B_2^3 - \left(2T_1A^2 + T_2B_1^2\right)B_2 = 0, \tag{128}$$

where, like in Equation (101), T_1 and T_2 are coefficients of the cross-interaction between the PW modes with angles, respectively, 45° and 90° between their wave vectors. According to Figure 6b, the stability conditions for the spatially uniform square-lattice and octagonal quasicrystalline patterns are, respectively, the following:

$$T_2 \leq 1, T_1 \geq T_2 + 1/2, \tag{129}$$

$$T_2 \leq 1, T_1 \leq T_2 + 1/2. \tag{130}$$

Accordingly, to secure the stability of the background square lattices and a possibility to have a broad transient layer between mutually rotated ones, which is filled by the effectively stable nearly-octagonal pattern, it is relevant to choose parameters belonging to the stability area (129), with values close to the stability boundary, $T_1 = T_2 + 1/2$. An appropriate choice is the following:

$$T_{1,2} \equiv 1 - \mu_{1,2},\ 0 < \mu_{1,2} \ll 1, \tag{131}$$

$$\text{with } m \equiv 2\mu_1/\mu_2 < 1. \tag{132}$$

Similar to what is considered above in Equations (63) and (64), Equation (128) obviously splits in two options, $B_2 = 0$ or the following one:

$$B_2^2 + 2T_1 A^2 + T_2 B_1^2 = 1. \tag{133}$$

In either case, Equations (126) and (127) simplify accordingly. The solutions corresponding to $B_2 = 0$ or to Equation (133) must be "dovetailed" at a stitch point, $x = x_0$, cf. Equation (65). An example of the so obtained solutions for amplitudes $A(x)$ and $B_{1,2}(x)$ is displayed in Figure 9b.

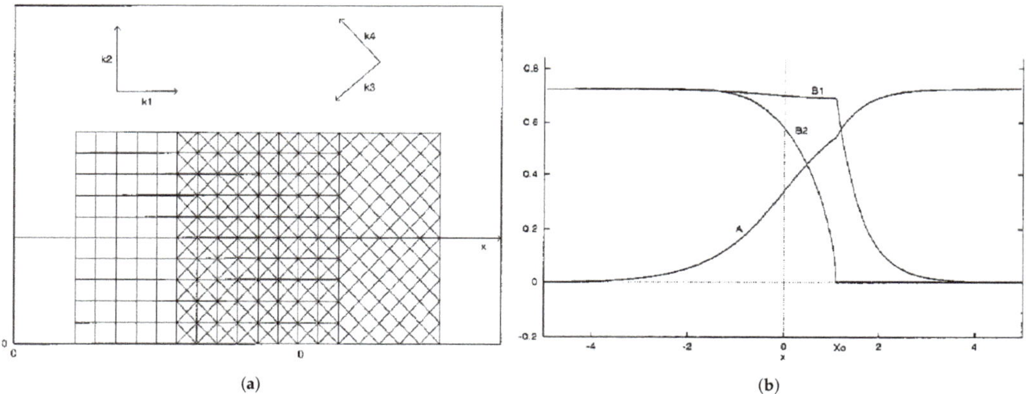

(a) (b)

Figure 9. (**a**) The scheme for building a broad stripe of the octagonal quasicrystalline state as a transient layer between semi-infinite domains filled by square-lattice patterns, mutually rotated by 45°. (**b**) An example of the corresponding solution for amplitudes $A(x)$ and $B_{1,2}(x)$. Parts of the solution corresponding to Equation (133) or to $B_2 = 0$ are connected at the stitch point $x = x_0$. Reprinted with permissions from Ref. [65].

3.4. Three-Dimensional Quasicrystals

A setting which makes it possible to predict a stable quasiperiodic pattern based on a set of four PW modes in the 3D space was put forward in Ref. [64]. It originates from the model of a lasing cavity, based on the standard system of coupled Maxwell–Bloch equations. The evolutional variable in this system is time, while the spatial structure is strongly anisotropic, as the field (Maxwell's) equation in the system contains only the first derivative, $\partial/\partial z$, with respect to the longitudinal coordinate, z, and the usual paraxial-diffraction operator, $i\left(\partial_x^2 + \partial_y^2\right)$, acting on the transverse coordinates, (x, y). As a result, at the lasing threshold components of 3D wave vectors carrying the PW modes,

$$\mathbf{K} = (\mathbf{k}, k_z),\ \mathbf{k} \equiv (k_x, k_y), \tag{134}$$

satisfy the following dispersion relation, which couples them to the wave's frequency Ω as follows:

$$\Omega = k^2 + k_z. \tag{135}$$

Eventually, above the lasing threshold the cubic nonlinearity of the Maxwell–Bloch system may produce a resonant quartet of 3D wave vectors, coupled by the following condition:

$$\mathbf{K}_1 + \mathbf{K}_2 = \mathbf{K}_3 + \mathbf{K}_4. \tag{136}$$

For comparison, in the 2D space the same relation (136), taken close to the threshold, i.e., for nearly equal length of the wave vectors, would imply that the four vectors form a rhombus, and the cubic interaction between the corresponding amplitudes, $u_{1,2,3,4}$, would be represented by usual nonresonant nonlinear terms, essentially the same as in Equation (101), with A_l replaced by u_l and $A_m^2 A_l$ replaced by the XPM terms, $|u_m|^2 u_l$. However, in the 3D setting, the resonance condition (136), combined with the dispersion relation (135), leads to a nontrivial possibility to add four-wave-mixing (FWM) cubic terms to the XPM ones, see below.

Substituting expression (134) for the 3D wave vector in Equations (136) and (135) leads to the following elementary exercise in planar geometry: find two pairs of 2D vectors, $(\mathbf{k}_1, \mathbf{k}_2)$ and $(\mathbf{k}_3, \mathbf{k}_4)$, satisfying the following conditions:

$$\mathbf{k}_1 + \mathbf{k}_2 = \mathbf{k}_3 + \mathbf{k}_4, \; k_1^2 + k_2^2 = k_3^2 + k_4^2. \tag{137}$$

An obvious solution of this exercise is plotted in Figure 10a.

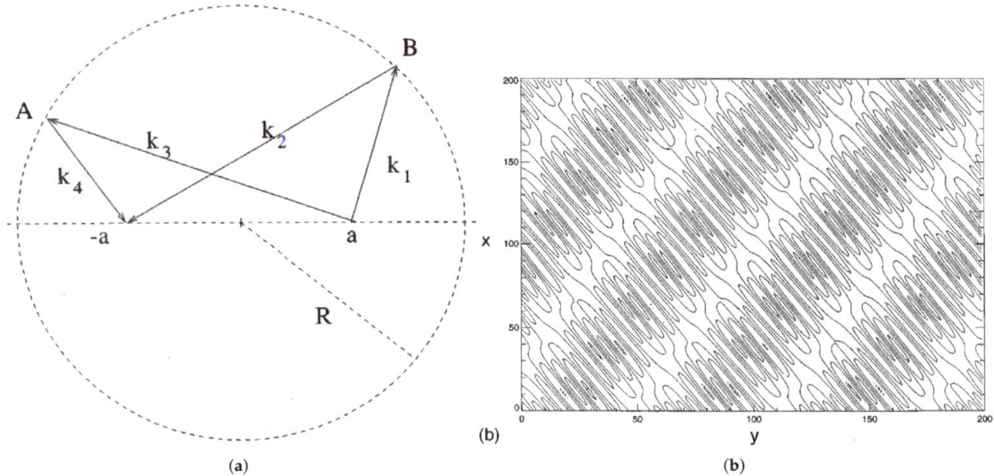

(a) (b)

Figure 10. (**a**) A set of four two-dimensional vectors $\mathbf{k}_{1,2,3,4}$ which solves equations (137). Here, A and B are two arbitrary points belonging to the circumference of arbitrary radius R, and $a < R$ is an arbitrary value of coordinate x. (**b**) An example of the three-dimensional quasiperiodic pattern, projected onto the (x, y) plane. Shown are contour plots of the corresponding distribution of the order parameter, $\mathrm{Re}\left[\sum_{l=1}^{4} u_l \cos(\mathbf{k}_l \cdot \mathbf{R})\right]$, where u_l are the complex amplitudes given by Equations (141)–(143). In this case, the phases are $\varphi_1 = \pi$, $\varphi_2 = -\pi/2$, $\varphi_3 = \pi/7$, while φ_4 is determined by Equation (142). The angle between vectors $\mathbf{k}_1 - \mathbf{k}_2$ and $\mathbf{k}_3 - \mathbf{k}_4$ is $\pi/5$. Figure is reprinted with permissions from Ref. [64].

Once a resonantly coupled quartet of four wave vectors is chosen, the respective system of evolution equations for the corresponding complex amplitudes is [64]

$$\begin{aligned} \frac{du_1}{dt} &= \gamma_0 u_1 - \left(|u_1|^2 + 2\sum_{l\neq 1}|u_l|^2\right)u_1 - 2u_2^* u_3 u_4, \\ \frac{du_2}{dt} &= \gamma_0 u_2 - \left(|u_2|^2 + 2\sum_{l\neq 2}|u_l|^2\right)u_2 - 2u_1^* u_3 u_4, \\ \frac{du_3}{dt} &= \gamma_0 u_3 - \left(|u_3|^2 + 2\sum_{l\neq 3}|u_l|^2\right)u_3 - 2u_1 u_2 u_4^*, \\ \frac{du_4}{dt} &= \gamma_0 u_4 - \left(|u_4|^2 + 2\sum_{l\neq 4}|u_l|^2\right)u_4 - 2u_1 u_2 u_3^*. \end{aligned} \tag{138}$$

In these equations, $\gamma_0 > 0$ is the linear gain, as above, and the last terms represent the four-wave-mixing (FWM) effect. Particular values of coefficients in front of nonlinear terms are standard ones which correspond to the XPM and FWM interactions in nonlinear optics [17], unlike general values of coefficients T_{l-m} in Equation (101). Similar to Equation (101), the system of Equations (138) admits the presentation in the form of $du_l/dt = -\,\partial L/\partial u_l^*$, with the Lyapunov function

$$L = -\gamma_0 \sum_l |u_l|^2 + \frac{1}{2}\sum_l |u_l|^4 + 2\sum_{l>m}|u_l|^2|u_m|^2 + 4\mathrm{Re}(u_1 u_2 u_3^* u_4^*). \tag{139}$$

Further analysis performed in Ref. [64] has produced two stable stationary solutions of Equation (138). First, this is a simple single-mode state (rolls), with

$$|u_1|^2 = \gamma_0,\ u_{2,3,4} = 0. \tag{140}$$

Next, dodecagonal quasicrystals with equal absolute values of all the four amplitudes are looked for as follows:

$$u_l = A\exp(i\varphi_l), \tag{141}$$

where the phases are locked so that

$$\varphi_1 + \varphi_2 - \varphi_3 - \varphi_4 = \pi, \tag{142}$$

and the squared absolute value of the amplitudes is

$$|u_l|^2 = \gamma_0/5, \tag{143}$$

cf. the rolls solution (140). Note that values of the Lyapunov function (139) for the rolls and 3D quasicrystal are as follows:

$$L_{\text{rolls}} = -\gamma_0^2/2,\ L_{\text{quasicryst}} = -2\gamma_0^2/5, \tag{144}$$

Hence, the rolls represent the ground state of the system, while the quasicrystal is a metastable state, as its value of L is slightly higher.

An example of the shape of the 3D quasiperiodic solution is displayed, in the projection onto plane (x, y), in Figure 10b. Additional examples can be found in Ref. [64].

Besides these solutions, Equations (138) give rise to another quasiperiodic state, with $\varphi_1 + \varphi_2 - \varphi_3 - \varphi_4 = 0$ and $|u_l|^2 = \gamma_0/9$ (cf. Equations (142) and (143)), but it is unstable. Two-mode solutions, e.g., ones with $|u_{1,2}|^2 = \gamma_0/3, u_{3,4} = 0$, also exist but are unstable [64].

4. Conclusions

The aim of this paper is to present a concise overview of two important topics in the theory of pattern formation in nonlinear dissipative media, viz., DWs (domain walls) and QP (quasiperiodic) patterns. The topics are selected as those important contributions to which were made in the works of Prof. Mikhail Tribelsky. Most of the results collected in this review, may be considered as "old" ones, as they were published ca. 30–35 [8,9,11,25,38,41,48,56,58–60,74,102] or 20 [57,63–65,75] years ago. Nevertheless, these results remain relevant in the context of ongoing theoretical and experimental studies in the ever expanding pattern-formation research area. This conclusion is upheld by the fact that the present paper includes a few novel exact analytical results, obtained as a relevant addition to the old theoretical findings concerning the DWs in systems of coupled real GL (Ginzburg–Landau) equations [10]. The new results, represented by Equations (54), (63)–(68), (72)–(75), and (87)–(90), produce exact solutions for symmetric DWs in the system of real GL equations, including linear mixing between the components; the solution for strongly asymmetric DWs in the case when the diffusion term is present only in one GL equation; the three-component composite state, including the DW in two components and a bright soliton in the third one; and the particular exact solution for DWs between waves governed by the real GL equations, including group-velocity terms with opposite signs.

The significance of the results presented in this brief review is enhanced by the fact that essentially the same coupled equations describe patterns of the DW and QP types not only in thermal convection, but also in nonlinear optics, BEC, and other physical systems. In particular, the pattern formation in BEC of cesium atoms under the action of a temporally periodic modulation of the nonlinearity (imposed by means of the Feshbach resonance [108]), similar to the Faraday instability, was recently experimentally demonstrated and theoretically modeled in the framework of amplitude equations similar to Equation (101) in Ref. [109]. Another novel realization of the pattern formation was proposed for a driven dissipative Bose–Hubbard lattice, which can be implemented in superconducting circuit arrays [110].

It is expected that theoretical and experimental studies along the directions outlined in this review have potential for further development, which will make it possible to add new findings to the above-mentioned well-established results.

Funding: This research was partly funded by the Israel Science Foundation through grant number 1286/17.

Data Availability Statement: Data sharing not applicable.

Acknowledgments: First of all, I would like to thank Michael Tribelsky for the collaboration, established long ago, which produced the essential results summarized in this article. I would also like to thank other colleagues in collaboration with whom these results were obtained: Alexander Nepomnyashchy, Jerry Moloney, Alan Newell, Natalia Komarova, Martin Van Hecke, and Horacio Rotstein. Valuable comments on the original version of this review were provided by Victor Kabanov, Ron Lifshitz, and Yury Yerin. I appreciate the help of Zhaopin Chen in producing Figures 4 and 5. As one of editors of the Special Issue of the journal *Physics* (published by MDPI), dedicated as a festschrift to the celebration of the 70th birthday of Michael Tribelsky, for which this article was written, I thank two other editors of the Special Issue, Andrey Miroshnichenko and Fernando Moreno, for a very efficient collaboration.

Conflicts of Interest: The author declares no conflict of interest.

References

1. Anisimov, S.I.; Tribel'skiĭ, M.I.; Épel'baum, Y.G. Instability of Plane Evaporation Boundary in Interaction of Laser Radiation with Matter. *Sov. Phys.—JETP* **1980**, *51*, 802–806. Available online: http://jetp.ras.ru/cgi-bin/dn/e_051_04_0802.pdf (accessed on 1 November 2021).
2. Bunkin, F.V.; Tribel'sky, M.I. Non-resonant interaction of high-power optical radiation with a liquid. *Sov. Physics Uspekhi* **1980**, *130*, 105–133. [CrossRef]
3. Luk'yanchuk, B.S.; Tribelsky, M.I. Anomalous light scattering by small particles. *Phys. Rev. Lett.* **2006**, *97*, 263902. [CrossRef]

4. Tribelsky, M.I.; Flach, S.; Miroshnichenko, A.E.; Gorbach, A.V.; Kivshar, Y.S. Light scattering by a finite obstacle and Fano resonances. *Phys. Rev. Lett.* **2008**, *100*, 043903. [CrossRef]
5. Tribelsky, M.I.; Geffrin, J.M.; Litman, A.; Eyraud, C.; Moreno, F. Small dielectric spheres with high refractive index as new multifunctional elements for optical devices. *Sci. Rep.* **2015**, *5*, 12288. [CrossRef]
6. Miroshnichenko, A.E.; Tribelsky, M.I. Giant in-particle field concentration and Fano resonances at light scattering by high-refractive-index particles. *Phys. Rev. A.* **2016**, *83*, 053837. [CrossRef]
7. Miroshnichenko, A.E.; Tribelsky, M.I. Ultimate absorption in light scattering by a finite obstacle. *Phys. Rev. Lett.* **2018**, *120*, 263902. [CrossRef]
8. Malomed, B.A.; Nepomnyashchy, A.A.; Tribelsky, M.I. Domain boundaries in convection patterns. *Phys. Rev. A* **1990**, *42*, 7244–7263. [CrossRef]
9. Malomed, B.A.; Nepomnyashchiĭ, A.A.; Tribel'skiĭ, M.I. Two-Dimensional Quasiperiodic Structures in Nonequilibrium Systems. *Sov. Phys.—JETP* **1989**, *69*, 388–396. Available online: http://jetp.ras.ru/cgi-bin/dn/e_069_02_0388.pdf (accessed on 1 November 2021).
10. Malomed, B.A. New Findings for the Old Problem: Exact Solutions for Domain Walls in Coupled Real Ginzburg-Landau Equations. To be published.
11. Malomed, B.A.; Tribelsky, M.I. Bifurcations in distributed kinetic systems with aperiodic instability. *Phys. D* **1984**, *14*, 67–87. [CrossRef]
12. Ginzburg, V.L.; Landau, L.D. On the theory of superconductivity. *Zh. Eksp. Teor. Fiz.* **1950**, *20*, 1064–1082. (In Russian). English Translation: *Collected Papers of L.D. Landau*; Ter-Haar, D.; Pergamon Press: Oxford, UK, 1965; pp. 546–568. [CrossRef]
13. Cross, M.C. Ingredients of a theory of convective textures close to onset. *Phys. Rev. A* **1982**, *25*, 1065–1076. [CrossRef]
14. Eckhaus, W. *Studies in Non-Linear Stability Theory*; Springer: New York, NY, USA, 1965. [CrossRef]
15. Aranson, I.S.; Kramer, L. The world of the complex Ginzburg-Landau equation. *Rev. Mod. Phys.* **2002**, *74*, 99–143. [CrossRef]
16. Malomed, B.A. Complex Ginzburg-Landau equation. In *Encyclopedia of Nonlinear Science*; Scott, A., Ed.; Routledge: New York, NY, USA, 2005; pp. 157–160.
17. Kivshar, Y.S.; Malomed, B.A. Dynamics of solitons in nearly integrable systems. *Rev. Mod. Phys.* **1989**, *61*, 763–915. [CrossRef]
18. Arecchi, F.T.; Boccaletti, S.; Ramazza, P. Pattern formation and competition in nonlinear optics. *Phys. Rep.* **1999**, *318*, 1–83. [CrossRef]
19. Rosanov, N.N. Transverse patterns in wide-aperture nonlinear optical systems. *Progr. Opt.* **1996**, *35*, 1–60. [CrossRef]
20. Rosanov, N.N. *Spatial Hysteresis and Optical Patterns*; Springer: Berlin/Heidelberg, Germany, 2002. [CrossRef]
21. Lega, J. Traveling hole solutions of the complex Ginzburg-Landau equation: A review. *Phys. D* **2001**, *152*, 269–287. [CrossRef]
22. Cross, M.C.; Hohenberg, P.C. Pattern-formation outside of equilibrium. *Rev. Mod. Phys.* **1993**, *65*, 851–1112. [CrossRef]
23. Ipsen, M.; Kramer, L.; Sorensen, P.G. Amplitude equations for description of chemical reaction-diffusion systems. *Phys. Rep.* **2000**, *337*, 193–235. [CrossRef]
24. Hoyle, R. *Pattern Formation: An Introduction to Methods*; Cambridge University Press: Cambridge, MA, USA, 2006. [CrossRef]
25. Malomed, B.A. Evolution of nonsoliton and "quasiclassical" wavetrains in nonlinear Schrödinger and Korteweg—de Vries equations with dissipative perturbations. *Phys. D* **1987**, *29*, 155–172. [CrossRef]
26. Sakaguchi, H. Motion of pulses and vortices in the cubic-quintic complex Ginzburg-Landau equation without viscosity. *Phys. D* **2005**, *210*, 138–148. [CrossRef]
27. Akhmediev, N.; Ankiewicz, A. (Eds.) *Dissipative Solitons: From Optics to Biology and Medicine*; Springer: Berlin/Heidelberg, Germany, 2008. [CrossRef]
28. Wise, F.W.; Chong, A.; Renninger, W.H. High-energy femtosecond fiber lasers based on pulse propagation at normal dispersion. *Laser Phot. Rev.* **2008**, *2*, 58–73. [CrossRef]
29. Ackemann, T.; Firth, W.J.; Oppo, G.L. Fundamentals and applications of spatial dissipative solitons in photonic devices. *Adv. At. Mol. Opt. Phys.* **2009**, *57*, 323–421. [CrossRef]
30. Leblond, H.; Mihalache, D. Models of few optical cycle solitons beyond the slowly varying envelope approximation. *Phys. Rep.* **2013**, *523*, 61–126. [CrossRef]
31. Song, Y.F.; Shi, X.J.; Wu, C.F.; Tang, D.Y.; Zhang, H. Recent progress of study on optical solitons in fiber lasers. *Appl. Phys. Rev.* **2019**, *6*, 0213139. [CrossRef]
32. Hocking, L.M.; Stewartson, K. On the nonlinear response of a marginally unstable plane parallel flow to a two-dimensional disturbance. *Proc. R. Soc. London Ser. A* **1972**, *326*, 289–313. [CrossRef]
33. Pereira, N.R.; Stenflo, L. Nonlinear Schrödinger equation including growth and damping. *Phys. Fluids* **1977**, *20*, 1733–1734. [CrossRef]
34. Malomed, B.A. Strong periodic amplification of solitons in a lossy optical fiber: Analytical results. *J. Opt. Soc. Am. B* **1994**, *11*, 1261–1266. [CrossRef]
35. Berntson, A.; Malomed, B.A. Dispersion-management with filtering. *Opt. Lett.* **1999**, *24*, 507–509. [CrossRef]
36. Bekki, N.; Nozaki, K. Formation of spatial patterns and holes in the generalized Ginzburg-Landau equation. *Phys. Lett. A* **1985**. *110*, 133–135. [CrossRef]
37. Petviashvili, V.I.; Sergeev, A.M. Spiral solitons in active media with an excitation threshold. *Dokl. Akad. Nauk SSSR* **1984**, *276*, 1380–1384. (In Russian). Available online: http://mi.mathnet.ru/dan46625 (accessed on 1 November 2021).

38. Fauve, S.; Thual, O. Solitary waves generated by subcritical instabilities in dissipative systems. *Phys. Rev. Lett.* **1990**, *64*, 282–284. [CrossRef]
39. van Saarloos, W.; Hohenberg, P.C. Pulses and fronts in the complex Ginzburg-Landau equation near a subcritical bifurcation. *Phys. Rev. Lett.* **1990**, *84*, 749–752. [CrossRef]
40. Hakim, V.; Jakobsen, P.; Pomeau, Y. Fronts vs. solitary waves in nonequilibrium systems. *Europhys. Lett.* **1990**, *11*, 19–24. [CrossRef]
41. Malomed, B.A.; Nepomnyashchy, A.A. Kinks and solitons in the generalized Ginzburg-Landau equation. *Phys. Rev. A* **1990**, *42*, 6009–6014. [CrossRef] [PubMed]
42. Kuramoto, Y.; Tsuzuki, T. Persistent propagation of concentration waves in dissipative media far from thermal equilibrium. *Progr. Theor. Phys.* **1976**, *55*, 356–369. [CrossRef]
43. Sivashinsky, G.I. Nonlinear analysis of hydrodynamic instability in laminar flames—I. Derivation of basic equations. *Acta Astronaut.* **1977**, *4*, 1177–1206. [CrossRef]
44. Kuramoto, Y. Diffusion-induced chaos in reaction systems. *Progr. Theor. Phys. Suppl.* **1978**, *64*, 346–367. [CrossRef]
45. Cladis, P.E.; Fradin, C.; Finn, P.L.; Brand, H.R. A novel route to defect turbulence in nematics. *Mol. Cryst. Liq. Cryst. Sci. Tech. A: Mol. Cryst. Liq. Cryst.* **1998**, *328*, 513–521. [CrossRef]
46. Manneville, P.; Pomeau, Y. A grain-boundary in cellular structures near the onset of convection. *Phil. Mag. A* **1983**, *48*, 607–621. [CrossRef]
47. Haragus, M.; Iooss, G. Bifurcation of symmetric domain walls for the Bénard-Rayleigh convection problem. *Arch. Ration. Mech. Anal.* **2021**, *239*, 733–781. [CrossRef]
48. Steinberg, V.; Ahlers, G.; Cannell, D.S. Pattern formation and wave-number selection by Rayleigh-Bénard convection in a cylindrical container. *Phys. Scr.* **1985**, *32*, 534–547. [CrossRef]
49. Rohrer, G.S. Grain boundary energy anisotropy: A review. *J. Mater. Sci.* **2011**, *46*, 5881–5895. [CrossRef]
50. Lim, H.; Lee, M.G.; Wagoner, R.H. Simulation of polycrystal deformation with grain and grain boundary effects. *Int. J. Plast.* **2011**, *27*, 1328–1354. [CrossRef]
51. Rudolph, P. Dislocation patterning and bunching in crystals and epitaxial layers—A review. *Cryst. Res. Tech.* **2017**, *52*, 1600171. [CrossRef]
52. Atxitia, U.; Hinzke, D.; Nowak, U. Fundamentals and applications of the Landau-Lifshitz-Bloch equation. *J. Phys. D Appl. Phys.* **2017**, *50*, 033003. [CrossRef]
53. Galkina, E.G.; Ivanov, B.A. Dynamic solitons in antiferromagnets. *Low Temp. Phys.* **2018**, *44*, 618–633. [CrossRef]
54. Yao, W.; Wu, B.; Liu, Y. Growth and grain boundaries in 2D materials. *ACS NANO* **2020**, *14*, 9320–9346. [CrossRef] [PubMed]
55. Yerin, Y.; Drechsler, S.-L. Phase solitons in a weakly coupled three-component superconductor. *Phys. Rev. B* **2021**, *104*, 014518. [CrossRef]
56. Malomed, B.A. Optical domain walls. *Phys. Rev. E* **1994**, *50*, 1565–1571. [CrossRef] [PubMed]
57. Trippenbach, M.; Góral, K.; Rzażewski, K.; Malomed. B.; Band, Y.B. Structure of binary Bose-Einstein condensates. *J. Phys. B At. Mol. Opt. Phys.* **2000**, *33*, 4017–4031. [CrossRef]
58. Malomed, B.A. Nonsteady waves in distributed dynamical systems. *Phys. D* **1983**, *8*, 353–359. [CrossRef]
59. Malomed, B.A. Stability and grain boundaries in the dispersive Newell-Whitehead-Siegel equation. *Phys. Scr.* **1997**, *57*, 115–117. [CrossRef]
60. Cross, M.C. Traveling and standing waves in binary-fluid convection in finite geometries. *Phys. Rev. Lett.* **1986**, *57*, 2935–2938. [CrossRef]
61. Cross, M.C. Structure of nonlinear traveling-wave states in finite geometries. *Phys. Rev. A* **1988**, *38*, 3593–3600. [CrossRef] [PubMed]
62. Coullet, P.; Frisch, T.; Plaza, F. Sources and sinks of wave patterns. *Phys. D* **1993**, *62*, 75–79. [CrossRef]
63. Voss, H.U.; Kolodner, P.; Abel, M.; Kurths, J. Amplitude equations from spatiotemporal binary-fluid convection data. *Phys. Rev. Lett.* **1999**, *83*, 3422–3425. [CrossRef]
64. Komarova, N.; Malomed, B.A.; Moloney, J.V.; Newell, A.C. Resonant quasiperiodic patterns in a three-dimensional lasing medium. *Phys. Rev. A* **1997**, *56*, 803–812. [CrossRef]
65. Rotstein, H.; Malomed, B.A. A quasicrystallic domain wall in nonlinear dissipative patterns. *Phys. Scr.* **2000**, *62*, 164–168.
66. Lugiato, A.A.; Lefever, R. Spatial dissipative structures in passive optical systems. *Phys. Rev. Lett.* **1987**, *58*, 2209–2211. [CrossRef]
67. Oppo, G.-L.; Brambilla, M.; Lugiato, L. Formation and evolution of roll patterns in optical parametric oscillators. *Phys. Rev. A* **1994**, *49*, 2028–2032. [CrossRef]
68. Chembo, Y.K.; Menyuk, C.R. Spatiotemporal Lugiato-Lefever formalism for Kerr-comb generation in whispering-gallery-mode resonators. *Phys. Rev. A* **2013**, *87*, 053852. [CrossRef]
69. Huang, S.W.; Yang, J.H.; Yang, S.H.; Yu, M.E.; Kwong, D.L.; Zelevinsky, T.; Jarrahi, M.; Wong, C.W. Globally stable microresonator Turing pattern formation for coherent high-power THz radiation on-chip. *Phys. Rev. X* **2017**, *7*, 041002. [CrossRef]
70. de Valcarcel, G.J.; Staliunas, K. Phase-bistable Kerr cavity solitons and patterns. *Phys. Rev. A* **2013**, *87*, 043802. [CrossRef]
71. Garbin, B.; Wang, Y.D.; Murdoch, S.G.; Oppo, G.L.; Coen, S.; Erkintalo, M. Experimental and numerical investigations of switching wave dynamics in a normally dispersive fibre ring resonator. *Eur. Phys. J. D* **2017**, *71*, 240. [CrossRef]

72. Mineev, V.P. The theory of the solution of two near-ideal Bose gases. *Sov. Phys.—JETP* **1974**, *40*, 132–136. Available online http://jetp.ras.ru/cgi-bin/dn/e_040_01_0132.pdf (accessed on 1 November 2021).
73. Busse, F.H. The stability of finite amplitude cellular convection and its relation to an extremum principle. *J. Fluid Mech.* **1967**, *30*, 625–649. [CrossRef]
74. Pomeau, Y. Front motion, metastability and subcritical bifurcations in hydrodynamics. *Phys. D* **1986**, *23*, 3–11. [CrossRef]
75. van Hecke, M.; Malomed, B.A. A domain wall between single-mode and bimodal states and its transition to dynamical behavior in inhomogeneous systems. *Phys. D* **1997**, *101*, 131–156. [CrossRef]
76. Kivshar, Y.S.; Agrawal, G.P. *Optical Solitons: From Fibers to Photonic Crystals*; Academic Press: Cambridge, MA, USA, 2003.
77. Skorobogatiy, M.; Yang, J. *Fundamentals of Photonic Crystal Guiding*; Cambridge University Press: Cambridge, MA, USA, 2009. [CrossRef]
78. Pitaevskii, L.P.; Stringari, S. *Bose-Einstein Condensation and Superfluidity*; Oxford University Press: Oxford, UK, 2016. [CrossRef]
79. Ballagh, R.J.; Burnett, K.; Scott, T.F. Theory of an output coupler for Bose-Einstein condensed atoms. *Phys. Rev. Lett.* **1997**, *78*, 1608–1611. [CrossRef]
80. Asghari, M.; White, I.H.; Penty, R.V. Wavelength conversion using semiconductor optical amplifiers. *J. Light. Tech.* **1997**, *15*, R3310–R3313. [CrossRef]
81. Kim, J.; Laemmlin, M.; Meuer, C.; Bimberg, D.; Eisenstein, G. Theoretical and experimental study of high-speed small-signal cross-gain modulation of quantum-dot semiconductor optical amplifiers. *IEEE J. Quant. Elect.* **2009**, *45*, 240–248. [CrossRef]
82. Merhasin, M.I.; Malomed, B.A.; Driben, R. Transition to miscibility in a binary Bose-Einstein condensate induced by linear coupling. *J. Phys. B At. Mol. Opt. Phys.* **2005**, *38*, 877–892. [CrossRef]
83. Alama, S.; Bronsard, L.; Contreras, A.; Pelinovsky, D.E. Domains walls in the coupled Gross-Pitaevskii equations. *Arch. Rat. Mech Appl.* **2015**, *215*, 579–615. [CrossRef]
84. Marzok, C.; Deh, B.; Courteille, P.W.; Zimmermann, C. Ultracold thermalization of ^{7}Li and ^{87}Rb. *Phys. Rev. A* **2007**, *76*, 052704. [CrossRef]
85. Alexandrov, A.S.; Kabanov, V.V. Excitations and phase segregation in a two-component Bose–Einstein condensate with an arbitrary interaction. *J. Phys. Condens. Matter* **2002**, *14*, L327–L332. [CrossRef]
86. Malomed, B.A. Domain wall between traveling waves. *Phys. Rev. E* **1994**, *50*, R3310–R3313. [CrossRef] [PubMed]
87. Kai, Y.; Yin, Z. Asymptotic analysis to domain walls between traveling waves modeled by real coupled Ginzburg-Landau equations. *Chaos Solitons Fractals* **2021**, *152*, 111266. [CrossRef]
88. Shechtman, D.; Blech, I.; Gratias, D.; Cahn, J. Metallic phase with long-range orientational order and no translational symmetry. *Phys. Rev. Lett.* **1984**, *53*, 1951–1953. [CrossRef]
89. Wang, N.; Chen, H.; Kuo, K. Two-dimensional quasicrystal with eightfold rotational symmetry. *Phys. Rev. Lett.* **1987**, *59*, 1010–1013. [CrossRef]
90. Barkan, K.; Diamant, H.; Lifshitz, R. Stability of quasicrystals composed of soft isotropic particles. *Phys. Rev.* B **2011**, *83*, 172201. [CrossRef]
91. Torquato, S. Hyperuniform states of matter. *Phys. Rep.* **2018**, *745*, 1–95. [CrossRef]
92. Steurer, W. Quasicrystals: What do we know? What do we want to know? What can we know? *Acta Crystallogr. A* **2018**, *74*, 1–11. [CrossRef]
93. Savitz, S.; Babadi, M.; Lifshitz, R. Multiple-scale structures: From Faraday waves to soft-matter quasicrystals. *IUCrJ* **2018**, *5*, 247–268. [CrossRef] [PubMed]
94. Skjaervo, S.H.; Marrows, C.H.; Stamps, R.L.; Leyderman, L.J. Advances in artificial spin ice. *Nat. Rev. Phys.* **2020**, *2*, 13–28. [CrossRef]
95. Lifshitz, R.; Petrich, D.M. Theoretical model for Faraday waves with multiple-frequency forcing. *Phys. Rev. Lett.* **1997**, *79*, 1261–1264. [CrossRef]
96. von Freymann, G.; Ledermann, A.; Thiel, M.; Staude, I.; Essig, S.; Busch, K.; Wegener, M. Three-dimensional nanostructures for photonics. *Adv. Funct. Mater.* **2010**, *20*, 1038–1052. [CrossRef]
97. Bellingeri, M.; Chiasera, A.; Kriegel, I.; Scotognella, F. Optical properties of periodic, quasi-periodic, and disordered one-dimensional photonic structures. *Opt. Mater.* **2017**, *72*, 403–421. [CrossRef]
98. Vardeny, Z.V.; Nahata, A.; Agrawal, A. Optics of photonic quasicrystals. *Nat. Photonics* **2013**, *7*, 177–187. [CrossRef]
99. Lu, L.; Joannopoulos, J.D.; Soljacic, M. Topological photonics. *Nat. Photonics* **2014**, *8*, 821–829. [CrossRef]
100. Steurer, W.; Sutter-Widmer, D. Photonic and phononic quasicrystals. *J. Phys. D: Appl. Phys.* **2007**, *40*, R229–R247. [CrossRef]
101. Malomed, B.A.; Tribelskiĭ, M.I. On the Stability of Stationary Weakly Overcritical Patterns in Convection and Allied Problems. *Sov. Phys.—JETP* **1987**, *65*, 305–310. Available online: http://jetp.ras.ru/cgi-bin/dn/e_065_02_0305.pdf (accessed on 1 November 2021).
102. Malomed, B.A.; Nepomnyashchy, A.A.; Tribelsky, M.I. Two-dimensional dissipative structures with a quasicrystallic symmetry. *Pis'ma Zh. Tekh. Fiz. (Sov. Phys. Tech. Phys. Lett.)* **1987**, *13*, 1165–1167.
103. Steurer, W. Twenty years of structure research on quasicrystals. Part I. Pentagonal, octagonal, decagonal and dodecagonal quasicrystals. *Z. für Krist.* **2004**, *219*, 391–446. [CrossRef]
104. Socolar, J.E.S.; Lubensky, T.C.; Steinhardt, P.J. Phonons, phasons, and dislocations in quasi-crystals. *Phys. Rev. B* **1986**, *34*, 3345–3360. [CrossRef]

105. Yamamoto, A. Crystallography of quasiperiodic crystals. *Acta Crystallogr. A* **1996**, *52*, 509–560. [CrossRef]
106. Freedman, B.; Lifshitz, R.; Fleischer, J.W.; Segev, M. Phason dynamics in nonlinear photonic quasicrystals. *Nat. Mater.* **2007**, *6*, 776–781. [CrossRef]
107. Iooss, G.; Joseph, D.D. *Elementary Stability Bifurcation Theory*; Springer: New York, NY, USA, 1980.
108. Chin, C.; Grimm, R.; Julienne, P.; Tiesinga, E. Feshbach resonances in ultracold gases. *Rev. Mod. Phys.* **2010**, *82*, 1225–1286. [CrossRef]
109. Zhang, Z.; Yao, K.-X.; Feng, L.; Hu, J.; Chin, C. Pattern formation in a driven Bose–Einstein condensate. *Nat. Phys.* **2020**, *16*, 652–656. [CrossRef]
110. Wang, Z.; Navarrete-Benlloch, C.; Cai, Z. Pattern formation and exotic order in driven-dissipative Bose-Hubbard systems. *Phys. Rev. Lett.* **2020**, *125*, 115301. [CrossRef] [PubMed]

Communication

Emergence of Many Mini-Circles from a Coffee Suspension with Mechanical Rotation

Hiroshi Ueno [1], Mayu Shono [1], Momoko Ogawa [1], Koichiro Sadakane [1,*] and Kenichi Yoshikawa [1,2,*]

1 Faculty of Life and Medical Sciences, Doshisha University, Tatara, Kyotanabe, Kyoto 610-0394, Japan; uaeneau@dmpl.doshisha.ac.jp (H.U.); smayu1828@gmail.com (M.S.); ctuf1025@mail4.doshisha.ac.jp (M.O.)

2 Center for Integrative Medicine and Physics, Institute for Advanced Study, Kyoto University, Kyoto 606-8501, Japan

* Correspondence: ksadakan@mail.doshisha.ac.jp (K.S.); keyoshik@mail.doshisha.ac.jp (K.Y.)

Abstract: Drying of an aqueous suspension containing fine granules leads to the formation of a circular pattern, i.e., the coffee-ring effect. Here, we report the effect of mechanical rotation with drying of an aqueous suspension containing a large amount of granular particles as in the Turkish coffee. It was found that wavy fragmented stripes, or a "waggly pattern", appear in the early stage of the drying process and a "polka-dot pattern" with many small circles is generated in the late stage. We discuss the mechanism of these patterns in terms of the kinetic effect on micro phase-segregation. We suggest that the waggly pattern is induced through a mechanism similar to spinodal decomposition, whereas polka-dot formation is accompanied by the enhanced segregation of a water-rich phase under mechanical rotation.

Keywords: coffee-ring; micro phase-segregation; transition of drying pattern

Citation: Ueno, H.; Shono, M.; Ogawa, M.; Sadakane, K.; Yoshikawa, K. Emergence of Many Mini-Circles from a Coffee Suspension with Mechanical Rotation. *Physics* **2021**, *3*, 8–16. https://doi.org/10.3390/physics3010003

Received: 4 December 2020
Accepted: 14 January 2021
Published: 22 January 2021

1. Introduction

The formation of a deposition pattern with the evaporation of a liquid containing nonvolatile particles has attracted considerable interest not only from a fundamental scientific aspects perspective [1–3], but also from an engineering point of view with respect to coating and patterning processes [4,5]. As a typical pattern, a so-called coffee-ring is caused by the transportation of solute particles toward a pinned contact line driven by Marangoni effect or spatial gradient of the surface tension, under a differential evaporation rate over the liquid/air surface [6–11]. In addition to the formation of a ring-like pattern [12], the generation of various kinds of morphologies, such as fractures, cracks, straight lines, spiral and dry parch, have been reported in the drying of droplets containing micro or nanoparticles [13–22]. Smart control of the positioning of nanoparticles by using photo-sensitive surfactant in drying droplets was also reported [23]. It has been shown that particles can be concentrated at the center of a droplet through spot-irradiation of its apex with a heating laser, by dismissing the coffee-ring pattern, which phenomenon was interpreted in terms of the reversal of intra-droplet flow induced by a thermal Marangoni effect [24,25]. A similar manner of particle deposition at the center of a droplet was observed when the solvent was changed from water to octane [26]. To suppress the coffee-ring effect, or the heterogeneous deposition of particles, various methodologies have been proposed, including the application of a surface acoustic wave [27], the imposition of electronic fields [28,29], heating of the solid substrate [30], and the addition of a surfactant [31,32]. In the present study, we performed a drying experiment by adopting an aqueous suspension containing fine coffee powder/granules, i.e., Turkish coffee, which is usually served without filtering and thus contains a relatively large amount of micro-particles. Drying this solution under a horizontal static condition results in the formation of a homogeneous granular layer without the formation of a coffee-ring. Interestingly, characteristic patterns of drying granules, such as multiple *wavy* segments and several mini-circles, are generated using a rotating dish under a tilting condition.

2. Materials and Methods

Roasted coffee beans were ground with a conical burr coffee grinder (product MSCS-2B, Hario Co. Ltd., Tokyo, Japan). Larger particles were sieved out of the ground powder with a sifter (grid size of 250 μm, Tokyo Screen Co., Ltd., Tokyo, Japan). In Figure 1 the experimental procedure in a schematic manner is shown, together with the photograph of the coffee powder (average diameter of 68 μm, and standard deviation of 23 μm) Aqueous solution was prepared by mixing 900 mg of ground coffee beans with 3 mL of ultrapure water (produced with Milli-Q water purification system, Millipore, Merck) The mixed solution was transferred onto a paper dish, of which the surface laminated with polyethylene terephthalate is hydrophobic and the diameter of the bottom planar part is 140 mm (RS-362, Dixie Japan Ltd., Tokyo, Japan). Then, the solution was mechanically homogenized with a vortex mixer (SI-0286, Scientific Industries Inc., Bohemia, NY, USA) In the present Communication, we report the experimental results under the conditions that the paper dish was fixed to a rotating dish with a tilting angle of θ = 45° and was rotated at 60 rpm by a direct current motor (mini-motor multi-ratio gearbox (12-speed), item 70190, Tamiya Ltd., Tokyo, Japan). As for the effect of tilting, we found that the coffee solution tends to flow out from the dish when θ is larger than 60°, whereas contrast of the generating pattern becomes relatively unclear when θ is smaller than 30°. Thus, we have carried out the experiment by taking the tilting angle as 45°. Under the condition θ = 45°, when the rotation rate is smaller than 30 rpm, the solution tends to flow downward outside the dish. When the rotation rate is larger than 100 rpm, the generated pattern tends to be inhomogeneous between the inner and outer regions of the dish, because of the relatively large magnitude of the oscillation on the centrifugal force. Based on the results of these preliminary experiments, we report the experimental results at the fixed values of the tilting angle at 45° and rotational rate at 60 rpm, in order to reveal the representable transition of the drying patters between wavy fragmented stripes and many mini-circles.

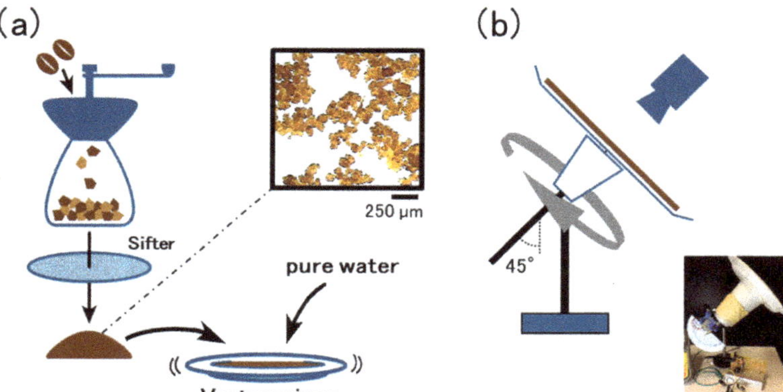

Figure 1. Experimental scheme. (**a**) Roasted coffee beans were ground with a conical burr grinder. Larger particles were sieved out of the ground powder with a sifter (grid size: 250 μm). The ground coffee was mixed with pure water on a paper dish (diameter of the horizontal circular area: 140 mm), and the solution was spread over the whole dish by vibration with a vortex mixer. (**b**) Experimental apparatus to rotate the tilted dish with the solution containing the coffee particles. The paper dish with the coffee solution was fixed to a rotating dish with a tilting angle of θ = 45°. The dish was rotated at 60 rpm. During rotation, the whole experimental apparatus was situated inside a control box with constant humidity (60%) and temperature (20 °C).

3. Results and Discussion

Figure 2 shows the drying patterns obtained from the coffee solution, by adopting (a–d) a solution containing coffee powder (see Figure 1a) and (e) filtered solution without powder. All of the pictures were taken for the completely dried states after standing still for 24 h with horizontal positioning. For the experiments shown in (a–d), we used a suspension with the solution of coffee powder on a paper dish, of which the surface is hydrophobic. Figure 2a shows the appearance of a pattern with many wavy shapes, which was obtained by fixing the solution on the tilted plate for 10 s in a stationary manner and then rotating it for 1 min. Hereafter, we call this morphology a "waggly" pattern. Figure 2b shows the drying pattern after 10 s of stationary tilting and then 20 min of plate rotation. The appearance of many mini-circles with a diameter of ~1 mm is observed, which we call "polka-dot" in this article. Here, it is to be noted that the waggly and polka-dot patterns appeared for the same experimental solution with different time-period of the dish rotation. Figure 2c shows a tree-like pattern which was generated under the stationary tilt condition for 1 min without rotation. Figure 2d shows a homogeneous layer of powder obtained by drying the coffee suspension under a horizontal arrangement. For comparison, Figure 2e shows a so-called coffee-ring, which was generated under horizontal drying of a droplet of coffee solution prepared through filtration. In both Figure 2d,e, 0.1 mL of coffee solution was deposited on the paper plate.

As shown in the experimental observations (Figure 2), it has become clear that drying under tilted rotation strongly affects the outcome; a waggly pattern appears first and then a polka-dot pattern develops. Next, we discuss the mechanism of the occurrence of the characteristic patterns. Under dish rotation, the coffee suspension is segregated into grain-rich and water-rich solutions as revealed in Figure 2. It would be expected that the underlying mechanism of the pattern formation observed for the suspension could be interpreted in terms of a kinetic effect in the first-order phase-transition. Thus, we will consider the appearance of the waggly and polka-dot patterns by adopting Cahn–Hilliard-type simple model equations [33–39]:

$$\frac{\partial \eta}{\partial t} = \nabla\left(M_c \nabla \frac{\delta F}{\delta \eta}\right), \tag{1}$$

where the free energy F exhibits two different contributions: bimodality with the order parameter and the interfacial energy. Here, M_c is a parameter of diffusivity and t is time.

$$F = \int \left(L\eta(1-\eta) + \frac{\alpha}{2}|\nabla \eta|^2\right) dv, \tag{2}$$

where L, α and dv are interaction parameter, gradient energy coefficient and differential volume, respectively. For simplicity, we chose the bimodal profile of the interaction energy as a function of η, corresponding to the water content in the solution containing coffee grains; $\eta = 1$ corresponds to pure water. We also neglected the contribution from the mixing entropy, since we are considering the segregation of relatively large particles of coffee grains. For the calculation of Equation (2), we tentatively adopted the parameters $L = 6.4 \times 10^3$ J/mol and $\alpha = 3.0 \times 10^{-3}$ Jm^2/mol, so as to obtain the pattern with usual spinodal decomposition. We may regard that $\eta = 0, 1$ correspond to the dense coffee grains and the clear solution, respectively. Strictly speaking, our experimental system is non-conservative, because of the evaporation of water to cause the spatial pattern. Thus, the usual Cahn–Hilliard equation does not hold in a strict manner for our experiments, especially for the experimental conditions with relatively large effect of the evaporation. However, the numerical results can still be expected to provide useful insight into the mechanism of pattern formation. Actually, for the initial stage of the drying process when the water content does not decrease so much, the kinetic equation based on Cahn–Hilliard model would represent the essential feature of the segregation. Since the order parameter η is dependent primarily on the relative concentration of the coffee grains, we may need to consider that the diffusivity M_c is sensitively dependent on η, in addition to the bimodal dependence (the term η $(1-\eta)$),

$$M_c = \left[\frac{D_0}{RT}\eta + \frac{D_1}{RT}(1-\eta)\right]\eta(1-\eta), \tag{3}$$

where D_0 and D_1 are the diffusion constants for the states with η = 0 and 1, respectively. In the following simulation, we used the universal gas constant R = 8.31 J/mol. We adopted the apparent diffusion constants $D_0 = 4.0 \times 10^{-10}$ m^2/s and $D_1 = 4.0 \times 10^{-8}$ m^2/s, by taking into account the effect of the smaller diffusivity of the grain rich solution. We adapted one-order larger value for the apparent diffusivity of water, D_1, as that of the pure water with stationary standing state [40], by considering the effect owe to the rhythmic change in gravitational field induced by the dish rotation. As for the diffusivity of the grain powder (the diameter is ca. 40 μm as estimated from the average diameter of 68 μm, as mentioned in Materials and Methods), it is expected that its diffusion constant is on the order of 10^{-5}–10^{-6} comparted to that of water for the usual Brownian motion under thermal equilibrium, as estimated from the Stokes–Einstein relationship. In addition, with the decrease of the water content, the diffusion of the coffee grain should become much lower. Thus, it is noted that the adapted value for D_1 is rather large compared to the intrinsic diffusivity under the fluctuation-dispersion relationship near thermal equilibricity. In other words, we perform the numerical modeling with the consideration of the effect induced by the external agitation, i.e., the periodic change of the gravitational field accompanied by the rotation of the tilted dish. Through such simple assumptions, we performed a numerical simulation using a two-dimensional system to shed light on the essential mechanism on the time-development of the generated pattern. It may be possible to include the effect of the periodic acceleration during dish rotation by tuning the effective temperature in the simulation. However, in the present study, for simplicity we used room temperature, T = 293 K. We carried out the numerical simulation by modifying the source code of Python available from the open access version [41], provided by the "Yamanaka Laboratoty" at Tokyo University of Agriculture and Technology, Japan. The grid spacing in the computation is taken as 1.0×10^{-3} m. The time width and step number are 0.01 s and 13,000, respectively; corresponding to a time-period of 130 s.

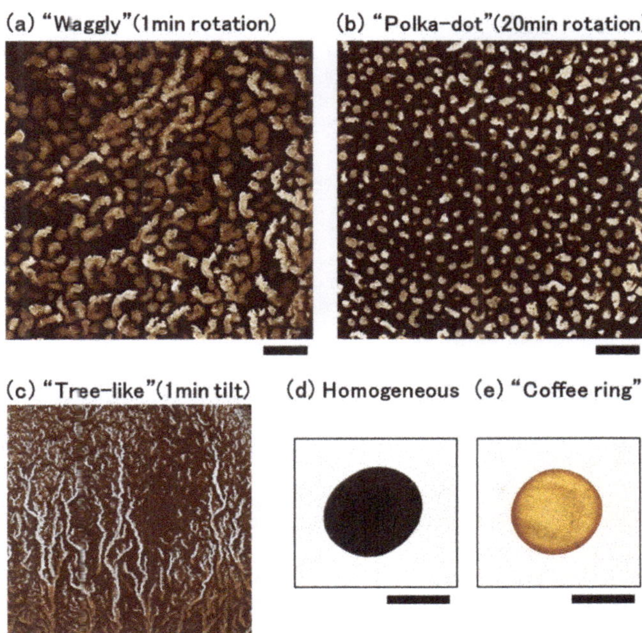

Figure 2. Generation of various characteristic patterns from a drying solution containing coffee powder under different conditions. The scale bars are 10 mm. (**a**) A waggly pattern with many wavy shapes formed when the paper dish was tilted while stationary for 10 s and then rotated for 1 min at a fixed angle of 45° (see Figure 1b). After the rotation, the dish was stood still horizontally for 24 h. The light brown and dark brown parts indicate water-rich and powder-rich regions, respectively. (**b**) Polka-dot pattern with many mini-circles generated from the coffee solution, with tilting without rotation for 10 s and then rotation for 20 min. After the rotation, the dish was stood still horizontally for 24 h. (**c**) Tree-like pattern caused by the downward flow of coffee solution when the plate was tilted at a fixed angle of 45° for 1 min without rotation, the dish was stood still horizontally for 24 h. (**d**) Homogeneous pattern formed by drying the coffee solution containing the powder, i.e., essentially the same solution as in (**a**–**c**). (**e**) Usual so-called coffee-ring formed by drying the filtered coffee solution with almost no grained powder.

Figure 3 exemplifies the segregation pattern generated after 130 s from the start of the segregation in the simulation. Figure 3a shows the appearance of wavy short-fragmented stripes, where the coloring of the segregation pattern is carried out with a threshold value of $\eta = 0.53$. This wavy pattern is familiar for phase segregation with spinodal decomposition [42] and apparently is similar to the waggly pattern observed in the early state (1 min rotation) of the drying process with vessel rotation as in Figure 2a. In contrast, Figure 3b shows the appearance of many mini-circles when the threshold is $\eta = 0.56$, corresponding to the polka-dot pattern observed in the late stage with rotation as in Figure 2b. Here, note that the apparent patterns change markedly depending on the threshold value for the same stage of the phase segregation kinetics. Figure 3c shows the spatial profile of the order parameter for the same region as in Figure 3a,b, revealing the existence of multiple domains with a larger η value along a wavy stripe. The appearance of multiple spots implies the occurrence of mini water-rich spots and such water-rich regions would prefer the formation of round shaped domain owe to the effect of surface tension. Thus, it is expected that such water-rich mini-domains tend to develop circular aqueous droplets during the longer drying process with rotation under tilting. The rate of water evaporation is expected to be faster in the grain-rich region (corresponding to the domain with smaller parameter η) compared to that from the relatively smooth surface of the mini water-rich region, which may induce the formation of mini water droplets with circular shapes by causing the polka-dot pattern.

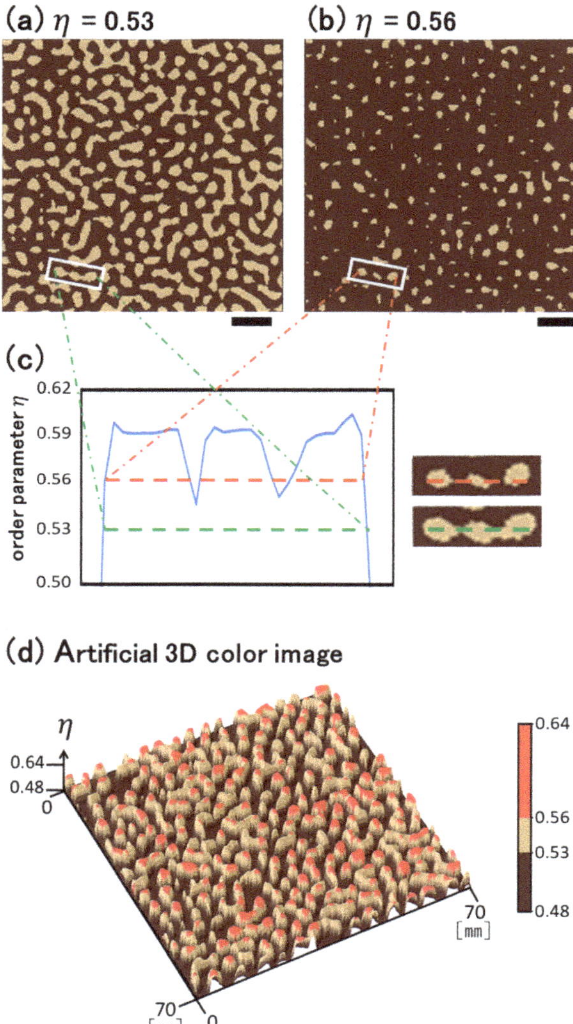

Figure 3. Segregation pattern obtained from numerical simulation for phase-segregation with the simple model equations (Equations (1)–(3)). The scale bars are 10 mm. (**a**,**b**): Spatial patterns with different threshold values of the parameter, η = 0.53 and 0.56, respectively, both of which correspond to the pattern generated after 130 s from the start of segregation. The bright parts in (**b**) show the region that is more water-rich than that in (**a**). (**c**): Order parameter along a section as indicated by a green bar in (**a**) and a red bar in (**b**), which are chosen from the spatial patterns in (**a**,**b**). (**d**): Artificial 3D color image on the same numerical simulation as in (**a**,**b**), revealing the existence of mini water-rich spots on the upper part (larger η value) of the waggly pattern.

4. Conclusions

We have reported the formation of waggly and polka-dot patterns for a drying solution containing fine coffee granules under tilted rotation. The results showed that mm-sized phase-segregation between the powder-rich and water-rich phases occurs for the drying solution with dish-rotation, whereas a homogeneous drying layer is generated without rotation. In relation to our observation, the appearance of various unique patterns from a coffee solution with a large amount of grain powder is known as a "fortune telling" pattern with Turkish coffee [43]. Inspired by such interesting pattern formation, we have performed the present study by introducing the effect of mechanical rotation of the plate. The appearance of the polka-dot pattern implies the realization of a uniform pattern with many mini-circles. It is noted that the time-development from waggly onto polka-dot pattern implies a kind of reverse process of coarse-graining. On the other hand, it is well known that coarsening or Ostwald ripening is the usual scenario in spinodal decomposition. Recently, it has been suggested that assemblies of self-propelled particles can cause reverse Ostwald ripening, i.e., reverse process of coarsening [44]. As similar phenomenon, the formation of spherical domains through the kinetics of spinodal decomposition was observed for a rubber-modified epoxy resin accompanied by a chemical reaction [45–48]. It is also noted that, from theoretical considerations, self-propelled particles are expected to undergo phase-separation [49–51], suggesting the occurrence of reverse process of coarsening during the development of phases separation. Thus, it is expected that the occurrence of the reverse-coarsening is generated under the far-equilibrium conditions through the violation of the fluctuation-dissipation relationship, or caused by the external mechanical agitation. In our experiment, the periodic change of the gravitational field should cause fluctuating translational motion of the segregating domains and such forcing effect may concern with the underlying mechanism on the specific phase-segregation of self-propelled particles under the violation of the fluctuation-dissipation relationship. The results of the present study as in Figure 2 suggest that the formation of many mini-circular pattern from waggly pattern, or reverse Ostwald ripening, can be generated for passive particles under external agitation of the mechanical dish rotation with a tilted state. Here, it is to be noted that, for the transition of the patterns, surface tension should play an important role in the formation of the circular domain as in the Polka-dot pattern through the decrease of the droplet surface area in the water-rich domains. In our 2D model simulation, we have not adapted these important effects in an apparent manner. It is highly expected that our results will stimulate experimental studies to examine the possible appearance of unique drying-induced patterns for solutions under various types of external mechanical agitation and also theoretical studies to clarify the detailed mechanism of the time-development from waggly pattern onto polka-dot pattern.

Author Contributions: Conceptualization and methodology, H.U.; investigation, H.U., M.S., M.O.; writing—original draft preparation, H.U., M.S., M.O.; writing—review and editing, K.S., K.Y.; Supervision, K.Y. All authors have read and agreed to the published version of the manuscript.

Funding: This work was partially supported by JSPS KAKENHI Grant Number JP20H01877 awarded to K.Y.

Data Availability Statement: Data available in a publicly accessible repository.

Acknowledgments: We are grateful to Mana Shimobayashi for her technical support with the experimental observation. H.U. acknowledges the support by Yoshida Scholarship Foundation. We thank S. Hagino for the technical advice concerning Turkish coffee.

Conflicts of Interest: The authors declare no conflict of interest.

References

1. Deegan, R.D.; Bakajin, O.; Dupont, T.F.; Huber, G.; Nagel, S.R. Capillary Flow as the Cause of Ring Stains from Dried Liquid Drops. *Nature* **1997**, *5*, 736. [CrossRef]
2. Carrithers, A.D.; Brown, M.J.; Rashed, M.Z.; Islam, S.; Velev, O.D.; Williams, S.J. Multiscale Self-Assembly of Distinctive Weblike Structures from Evaporated Drops of Dilute American Whiskeys. *ACS Nano* **2020**, *14*, 5417–5425. [CrossRef] [PubMed]
3. Thiele, U.; Todorova, D.V.; Lopez, H. Gradient Dynamics Description for Films of Mixtures and Suspensions: Dewetting Triggered by Coupled Film Height and Concentration Fluctuations. *Phys. Rev. Lett.* **2013**, *111*, 117801. [CrossRef] [PubMed]
4. Park, J.; Moon, J. Control of Colloidal Particle Deposit Patterns within Picoliter Droplets Ejected by Ink-Jet Printing. *Langmuir* **2006**, *22*, 3506–3513. [CrossRef]
5. Kuang, M.; Wang, L.; Song, Y. Controllable Printing Droplets for High-Resolution Patterns. *Adv. Mater.* **2014**, *26*, 6950–6958. [CrossRef]
6. Deegan, R.D.; Bakajin, O.; Dupont, T.F.; Huber, G.; Nagel, S.R.; Witten, T.A. Contact Line Deposits in an Evaporating Drop. *Phys. Rev. E* **2000**, *62*, 756–765. [CrossRef]
7. Hampton, M.A.; Nguyen, T.A.H.; Nguyen, A.V.; Xu, Z.P.; Huang, L.; Rudolph, V. Influence of Surface Orientation on the Organization of Nanoparticles in Drying Nanofluid Droplets. *J. Colloid Interface Sci.* **2012**, *377*, 456–462. [CrossRef] [PubMed]
8. Weon, B.M.; Je, J.H. Fingering inside the Coffee Ring. *Phys. Rev. E* **2013**, *87*, 013003. [CrossRef]
9. Li, Y.; Diddens, C.; Segers, T.; Wijshoff, H.; Versluis, M.; Lohse, D. Evaporating Droplets on Oil-Wetted Surfaces: Suppression of the Coffee-Stain Effect. *Proc. Natl. Acad. Sci. USA* **2020**, *117*, 16756–16763. [CrossRef]
10. Parsa, M.; Harmand, S.; Sefiane, K. Mechanisms of Pattern Formation from Dried Sessile Drops. *Adv. Colloid Interface Sci.* **2018**, *254*, 22–47. [CrossRef]
11. Still, T.; Yunker, P.J.; Yodh, A.G. Surfactant-Induced Marangoni Eddies Alter the Coffee-Rings of Evaporating Colloidal Drops. *Langmuir* **2012**, *28*, 4984–4988. [CrossRef] [PubMed]
12. Hu, G.; Yang, L.; Yang, Z.; Wang, Y.; Jin, X.; Dai, J.; Wu, Q.; Liu, S.; Zhu, X.; Wang, X.; et al. A General Ink Formulation of 2D Crystals for Wafer-Scale Inkjet Printing. *Sci. Adv.* **2020**, *6*, eaba5029. [CrossRef]
13. Parisse, F.; Allain, C. Drying of Colloidal Suspension Droplets: Experimental Study and Profile Renormalization. *Langmuir* **1997**, *13*, 3598–3602. [CrossRef]
14. Sugiyama, Y.; Larsen, R.J.; Kim, J.-W.; Weitz, D.A. Buckling and Crumpling of Drying Droplets of Colloid–Polymer Suspensions. *Langmuir* **2006**, *22*, 6024–6030. [CrossRef] [PubMed]
15. Sen, D.; Melo, J.S.; Bahadur, J.; Mazumder, S.; Bhattacharya, S.; Ghosh, G.; Dutta, D.; D'Souza, S.F. Buckling-Driven Morphological Transformation of Droplets of a Mixed Colloidal Suspension during Evaporation-Induced Self-Assembly by Spray Drying. *Euro. Phys. J. E* **2010**, *31*, 393–402. [CrossRef]
16. Bück, A.; Peglow, M.; Naumann, M.; Tsotsas, E. Population Balance Model for Drying of Droplets Containing Aggregating Nanoparticles. *AIChE J.* **2012**, *58*, 3318–3328. [CrossRef]
17. Goehring, L.; Clegg, W.J.; Routh, A.F. Plasticity and Fracture in Drying Colloidal Films. *Phys. Rev. Lett.* **2013**, *110*, 024301. [CrossRef]
18. Li, W.; Lan, D.; Wang, Y. Dewetting-Mediated Pattern Formation inside the Coffee Ring. *Phys. Rev. E* **2017**, *95*, 042607. [CrossRef]
19. Winkler, A.; Virnau, P.; Binder, K.; Winkler, R.G.; Gompper, G. Hydrodynamic Mechanisms of Spinodal Decomposition in Confined Colloid-Polymer Mixtures: A Multiparticle Collision Dynamics Study. *J. Chem. Phys.* **2013**, *138*, 054901. [CrossRef]
20. Varanakkottu, S.N.; Anyfantakis, M.; Morel, M.; Rudiuk, S.; Baigl, D. Light-Directed Particle Patterning by Evaporative Optical Marangoni Assembly. *Nano Lett.* **2016**, *16*, 644–650. [CrossRef]
21. Joksimovic, R.; Watanabe, S.; Riemer, S.; Gradzielski, M.; Yoshikawa, K. Self-Organized Patterning through the Dynamic Segregation of DNA and Silica Nanoparticles. *Sci. Rep.* **2014**, *4*, 3660. [CrossRef] [PubMed]
22. Mae, K.; Toyama, H.; Nawa-Okita, E.; Yamamoto, D.; Chen, Y.-J.; Yoshikawa, K.; Toshimitsu, F.; Nakashima, N.; Matsuda, K.; Shioi, A. Self-Organized Micro-Spiral of Single-Walled Carbon Nanotubes. *Sci. Rep.* **2017**, *7*, 5267. [CrossRef] [PubMed]
23. Anyfantakis, M.; Varanakkottu, S.N.; Rudiuk, S.; Morel, M.; Baigl, D. Evaporative Optical Marangoni Assembly: Tailoring the Three-Dimensional Morphology of Individual Deposits of Nanoparticles from Sessile Drops. *ACS Appl. Matter Interfaces* **2017**, *9*, 37435–37445. [CrossRef] [PubMed]
24. Yen, T.M.; Fu, X.; Wei, T.; Nayak, R.U.; Shi, Y.; Lo, Y.-H. Reversing Coffee-Ring Effect by Laser-Induced Differential Evaporation. *Sci. Rep.* **2018**, *8*, 3157. [CrossRef] [PubMed]
25. Mouhamad, Y. Dynamics and Phase Separation during Spin Casting of Polymer Films. Ph.D. Thesis, University of Sheffield, Sheffield, UK, 2014.
26. Hu, H.; Larson, R.G. Marangoni Effect Reverses Coffee-Ring Depositions. *J. Phys. Chem. B* **2006**, *110*, 7090–7094. [CrossRef]
27. Mampallil, D.; Reboud, J.; Wilson, R.; Wylie, D.; Klug, D.R.; Cooper, J.M. Acoustic Suppression of the Coffee-Ring Effect. *Soft Matter* **2015**, *11*, 7207–7213. [CrossRef]
28. Kim, S.J.; Kang, K.H.; Lee, J.-G.; Kang, I.S.; Yoon, B.J. Control of Particle-Deposition Pattern in a Sessile Droplet by Using Radial Electroosmotic Flow. *Anal. Chem.* **2006**, *78*, 5192–5197. [CrossRef]
29. Wray, A.W.; Papageorgiou, D.T.; Craster, R.V.; Sefiane, K.; Matar, O.K. Electrostatic Suppression of the "Coffee Stain Effect". *Langmuir* **2014**, *30*, 5849–5858. [CrossRef]
30. Li, Y.; Lv, C.; Li, Z.; Quéré, D.; Zheng, Q. From Coffee Rings to Coffee Eyes. *Soft Matter* **2015**, *11*, 4669–4673. [CrossRef]
31. Anyfantakis, M.; Geng, Z.; Morel, M.; Rudiuk, S.; Baigl, D. Modulation of the Coffee-Ring Effect in Particle/Surfactant Mixtures: The Importance of Particle–Interface Interactions. *Langmuir* **2015**, *31*, 4113–4120. [CrossRef]

32. Chao, Y.; Hung, L.T.; Feng, J.; Yuan, H.; Pan, Y.; Guo, W.; Zhang, Y.; Shum, H.C. Flower-like Droplets Obtained by Self-Emulsification of a Phase-Separating (SEPS) Aqueous Film. *Soft Matter* **2020**, *16*, 6050–6055. [CrossRef] [PubMed]
33. Cahn, J.W.; Hilliard, J.E. Free Energy of a Nonuniform System. I. Interfacial Free Energy. *J. Chem. Phys.* **1958**, *28*, 258–267. [CrossRef]
34. Yamanaka, A.; Aoki, T.; Ogawa, S.; Takaki, T. GPU-Accelerated Phase-Field Simulation of Dendritic Solidification in a Binary Alloy. *J. Cryst. Growth* **2011**, *318*, 40–45. [CrossRef]
35. Lee, D.; Huh, J.-Y.; Jeong, D.; Shin, J.; Yun, A.; Kim, J. Physical, Mathematical, and Numerical Derivations of the Cahn–Hilliard Equation. *Comput. Mat. Sci.* **2014**, *81*, 216–225. [CrossRef]
36. Liu, J.; Dedè, L.; Evans, J.A.; Borden, M.J.; Hughes, T.J.R. Isogeometric Analysis of the Advective Cahn–Hilliard Equation: Spinodal Decomposition under Shear Flow. *J. Comp. Phys.* **2013**, *242*, 321–350. [CrossRef]
37. Alizadeh Pahlavan, A.; Cueto-Felgueroso, L.; Hosoi, A.E.; McKinley, G.H.; Juanes, R. Thin Films in Partial Wetting: Stability, Dewetting and Coarsening. *J. Fluid Mech.* **2018**, *845*, 642–681. [CrossRef]
38. Li, Y.; Pang, Y.; Liu, W.; Wu, X.; Hou, Z. Effect of Diffusivity on the Pseudospinodal Decomposition of the Γ' Phase in a Ni-Al Alloy. *J. Phase Equilib. Diffus.* **2016**, *37*, 261–268. [CrossRef]
39. Tran Duc, V.-N.; Chan, P.K. Using the Cahn–Hilliard Theory in Metastable Binary Solutions. *ChemEngineering* **2019**, *3*, 75. [CrossRef]
40. Holz, M.; Heil, S.R.; Sacco, A. Temperature-Dependent Self-Diffusion Coefficients of Water and Six Selected Molecular Liquids for Calibration in Accurate 1H NMR PFG Measurements. *Phys. Chem. Chem. Phys.* **2000**, *2*, 4740–4742. [CrossRef]
41. Two-Dimensional Phase-Field Model for Conserved Order Parameter (Cahn-Hilliard Equation). Available online: http://web.tuat.ac.jp/~{}yamanaka/pcoms2019/Cahn-Hilliard-2d.html (accessed on 31 October 2020).
42. Elder, K.; Rogers, T.; Desai, R. Early Stages of Spinodal Decomposition for the Cahn-Hilliard-Cook Model of Phase Separation. *Phy. Rev. B* **1988**, *38*, 4725–4739. [CrossRef]
43. Nicolaou, L. *The Art of Coffee Cup Reading*; Zeus Publications: Mermaid Waters, Australia, 2015.
44. Tjhung, E.; Nardini, C.; Cates, M.E. Cluster Phases and Bubbly Phase Separation in Active Fluids: Reversal of the Ostwald Process. *Phys. Rev. X* **2018**, *8*, 031080. [CrossRef]
45. Yamanaka, K.; Takagi, Y.; Inoue, T. Reaction-Induced Phase Separation in Rubber-Modified Epoxy Resins. *Polymer* **1989**, *30*, 1839–1844. [CrossRef]
46. Wang, W.; Liu, Q.-X.; Jin, Z. Spatiotemporal Complexity of a Ratio-Dependent Predator-Prey System. *Phys. Rev. E* **2007**, *75*, 051913. [CrossRef] [PubMed]
47. Das, P.; Jaiswal, P.K.; Puri, S. Surface-Directed Spinodal Decomposition on Chemically Patterned Substrates. *Phys. Rev. E* **2020**, *102*, 012803. [CrossRef] [PubMed]
48. Posada, E.; López-Salas, N.; Carriazo, D.; Muñoz-Márquez, M.A.; Ania, C.O.; Jiménez-Riobóo, R.J.; Gutiérrez, M.C.; Ferrer, M.L.; del Monte, F. Predicting the Suitability of Aqueous Solutions of Deep Eutectic Solvents for Preparation of Co-Continuous Porous Carbons via Spinodal Decomposition Processes. *Carbon* **2017**, *123*, 536–547. [CrossRef]
49. Stenhammar, J.; Tiribocchi, A.; Allen, R.J.; Marenduzzo, D.; Cates, M.E. Continuum Theory of Phase Separation Kinetics for Active Brownian Particles. *Phys. Rev. Lett.* **2013**, *111*, 145702. [CrossRef]
50. Stenhammar, J.; Marenduzzo, D.; Allen, R.J.; Cates, M.E. Phase Behaviour of Active Brownian Particles: The Role of Dimensionality. *Soft Matter* **2014**, *10*, 1489–1499. [CrossRef]
51. Cates, M.E.; Tailleur, J. Motility-Induced Phase Separation. *Annu. Rev. Cond. Matter Phys.* **2015**, *6*, 219–244. [CrossRef]

Article

Design of Switchable *On/Off* Subpixels for Primary Color Generation Based on Molybdenum Oxide Gratings

Gonzalo Santos [1], Francisco González [1], Dolores Ortiz [1], José María Saiz [1], Maria Losurdo [2], Yael Gutiérrez [2] and Fernando Moreno [1,*]

1 Department of Applied Physics, Universidad de Cantabria, Avda. Los Castros s/n, 39005 Santander, Spain; gonzalo.santos@unican.es (G.S.); gonzaleff@unican.es (F.G.); ortizd@unican.es (D.O.); saizvj@unican.es (J.M.S.)

2 Institute of Nanotechnology, CNR-NANOTEC, Via Orabona 4, 70126 Bari, Italy; losurdo.maria@gmail.com (M.L.); yael.gutierrezvela@nanotec.cnr.it (Y.G.)

* Correspondence: morenof@unican.es

Abstract: Structural color emerges from the interaction of light with structured matter when its dimension is comparable to the incident wavelength. The reflected color can be switched by controlling such interaction with materials whose properties can be changed through external stimuli such as electrical, optical, or thermal excitation. In this research, a molybdenum oxide (MoO_x) reflective grating to get a switchable *on/off* subpixel is designed and analyzed. The design is based on subpixel *on* and *off* states that could be controlled through the oxidation degree of MoO_x. A suitable combination of three of these subpixels, optimized to get a control of primary colors, red, green, and blue, can lead to a pixel which can cover a wide range of colors in the color space for reflective display applications.

Keywords: color reflective displays; phase-change materials; structural color

Citation: Santos, G.; González, F.; Ortiz, D.; Saiz, J.M.; Losurdo, M.; Gutiérrez, Y.; Moreno, F. Design of Switchable *On/Off* Subpixels for Primary Color Generation Based on Molybdenum Oxide Gratings. *Physics* **2021**, 3, 655–663. https://doi.org/10.3390/physics3030038

Received: 2 June 2021
Accepted: 9 August 2021
Published: 12 August 2021

1. Introduction

For centuries, color has been a quite interesting topic for the scientific community [1,2]. The first systematic study was made by Newton when he performed his classical experiment, i.e., analysis and synthesis of light with a glass prism. Newton stated that the spectrum was constituted by seven colors: red, orange, yellow, green, blue, indigo, and violet. However, most colors in nature are not spectrally pure or able to fit in a small region of the spectrum, since they are often a result of a combination of phenomena. The color of a radiation depends not only on the reflectance—or transmittance—of the last object it went through, but also on the kind of illuminant and the photopic curve of the observer. Although a wide range of magnitudes can be used to characterize color, the best attributes according to the International Commission on Illumination (CIE) are chromaticity, brightness, and contrast [3].

Structural color is one of the most common manifestations of color in nature [4]. It is based on the selective light reflection depending on the interaction between light and structured matter (typically at nano- and microscale). This is the main pigmentary difference in which color is originated by the absorption of the electrons present in the pigment [5].

The most common mechanisms for obtaining structural colors are based on interference, diffraction, scattering, and photonic crystals [6]. Film interference can be considered as a typical Fabry–Perot (FP) effect. In such systems, light undergoes multiple reflections. When the optical path difference between two reflected rays is an odd multiple of a half of the incident wavelength, constructive interference takes place in specific spectral intervals and vivid colors can be generated. Many reflective displays are based on this phenomenon [7–11]. A typical configuration is a metal–insulator–metal (MIM) stack. A thin metal layer is used to control the spectral width of the reflectance peaks, a dielectric spacer tunes the resonant wavelength in the cavity, and a second metal layer works as a bottom mirror. However, if one looks for monochromatic intervals, there is an important drawback; these types of FP configurations generate reflectance resonances with a broad spectral width, and more than one resonance (second-order or even higher) is often generated, leading to undesirable resonance peaks in the visible range. This can be overcome by using diffraction gratings. They allow to generate narrow and isolated peaks in the visible part of the electromagnetic spectrum.

The diffraction effect by a grating is comparable to the multiple interference generated in FP cavities. The grating efficiency is determined by the period, functional shape of the profile, depth of the periodic structure, materials, angle of incidence and observation direction, wavelength, and polarization. Wood anomalies are known concept in the diffraction by periodic gratings [12]. The adjective "anomalies" is because, when discovered by R. Wood in 1902 [13], there was no clear explanation for the observation of narrow reflectance peaks in diffraction gratings. For metals, this is now explained because of surface plasmon excitation [12]. For dielectric gratings, the effect is due to the coupling of the propagating diffracted rays to the modes of the waveguide underneath, leading to guided mode resonances (GMR) [14,15]. The spectral width of a GMR is usually quite narrow; thus, as later shown, this is a favorable feature in terms of obtaining a wide color gamut. Although diffraction gratings have been proposed previously for their use on reflective displays [16–20], most of the gratings cannot control the reflectance for a fixed geometry. In this situation, only "static" colors are generated, which limits their potential as actively tunable color devices. A more recent study proposed a dielectric grating based on ITO to obtain an active color display by changing its permittivity with electrically tunable electron densities [21]. An alternative solution to the one proposed in [21] is the use of phase-change materials (PCM).

The most extended PCMs are chalcogenide materials. These are compounds of elements of the chalcogen group (sulfur, selenium, and tellurium) bound to network formers such as arsenic, germanium, antimony, and gallium [22]. They are known as the GST family due to the chalcogenide $Ge_2Sb_2Te_5$, which has revolutionized the blooming field of phase-change photonics. The peculiarity of these materials is that they can be switched between their crystalline and amorphous phases by controlled electrical, optical, or thermal excitation [23,24]. This process leads to a modulation of their optical and electrical properties. Another known PCM is vanadium dioxide (VO_2). VO_2 is a strongly correlated material that is dielectric at room temperature and becomes metallic in the infrared spectrum when heated at around 340 K [25–27]. On the other hand, molybdenum oxide (MoO_x) presents a metal–semiconductor transition in the visible (Vis) spectrum, by changing its oxidation degree from MoO_2 (metal) to MoO_3 (semiconductor), which makes it suitable for applications in this range such as reflective displays among others [28]. Very recently, a switchable pixel based on an FP configuration with MoO_x for color reflective displays was presented [29].

In this research, the design of a dielectric grating based on molybdenum oxide is proposed as a switchable *on/off* pixel for a color display. MoO_x can be considered a nonvolatile phase-change material, i.e., it does not require a constant energy supply of energy to keep the switched state. It can exist as MoO_2 or MoO_3 and has a wide variety of nonstoichiometric oxides. The change in the oxygen content strongly affects the band structure and, consequently, its optical behavior. In MoO_3, O *2p* orbitals give rise to the highest occupied states, wherein electrons are fully localized around the O atoms, giving a semiconducting behavior. However, the Fermi level of MoO_2 is composed of O *3d* orbitals that present the characteristics of a metal [30]. Therefore, by changing the oxidation state from MoO_3 to MoO_2 a semiconductor–conductor–metal transition is triggered, allowing a modulation of the light–matter interaction. Interestingly, the literature shows that the intervalence charge-transfer modulation within diverse valence states of Mo, going from Mo^{6+} (MoO_3) to Mo^{5+} and finally to Mo^{4+} (MoO_2), can occur thermally by annealing at

400 °C in air (oxidation of MoO_2 to Mo_3) [31,32] or in a reduction environment (reduction from MoO_3 to MoO_2), e.g., using a gas such as hydrogen or propane (which is a source of hydrogen), [32,33] involving the thermal-activated adsorption/desorption of oxygen.

As for oxidation [31,32], the feasibility of oxidation even from MoS_2 to MoO_2 and MoO_3 in air has also been shown by green laser irradiation on a millisecond time scale [34].

This annealing and change of oxidation state can be operated starting from deposited MoO_3 or MoO_2, which can be obtained (i) by controlling the stoichiometry during the growth by the oxygen partial pressure, or (ii) by post-growth processing which includes ion bombardment, which results in the preferential loss of bridging oxygen atoms and oxygen plasmas.

Noteworthy, this innovative way of modulating the oxygen content in oxides has been recently reported for the phase-change material VO_2 [35].

Although work is in progress for the practical implementation of this approach to MoO_x, the oxidation/hydrogenation approach has already been implemented in dynamic color devices [36], even using another material (magnesium, Mg), moving it between the two states of oxidation to MgO and hydrogenation to MgH_2.

According to this concept of oxidation/reduction applied to MoO_x, MoO_3 acts as a lossless transparent dielectric in the visible range (E_{gap} = 3 eV depending on the crystallinity), allowing the coupling of narrow modes of the GMR kind and, therefore, opening the possibility of getting vivid colors in the reflectance. On the contrary, MoO_2 absorbs the visible spectrum range and Wood anomalies cannot be generated, reflecting a pale color.

For an accurate description of the proposed color display, this paper is divided into various sections. Section 2 is devoted to describing the device design, Section 3 contains details about the numerical simulation method, Section 4 develops the working principle and Sections 5 and 6 contain, respectively, the main results and conclusions of this research.

2. Pixel Model

In this research, the reflective display pixel constituted three subpixels, each one associated to a primary color, red (R), green (G), and blue (B). Their reflection properties can be controlled through the optical properties of the MoO_x material according to what has been described previously. Each subpixel is a reflective diffraction grating based on periodic MoO_x ribs over a silicon dioxide (SiO_2) substrate, as shown in Figure 1 (top). A microheater [37] can be located under the substrate to control the annealing process under an oxygen or hydrogen atmosphere.

As MoO_x is the active tunable material of the reflective display, *on* and *off* states can be generated and controlled depending on its stoichiometry. In this work, two measurements of oxygen contents were considered: x = 2.9 (*on*) and x = 2.1 (*off*). The refractive index of molybdenum oxide for both oxygen contents is also shown in Figure 1 (bottom left and bottom right, respectively). The imaginary part of the refractive index, k, is almost zero in the *on* mode and greater than one in the *off* state. In the *on* state, light can travel through it, allowing the generation of a GMR (see Section 4 for more details), which in turn produces vivid colors. However, in the *off* state, most of the light is absorbed. In this case, resonances are not produced, and a pale unsaturated color is reflected. These optical constants for both oxygen contents were obtained from the literature [38].

In general, the main parameters for the characterization of a diffraction grating are the duty cycle, D, the height, d, of the ribs, the period, P, the polarization (perpendicular to rib direction, p-polarization), and the incident angle, θ. The duty cycle D can be considered as the ratio between the width of the rib, w, and the period P. The first three parameters can be varied for the optimization of the device. The optimization is based on the achievement of the best primary colors, R, G, and B. Each primary color corresponds to a different grating (subpixel) with different parameters. Their suitable combination gives rise to the desired color.

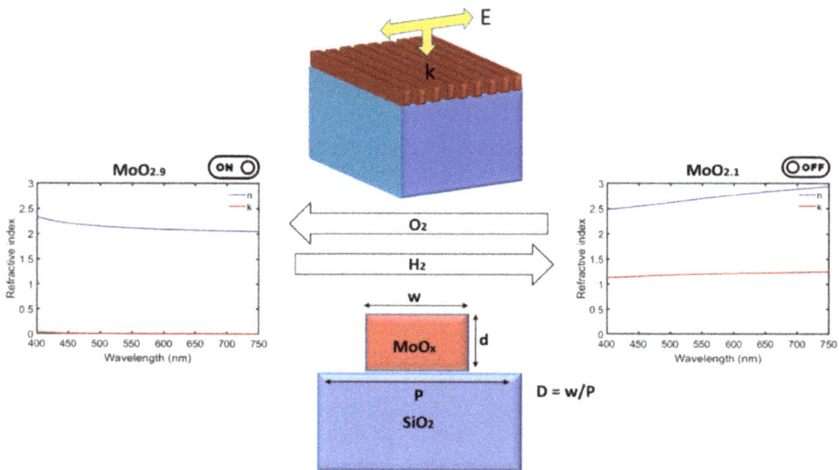

Figure 1. Scheme of the reflected display unit (subpixel associated with a primary color). This design is based on a periodic structure of ribs made of molybdenum oxide (MoO_x) over a SiO_2 substrate. Depending on the amount of oxygen content, MoO_x is either a transparent medium to visible radiation (*on* state) or an absorbing one (*off* state) (see the extinction coefficient *k* of the refractive index [38]). The transition between both states can be triggered by controlling the amount of H_2 or O_2 (see [35,36]).

All results were simulated considering normal incidence and light polarization perpendicular to the rib direction. For simplicity, this polarization was chosen in order to ignore the length of the ribs, allowing the study of the system under a 2D geometrical configuration without losing physical information.

3. Numerical Simulation Method

Spectral reflectivity was calculated using finite-difference time-domain (FDTD) simulation (LUMERICAL). This is a numerical analysis technique used for modeling computational electrodynamics (finding approximate solutions to a system of coupled differential equations by time discretization). Nonuniform mesh settings were used in these simulations, and the source used was always a plane wave. Periodicity boundary conditions were used to simulate an infinite number of MoO_x ribs over a silicon dioxide substrate.

All color simulations within this work assume a standard D65 illuminant, corresponding to average daylight, and a CIE standard observer (2°), representing mean human spectral sensitivity to visible spectrum range under 2° field observation [39,40]. As sunlight is not polarized and this study was performed for *p*-polarization, some polarizing element should be used in the actual device for the accurate generation of the colors described in this work.

The most common color space to characterize the color generated by reflective displays, as cited in the introduction, is CIE1931. In this space, color is defined by its tristimulus values (x, y, z) in the chromaticity diagram. In this diagram, it is possible to compare the simulated red, green, and blue colors generated by the device with the standard ones (sRGB).

4. Working Principle: Wood Anomalies and the Guided Mode Resonance

The color reflected by the designed subpixel device in the *on* mode is the result of the excitation of guided waves by the grating, also known as Wood anomalies [14]. This high-reflectance phenomenon is based on coupling light propagating in free space to the grating, leading to GMRs [41]. To better understand this phenomenon, spectral reflectivity and the electric and magnetic fields in the near-field regime of the proposed device were analyzed for a given subpixel case. To simulate an example, diffraction grating parameters were fixed to d = 150 nm, P = 340 nm, and D = 0.6 (normal incidence and p-polarization were also assumed).

In Figure 2a,b, the grating spectral reflectivity is shown for the *on* and *off* modes, respectively. In the *on* mode, a narrow reflectance peak appears at a wavelength around 517 nm. There is a single resonance because a mode is excited and guided by the subwavelength grating. On the contrary, the spectral reflectivity in the *off* state is a flat curve due to the absorbance predominance of $MoO_{2.1}$, making impossible the generation of GMR.

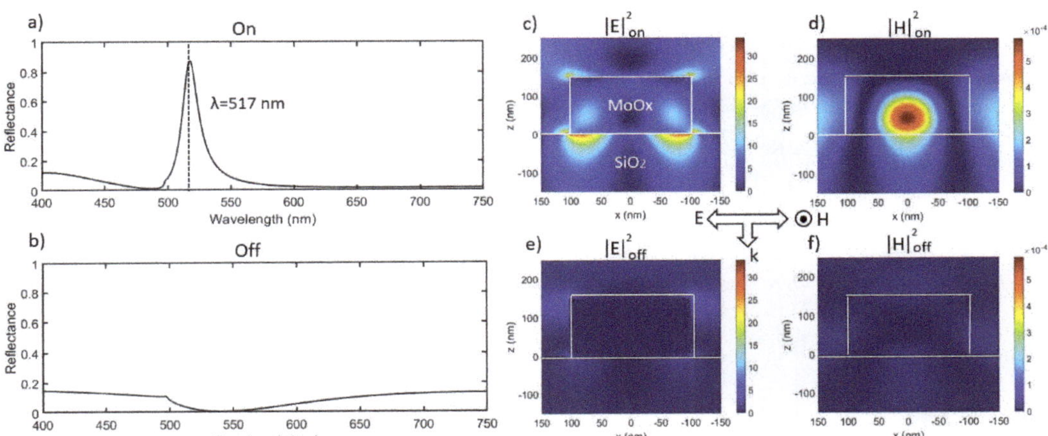

Figure 2. Reflectance of the proposed subpixel for the grating parameters d = 150 nm, p = 340 nm, and D = 0.6 for the (**a**) *on* and (**b**) *off* states, respectively. For the transparent version of the MoO_x ribs, the reflection of the grating peaks occurs at λ = 517 nm. Square module of the electric field in near-field regime when the incident wavelength is λ = 517 nm (spectral position of the reflectance maximum) for the (**c**) *on* and (**e**) *off* states, respectively. Square modulus of the magnetic field in the near-field regime when the incident wavelength is λ = 517 nm for the (**d**) *on* and (**f**) *off* states, respectively. The electric field is orthogonal to the grating ribs and the magnetic field is parallel.

The modules of the electric and magnetic fields for the *on* state and λ = 517 nm are represented in Figure 2c,d, respectively, where guided mode resonances can clearly be observed. Such resonances come from a coupling between nonhomogeneous diffraction orders and the eigenmodes of the grating (Wood anomalies) [41]. As a result, the electric dipole resonance is produced at the MoO_x–SiO_2 interface (the two hotspots in Figure 2c). The coupling of all the dipoles produced in each rib is attributed to the narrow and high reflectance peak. Depending on the number of interacting ribs (i.e., number of interacting dipoles), the electromagnetic response is different. A higher number of ribs leads to more efficient coupling. To assess the importance of this issue, the same simulation shown in Figure 2a was performed but now considering a finite number of ribs (Figure 3a). As this number was increased, the reflectance was more similar to that simulated by periodic boundary conditions. Figure 3b shows that a reflectance stationary regime was reached from a number of approximately 150 ribs.

Analogous simulations are presented in Figure 2e,f for the *off* state. No modes were excited due to the high value of the extinction coefficient k of $MoO_{2.1}$.

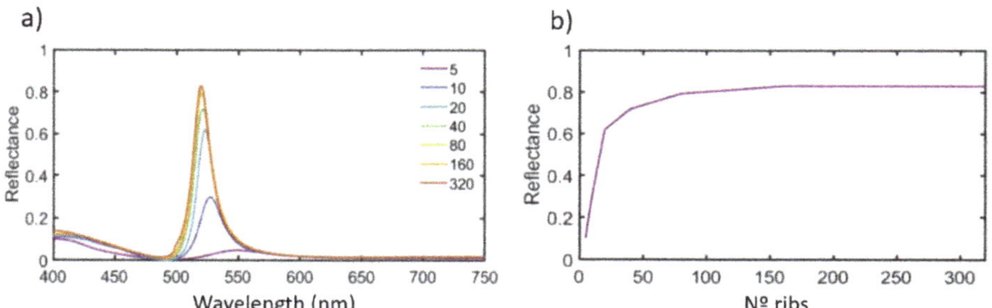

Figure 3. (**a**) Spectral reflectivity of the proposed device fixing grating parameters to d = 150 nm, P = 340 nm, and D = 0.6 (*on* state) for different numbers of ribs. (**b**) Maximum reflectance of the resonances generated by changing the number of ribs.

5. Results

The results are based on the optimization of each reflective grating subpixel to generate the best primary colors. For normal incidence and p-polarization, the way to control the GMR is by varying the duty cycle, the period, or the rib height. Although color does not only depend on spectral reflectivity, for RGB optimization, those magnitudes (D, P, d) were analyzed to generate resonances at 460 nm (blue), 520 nm (green), and 620 nm (red). Moreover, a good contrast between the *on* and *off* colors should be addressed. The grating subpixel parameters used for color primary generation are shown in Table 1.

Table 1. Grating subpixel parameters (height d, period P, and duty cycle D) for the generation of red, green, and blue colors.

	R Subpixel	G Subpixel	B Subpixel
d (nm)	230	125	150
P (nm)	400	340	280
D	0.55	0.8	0.6

The corresponding spectral reflectivities are shown in Figure 4a,b for the *on* and the *off* states, respectively. The height, the bandwidth, and the spectral position of those resonances delimited the quality of the generated color. For small bandwidths, monochromatic colors were obtained. However, a lower bandwidth led to less light being reflected. Therefore, the luminosity of the color was too low, generating a very dark color. For this reason, an equilibrium should be required. The resulting colors of these resonances are represented in Figure 4c,d for the *on* and the *off* states respectively, and both are plotted in CIE1931 in Figure 4e. *On* colors (white points) are close to standard RGB coordinates (triangle vertices). However, *off* colors (black points) are far from those vertices and close to each other, revealing a pale and similar color. A large color gamut can be obtained through this system.

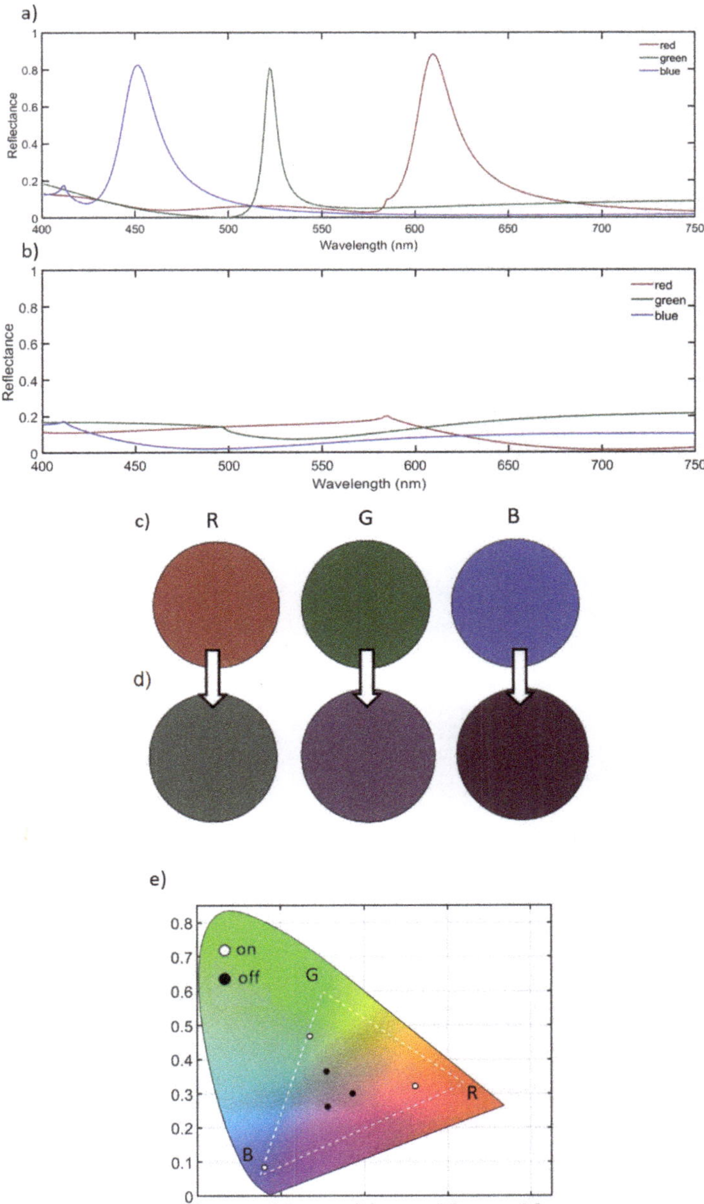

Figure 4. Spectral reflectance of the designed subpixel devices by considering the grating parameters shown in Table 1 for the *on* (**a**) and *off* (**b**) states, respectively. *On* (**c**) and *off* (**d**) colors generated by the reflective display pixel, respectively. (**e**) Representation of those colors in CIE1931 space. The triangle represents RGB standard coordinates, white points (*on* simulated colors), and black points (*off* simulated colors).

6. Conclusions

In this paper, a switchable *on/off* color reflective pixel model based on resonant effects by reflective subwavelength diffraction gratings (subpixel) was designed. The pixel model consisted of a suitable combination of three of those subpixels, each one optimized to generate a primary color, red (R), green (G), or blue (B). The excited resonances in each subpixel can be considered as the collective effect of the electric dipole modes generated in each grating rib for resonant wavelengths. Consequently, the generated color can be tuned spectrally by changing the height, period, and duty cycle of the grating of each subpixel. *On/off* modes can be generated in each subpixel due to the change in oxidation state of molybdenum oxide (MoO_x) from MoO_2 to MoO_3. This leads to the generation of a wide gamut of colors close to standard sRGB ones for the *on* mode and a pale color for the respective *off* state. Compared to other reflective devices based on the Fabry–Perot phenomenon, narrower reflectance peaks can be obtained due to the characteristics of the excited grating resonances, which correspond to the Wood anomalies. This allows the reflection of monochromatic colors and the generation of a large color gamut for applications in color reflective displays.

Author Contributions: Conceptualization and methodology, G.S., F.M. and Y.G.; software, G.S and Y.G.; formal analysis, all.; resources, M.L. and F.M.; writing—original draft preparation, G.S. writing—review and editing, G.S., D.O., J.M.S. and F.M.; supervision, F.G., M.L. and F.M. All authors have read and agreed to the published version of the manuscript.

Funding: The authors received funding from the European Union's Horizon 2020 research and innovation program under grant agreement No. 899598—PHEMTRONICS.

Conflicts of Interest: The authors declare no conflict of interest.

References

1. Shevell, S.K. (Ed.) *The Science of Color*; Optical Society of America/Elsevier Science: Oxford, UK, 2003. Available online https://www.sciencedirect.com/book/9780444512512/the-science-of-color (accessed on 9 August 2021).
2. Crone, R.A. *A History of Color: The Evolution of Theories of Light and Color*; Springer Science & Business Media: Dordrecht, Germany 1999. [CrossRef]
3. Schanda, J. (Ed.) *Colorimetry: Understanding the CIE System*; John Wiley & Sons, Inc.: Hoboken, NJ, USA, 2007. [CrossRef]
4. Sun, J.; Bhushan, B.; Tong, J. Structural coloration in nature. *RSC Adv.* **2013**, *3*, 14862–14889. [CrossRef]
5. Gürses, A.; Açıkyıldız, M.; Güneş, K.; Gürses, M.S. *Dyes and Pigments: Their Structure and Properties*; Springer: Cham, Switzerland, 2016; pp. 13–29. [CrossRef]
6. Kinoshita, S.; Yoshioka, S. Structural colors in nature: The role of regularity and irregularity in the structure. *ChemPhysChem* **2005**, *6*, 1442–1459. [CrossRef] [PubMed]
7. Sung, C.; Han, J.; Song, J.; Ah, C.S.; Cho, S.M.; Kim, T.Y. Reflective-type transparent/colored mirror switchable device using reversible electrodeposition with Fabry–Perot interferometer. *Adv. Mater. Technol.* **2020**, *5*, 2000367. [CrossRef]
8. Lin, Z.; Long, Y.; Zhu, X.; Dai, P.; Liu, F.; Zheng, M.; Zhou, Y.; Duan, H. Extending the color of ultra-thin gold films to blue region via Fabry-Pérot-Cavity-Resonance-Enhanced reflection. *Optik* **2019**, *178*, 992–998. [CrossRef]
9. Cho, S.M.; Cheon, S.H.; Kim, T.Y.; Ah, C.S.; Song, J.; Ryu, H.; Chu, H.Y. Design and fabrication of integrated Fabry-Perot type color reflector for reflective displays. *J. Nanosci. Nanotechnol.* **2016**, *16*, 5038–5043. [CrossRef]
10. Yang, Z.; Zhou, Y.; Chen, Y.; Wang, Y.; Dai, P.; Zhang, Z.; Duan, H. Reflective color filters and monolithic color printing based on asymmetric Fabry–Perot cavities using nickel as a broadband absorber. *Adv. Opt. Mater.* **2016**, *4*, 1196–1202. [CrossRef]
11. Zhao, J.; Qiu, M.; Yu, X.; Yang, X.; Jin, W.; Lei, D.; Yu, Y. Defining deep-subwavelength-resolution, wide-color-gamut, and large-viewing-angle flexible subtractive colors with an ultrathin asymmetric Fabry–Perot lossy cavity. *Adv. Opt. Mater.* **2019**, *7*, 1900646. [CrossRef]
12. Maystre, D. Theory of Wood's Anomalies. In *Plasmonics*; Enoch, S., Bonod, N., Eds.; Springer: Berlin/Heidelberg, Germany, 2012; pp. 39–83. [CrossRef]
13. Wood, R.W. On a remarkable case of uneven distribution of light in a diffraction grating spectrum. *Proc. Phys. Soc. Lond.* **1901**, *18*, 269–275. [CrossRef]
14. Wang, S.S.; Magnusson, R.J.A.O. Theory and applications of guided-mode resonance filters. *Appl. Opt.* **1993**, *32*, 2606–2613. [CrossRef]
15. Hessel, A.; Oliner, A.A. A new theory of Wood's anomalies on optical gratings. *Appl. Opt.* **1965**, *4*, 1275–1297. [CrossRef]
16. Chen, Y.; Liu, W. Design and analysis of multilayered structures with metal–dielectric gratings for reflection resonance and color generation. *Opt. Lett.* **2012**, *37*, 4–6. [CrossRef] [PubMed]

17. Lee, H.S.; Yoon, Y.T.; Lee, S.S.; Kim, S.H.; Lee, K.D. Color filter based on a subwavelength patterned metal grating. *Opt. Express* **2007**, *15*, 15457–15463. [CrossRef] [PubMed]
18. Cheong, B.-H.; Prudnikov, O.; Cho, E.; Kim, H.-S.; Yu, J.; Cho, Y.-S.; Choi, H.-Y.; Shin, S.T. High angular tolerant color filter using subwavelength grating. *Appl. Phys. Lett.* **2009**, *94*, 213104. [CrossRef]
19. Uddin, M.J.; Magnusson, R. Efficient guided-mode-resonant tunable color filters. *IEEE Photonics Technol. Lett.* **2012**, *24*, 1552–1554. [CrossRef]
20. Uddin, M.J.; Magnusson, R. Guided-mode resonant color filter array for reflective displays. In Proceedings of the 2013 IEEE Photonics Conference, Bellevue, WA, USA, 8–12 September 2013; IEEE: Piscataway, NJ, USA, 2013; pp. 28–29. [CrossRef]
21. Wang, W.; Guan, Z.; Xu, H. A high speed electrically switching reflective structural color display with large color gamut. *Nanoscale* **2021**, *13*, 1164–1171. [CrossRef]
22. Wuttig, M.; Bhaskaran, H.; Taubner, T. Phase-change materials for non-volatile photonic applications. *Nat. Photonics* **2017**, *11*, 465–476. [CrossRef]
23. Vassalini, I.; Alessandri, I.; de Ceglia, D. Stimuli-responsive phase change materials: Optical and optoelectronic applications. *Materials* **2021**, *14*, 3396. [CrossRef] [PubMed]
24. Gong, Z.; Yang, F.; Wang, L.; Chen, R.; Wu, J.; Grigoropoulos, C.P.; Yao, J. Phase change materials in photonic devices. *J. Appl. Phys.* **2021**, *129*, 030902. [CrossRef]
25. Zylbersztejn, A.; Mott, N.F. Metal-insulator transition in vanadium dioxide. *Phys. Rev. B* **1975**, *11*, 4383. [CrossRef]
26. Mott, N.F. Metal-insulator transition. *Rev. Mod. Phys.* **1968**, *40*, 677. [CrossRef]
27. Qazilbash, M.M.; Brehm, M.; Chae, B.-G.; Ho, P.-C.; Andreev, G.O.; Kim, B.-J.; Yun, S.J.; Balatsky, A.V.; Maple, M.B.; Keilmann, F.; et al. Mott transition in VO_2 revealed by infrared spectroscopy and nano-imaging. *Science* **2007**, *318*, 1750–1753. [CrossRef]
28. Zhang, Q.; Li, X.; Ma, Q.; Zhang, Q.; Bai, H.; Yi, W.; Liu, J.; Han, J.; Xi, G. A metallic molybdenum dioxide with high stability for surface enhanced Raman spectroscopy. *Nat. Commun.* **2017**, *8*, 14903. [CrossRef]
29. Santos, G.; González, F.; Ortiz, D.; Saiz, J.M.; Losurdo, M.; Moreno, F.; Gutiérrez, Y. Dynamic reflective color pixels based on molybdenum oxide. *Opt. Express* **2021**, *29*, 19417–19426. [CrossRef]
30. Imada, M.; Fujimori, A.; Tokura, Y. Metal-insulator transitions. *Rev. Mod. Phys.* **1998**, *70*, 1039. [CrossRef]
31. Camacho-López, M.A.; Escobar-Alarcón, L.; Picquart, M.; Arroyo, R.; Córdoba, G.; Haro-Poniatowski, E. Micro-Raman study of the m-MoO_2 to α-MoO_3 transformation induced by cw-laser irradiation. *Opt. Mater.* **2011**, *33*, 480–484. [CrossRef]
32. Ressler, T.; Wienold, J.; Jentoft, R.E.; Neisius, T. Bulk structural investigation of the reduction of MoO_3 with propene and the oxidation of MoO_2 with oxygen. *J. Catal.* **2002**, *210*, 67–83. [CrossRef]
33. Dang, J.; Zhang, G.-H.; Chou, K.-C.; Reddy, R.G.; He, Y.; Sun, Y. Kinetics and mechanism of hydrogen reduction of MoO_3 to MoO_2. *Int. J. Refract. Met. Hard Mater.* **2013**, *41*, 216–223. [CrossRef]
34. Austin, D.; Gliebe, K.; Muratore, C.; Boyer, B.; Fisher, T.S.; Beagle, L.K.; Benton, A.; Look, P.; Moore, D.; Ringe, E.; et al. Laser writing of electronic circuitry in thin film molybdenum disulfide: A transformative manufacturing approach. *Mater. Today* **2021**, *43*, 17–26. [CrossRef]
35. Duan, X.; White, S.T.; Cui, Y.; Neubrech, F.; Gao, Y.; Haglund, R.F.; Liu, N. Reconfigurable multistate optical systems enabled by VO_2 phase transitions. *ACS Photonics* **2020**, *7*, 2958–2965. [CrossRef]
36. Duan, X.; Kamin, S.; Liu, N. Dynamic plasmonic colour display. *Nat. Commun.* **2017**, *8*, 14606. [CrossRef] [PubMed]
37. Broughton, B.; Bandhu, L.; Talagrand, C.; Garcia-Castillo, S.; Yang, M.; Bhaskaran, H.; Hosseini, P. 38-4: Solid-state reflective displays (SRD®) utilizing ultrathin phase-change materials. *SID Symp. Digest Techn. Papers* **2017**, *48*, 546–549. [CrossRef]
38. Smith, G.B.; Golestan, D.; Gentle, A.R. The insulator to correlated metal phase transition in molybdenum oxides. *Appl. Phys. Lett.* **2013**, *103*, 051119. [CrossRef]
39. Capilla, P.; Pujol, J. *Fundamentos de Colorimetría*; Universitat de València: Valencia, Spain, 2002.
40. Malacara, D. *Color Vision and Colorimetry: Theory and Applications*; SPIE: Bellingham, WA, USA, 2011. [CrossRef]
41. Sarrazin, M.; Vigneron, J.-P.; Vigoureux, J.-M. Role of Wood anomalies in optical properties of thin metallic films with a bidimensional array of subwavelength holes. *Phys. Rev. B* **2003**, *67*, 085415. [CrossRef]

 physics

MDPI

Article

Scaling Conjecture Regarding the Number of Unknots among Polygons of $N \gg 1$ Edges

Alexander Y. Grosberg

Center for Soft Matter Research, Department of Physics, New York University, 726 Broadway, New York, NY 10003, USA; aygl@nyu.edu

Abstract: The conjecture is made based on a plausible, but not rigorous argument, suggesting that the unknot probability for a randomly generated self-avoiding polygon of $N \gg 1$ edges has only logarithmic, and not power law corrections to the known leading exponential law: $P_{\text{unknot}}(N) \sim \exp[-N/N_0 + o(\ln N)]$ with N_0 being referred to as the random knotting length. This conjecture is consistent with the numerical result of 2010 by Baiesi, Orlandini, and Stella.

Keywords: polymers; knots; unknot probability

Citation: Grosberg, A.Y. Scaling Conjecture Regarding the Number of Unknots among Polygons of $N \gg 1$ Edges. *Physics* **2021**, 3, 664–668. https://doi.org/10.3390/physics3030039

Received: 22 June 2021
Accepted: 30 July 2021
Published: 12 August 2021

1. Introduction and Problem Formulation

Randomly generated self-avoiding polygons represent an interesting object for mathematical physics, for several reasons. First, such polygons can serve as a zeroth approximation model for ring polymers. Different realizations, or members of the ensemble, of random polygons mimic different spatial arrangements of polymers, sampled via thermal fluctuations; importantly, ring polymers are currently the subject a great deal of interest, as evidenced, for example, by recent papers [1–5]. Despite this multitude of studies, the fundamentals of the statistical mechanics of topologically constrained polymers remain insufficiently understood. Second, random polygons—especially those comprised of the edges of a lattice, e.g., a cubic lattice—allow for concise mathematical formulation of the problems of interest, which, for an off-lattice model, is difficult even to formulate, let alone solve. Of course, this situation is by no means unique; other problems are also frequently more easily addressed using lattice models. Specifically, here, the problem in question is that of the knot entropy; see [6] for a general discussion. Indeed, this quantity is easy to define for the lattice polygon. Let $\Omega_{\text{unknot}}(N)$ be the total number of distinct rooted, i.e., with one point fixed, polygons with N edges, which are topologically equivalent to the trivial knot (an unknot or a circle). Since lattice polygons are considered, $\Omega_{\text{unknot}}(N)$ is a well-defined finite number. By definition, then, $\ln \Omega_{\text{unknot}}(N)$ is the entropy of an unknot.

Let us emphasize that this paper deals only with loops made by the closing of one single line, like letter O, not like letter θ or sign ∞, etc.; the loop may be embedded in various ways in three dimensions, forming an unknot or knots of different topologies, but the loop itself remains a simple O, albeit a "lattice O".

The number of unknots must be compared with the total number of distinct rooted polygons of N edges in three-dimensional space; let us call it $\Omega_{\text{loop}}(N)$. Then, $\ln \Omega_{\text{loop}}(N)$ is the entropy of the full ensemble of loops of all knot types. In terms of these quantities, the probability of finding an unknot among the randomly (and uniformly) generated polygons is:

$$P_{\text{unknot}}(N) = \Omega_{\text{unknot}}(N)/\Omega_{\text{loop}}(N)\,. \tag{1}$$

Clearly, $\ln P_{\text{unknot}}$ represents the corresponding change of entropy. Under the conditions of thermodynamic equilibrium, $\ln P_{\text{unknot}}$ is related to the minimal amount of mechanical work, needed to untie all the knots.

Some statements are rigorously established about these quantities; see review [7] and references therein. In particular, it is known that there exists a limit,

$$\lim_{N\to\infty} \frac{\ln P_{\text{unknot}}(N)}{N} = -\frac{1}{N_0} \, . \tag{2}$$

In other words,

$$P_{\text{unknot}}(N)|_{N\gg 1} \simeq \text{const} \cdot \exp[-N/N_0] \, . \tag{3}$$

The quantity N_0 is sometimes referred to as the random knotting length: for $N < N_0$ most polymers are unknots, while, for $N > N_0$, unknots are exceedingly rare. The exponential behavior of unknotting probability (3) is also proven for random off-lattice polygons and established numerically for a number of other models [7], albeit with very different values of N_0, ranging from a few hundred to a few million. Remarkably, no analytical method is known to find this quantity for any model.

The subject of the present note is the question—how does P_{unknot} approach its exponential asymptotic? In other words, how does the difference, $\ln P_{\text{unknot}}(N)/N + 1/N_0$ (note that $\ln P_{\text{unknot}}(N) < 0$), behave at large N, or how does this difference tend towards zero? The question is about the tail of probability $P_{\text{unknot}}(N)$ and whether it is similar to other subtle probability distributions known in various branches of physics; see, e.g., [8].

2. Developing the Argument

Let us start with $\Omega_{\text{linear}}(N)$—the number of distinct self-avoiding "open polygons" of N edges in three-dimensional space starting from, i.e., rooted in the origin (the open polygon is simply a broken line, with non-connected ends). This quantity was carefully studied in the theory of self-avoiding walks (see, e.g., [9], as a classical source), and it is known to behave as

$$\Omega_{\text{linear}}(N)|_{N\gg 1} \simeq \text{const} \cdot z^N \cdot N^{\gamma-1} \, , \tag{4}$$

where γ is a critical exponent which is universal, unlike the growth constant z, which is not universal. Therefore, z depends, for example, on the lattice type, while γ does not. The numerical value of γ was studied with great attention both analytically by renormalization group and ε-expansion [10], and by high-precision Monte Carlo [11,12]: the result was $\gamma \approx 1.16$.

Based on the knowledge of $\Omega_{\text{linear}}(N)$, one can deduce the estimate of $\Omega_{\text{loop}}(N)$. This deduction is known [9], and, here, for the purpose of subsequent generalization, let us repeat the derivation using the scaling argument, originally due to Khokhlov [13] and later developed by Duplantier [14]. This argument views transformation from a linear chain to a loop as a chemical reaction between chain ends. The argument suggests that the probability of two ends of a linear chain, meeting together in space, is of the same order as the conditional probability of the ends of two separate chains meeting in space, conditioned on the fact that these two chains share the same volume $R^3 \sim N^{3\nu}$. Here, $\nu \approx 0.588$ is a usual "metric" or Flory critical exponent, while R is the mean squared average gyration radius of the chain of length N. This argument yields the following estimate for $\Omega_{\text{loop}}(N)$:

$$\frac{\Omega_{\text{loop}}(N)}{\Omega_{\text{linear}}(N)} \sim \frac{\Omega_{\text{linear}}(2N)}{[\Omega_{\text{linear}}(N)] \times [R^3 \Omega_{\text{linear}}(N)]} \, . \tag{5}$$

This relation can also be explained in a different way. Equation (5) represents the statement that two different probabilities are of the same order, i.e., they scale with the same power of N. The left-hand side of Equation (5) is the probability that a randomly chosen linear chain of N monomers can be closed due to two ends being next to each other by pure chance. The right-hand side of Equation (5) estimates the probability, dealing with two linear chains of the same length N being co-localized in the same volume $\sim R^3$; for these two chains, the right-hand side of Equation (5) indicates the probability that the end of one of the chains is found next to the end of the other. Indeed, the numerator of the right-hand side of Equation (5) represents the number of states of one linear chain of combined length $2N$ that is the same as two separate chains with the ends of these chains forced to be next to each other. The first factor in the denominator is the number of states for one half-chain, while the second factor is enhanced by a factor R^3 as soon as the second chain can be rooted in any place within volume R^3 around the root of the first chain (the monomer size is taken to be unity). Assembling all this together, one arrives at:

$$\Omega_{\mathrm{loop}}(N)\Big|_{N\gg 1} \simeq \mathrm{const}\cdot z^N \cdot N^{-3\nu} \,. \tag{6}$$

This result also follows straight from Equations (1.10) and (1.11) of Ref. [14], and is known for self-avoiding polymers, as stated in textbooks, see, e.g., [9]. The most important property of the result (6) is that it does not involve the index γ, which cancels away from the "chemical equilibrium" condition (5). This cancellation of γ has an important physical interpretation: γ describes the situation around chain ends, as monomers close to the ends find themselves in a different kind of environment compared to internal monomers close to the middle of the chain. Since the loop does not have any ends, there is no effect to be described by γ.

The estimate (6) is accurate in terms of the power, so we can rewrite it as:

$$\Omega_{\mathrm{loop}}(N)\Big|_{N\gg 1} \simeq \mathrm{const}\cdot e^{N\ln z - 3\nu\ln N + o(\ln N)} \,. \tag{7}$$

Thus, corrections in the exponential are much smaller than $\ln N$.

The next step in building the argument is yet another mathematically proven statement [7] that the number of N-step self-avoiding unknots, $\Omega_{\mathrm{unknot}}(N)$, behaves such that there exists a limit,

$$\lim_{N\to\infty}\frac{\ln\Omega_{\mathrm{unknot}}(N)}{N} = z_0 < z \,, \tag{8}$$

or

$$\Omega_{\mathrm{unknot}}(N)|_{N\gg 1} \simeq z_0^{N+o(N)} \,. \tag{9}$$

At the same time, there is a scaling prediction [15], supported by a significant amount of numerical evidence [16–18], suggesting that a trivial knot loop, in terms of its overall size (e.g., gyration radius), is controlled by the same index $\nu \approx 0.588$, which describes the self-avoiding walks. Although there is a counter-argument pointing to the limited depth of analogy between trivial knots and self-avoiding loops [19], one can try to take this analogy one step further and conjecture that the number of unknots has the same scaling as the number of self-avoiding loops (6), but with a modified growth constant:

$$\begin{aligned}\Omega_{\mathrm{unknot}}(N)|_{N\gg 1} &\simeq \mathrm{const}\cdot z_0^N \cdot N^{-3\nu} \\ &\simeq \mathrm{const}\cdot e^{N\ln z_0 - 3\nu\ln N + o(\ln N)} \,.\end{aligned} \tag{10}$$

In other words, the above argument yields the conjecture that the cancellation of the index γ, as in Equation (5), occurs for knot-avoiding loops—just as it is proven to do for self-avoiding loops. From physics point of view, this is justified by the fact that γ is supposed to characterize chain ends, while an unknot has no ends. Since this is a non-rigorous conjecture, it is important to stress where the argument may have limitations. In this regard, cancellation of the index γ is the most essential point where the conjecture is justified only by a physical argument and not by mathematics.

Another important point is also that the power ν needs to be of the same value in both relation (6) and relation (10). If this conjecture is correct, then the probability of unknot, $P_{\mathrm{unknot}}(N) = \Omega_{\mathrm{unknot}}(N)/\Omega_{\mathrm{loop}}(N)$, is predicted to have the following asymptotic behavior:

$$P_{\mathrm{unknot}}(N)|_{N\gg 1} \simeq \mathrm{const}\cdot e^{N(\ln z_0 - \ln z) + o(\ln N)} \,. \tag{11}$$

To reiterate, the non-rigorous conjecture, found here, is, first, motivated by the analogy between knots-avoiding and self-avoiding—both described by the same metric exponent ν, and, second, by the fact that there is no index γ or its analogs since there are no ends in the loop.

3. Concluding Remarks

Equation (11) represents the result of the present paper. Of course, this equaion indicates that the random knotting chain length can be expressed as $1/N_0 = \ln(z/z_0)$. However, this is not a significant result because the growth constant z_0 (or even z) are not simple quantities to compute theoretically or to measure experimentally; essentially, z_0 and N_0 contain the same information. The real non-trivial statement is that there is $o(\ln N)$ instead of $o(N)$ in the exponential. In other words, the conjecture suggests that there is no power-law correction factor to the main exponential trend in unknot probability. The correction, of course, exists, but it is at most logarithmic. This can be contrasted with the fact that the probabilities of various non-trivial knots are routinely fitted to expressions like $N^{\mu} \exp(-N/N_0)$ (with N_0 as for trivial knots); see, e.g., [20–22]. In these terms, the result of the present paper is that for the trivial knots, $\mu = 0$ exactly.

The questions of critical exponents, related to the entropy of random polygons, were examined numerically, in quite some detail in the series of studies by the Italian group of E. Orlandini and co-authors [21–23]. In particular, Ref. [23] presented the most accurate study to date of (in the present notation) the exponent μ and it was found that, within the numerical accuracy of the Monte Carlo simulations made there, the result for an unknot was so small that it was not distinguishable from $\mu = 0$. In this sense, the result of the present study can be viewed as a confirmation or rather an explanation of the numerical observation made about a decade ago.

Does this result have practical implications beyond mathematical curiosity? In general, random knots are a fact of life in case of a DNA plasmid and a number of other biological contexts; see, e.g., [24]. Historically, in the first study [25] on random polymer topology, the main surprise was to observe that the probability of non-trivial knots, i.e., $1 - P_{\text{unknot}}(N)$, although still rather small at the tested range of N values, is, nevertheless, an increasing function of N. In this sense, the main observation is that for long polymers, $P_{\text{unknot}}(N)$ is small. This is, of course, consistent with the statement of the mathematical theorem (2), except the latter deals with the mathematical limit of $N \to \infty$, while, in practice, the exponential dependence of $P_{\text{unknot}}(N)$ on the chain length, N, seems to be consistent with observations, even at modest values of N being certainly smaller than random knotting length N_0; see, e.g., [26]. In this sense, the statement of the absence of power law corrections, made in the present paper, may have quite some practical implications.

Funding: The work of AYG was partially supported by the MRSEC Program of the National Science Foundation under Award Number DMR-1420073. This work was performed at the Aspen Center for Physics, which is supported by National Science Foundation Grant Number PHY-1607611.

Acknowledgments: The author acknowledges the fruitful atmosphere of the Aspen Center for Physics where this work was conceived and performed.

Conflicts of Interest: The author declares no conflict of interest.

References

1. Smrek, J.; Chubak, I.; Likos, C.N.; Kremer, K. Active topological glass. *Nat. Commun.* **2020**, *11*, 26–36. [CrossRef]
2. O'Connor, T.C.; Ge, T.; Rubinstein, M.; Grest, G.S. Topological linking drives anomalous thickening of ring polymers in weak extensional flows. *Phys. Rev. Lett.* **2020**, *124*, 027801. [CrossRef]
3. Kruteva, M.; Allgaier, J.; Monkenbusch, M.; Porcar, L.; Richter, D. Self-similar polymer ring conformations based on elementary loops: A direct observation by SANS. *ACS Macro Lett.* **2020**, *9*, 507–511. [CrossRef]
4. Liebetreu, M.; Likos, C.N. Shear-induced stack orientation and breakup in cluster glasses of ring polymers. *ACS Appl. Polym. Mater.* **2020**, *2*, 3505–3517. [CrossRef]

5. Parisi, D.; Costanzo, S.; Jeong, Y.; Ahn, J.; Chang, T.; Vlassopoulos, D.; Halverson, J.D.; Kremer, K.; Ge, T.; Rubinstein, M.; Grest, G.S.; Srinin, W.; Grosberg, A.Y. Nonlinear shear rheology of entangled polymer rings. *Macromolecules* **2021**, *54*, 2811–2827. [CrossRef]
6. Grosberg, A.Y.; Nechaev, S.K. Polymer topology. *Adv. Polym. Sci.* **1993**, *106*, 1–30.
7. Orlandini, E.; Whittington, S.G. Statistical topology of closed curves: Some applications in polymer physics. *Rev. Mod. Phys.* **2007**, *79*, 611–642. [CrossRef]
8. Tribelsky, M.I.; Grosberg, A.Y. Laser heating of a transparent medium containing random absorbing inhomogeneities. *Sov. Phys. JETP* **1976**, *41*, 524–526.
9. De Gennes, P.G. *Scaling Concepts in Polymer Physics*; Cornell University Press: Ithaca, NY, USA, 1979.
10. Le Guillou, J.C.; Zinn-Justin, J. Accurate critical exponents from field theory. *J. Phys.* **1989**, *50*, 1365–1370. doi: 10.1051/jphyslet:019850046040013700. [CrossRef]
11. Caracciolo, S.; Causo, M.S.; Pelissetto, A. High-precision determination of the critical exponent γ for self-avoiding walks. *Phys. Rev. E* **1998**, *57*, R1215–R1218. [CrossRef]
12. Clisby, N. Scale-free Monte Carlo method for calculating the critical exponent γ of self-avoiding walks. *J. Phys. A Math. Theor.* **2017**, *50*, 264003. [CrossRef]
13. Khokhlov, A.R. Influence of excluded volume effect on the rates of chemically controlled polymer-polymer reactions. *Makromol. Chem. Rapid Commun.* **1981**, *2*, 633–636. [CrossRef]
14. Duplantier, B. Statistical mechanics of polymer networks of any topology. *J. Stat. Phys.* **1989**, *54*, 581–680. [CrossRef]
15. Grosberg, A.Y. Critical exponents for random knots. *Phys. Rev. Lett.* **2000**, *85*, 3858 – 3861. [CrossRef] [PubMed]
16. Dobay, A.; Dubochet, J.; Millett, K.; Sottas, P.E.; Stasiak, A. Scaling behavior of random knots. *Proc. Natl. Acad. Sci. USA* **2003**, *100*, 5611–5615. [CrossRef] [PubMed]
17. Matsuda, H.; Yao, A.; Tsukahara, H.; Deguchi, T.; Furuta, K.; Inami, T. Average size of random polygons with fixed knot topology. *Phys. Rev. E* **2003**, *68*, 011102. [CrossRef] [PubMed]
18. Moore, N.T.; Lua, R.C.; Grosberg, A.Y. Topologically driven swelling of a polymer loop. *Proc. Natl. Acad. Sci. USA* **2004**, *101*, 13431–13435. [CrossRef]
19. Moore, N.T.; Grosberg, A.Y. Limits of analogy between self-avoidance and topology-driven swelling of polymer loops. *Phys. Rev. E* **2005**, *72*, 061803. [CrossRef]
20. Deguchi, T.; Uehara, E. Statistical and dynamical properties of topological polymers with graphs and ring polymers with knots. *Polymers* **2017**, *9*, 252. [CrossRef]
21. Orlandini, E.; Tesi, M.; Janse van Rensburg, E.; Whittington, S. Entropic exponents of lattice polygons with specified knot type. *J. Phys. Math. Gen.* **1996**, *29*, L299–L303. [CrossRef]
22. Orlandini, E.; Tesi, M.; Janse van Rensburg, E.; Whittington, S.G. Asymptotics of knotted lattice polygons. *J. Phys. A Math. Gen.* **1998**. *31*, 5953–5967. [CrossRef]
23. Baiesi, M.; Orlandini, E.; Stella, A.L. The entropic cost to tie a knot. *J. Stat. Mech. Theory Exp.* **2010**, *2010*, P06012. [CrossRef]
24. Vologodskii, A.V. *Biophysics of DNA*; Cambridge University Press: Cambridge, UK, 2015. [CrossRef]
25. Frank-Kamenetskii, M.D.; Lukashin, A.V.; Vologodskii, A.V. Statistical mechanics and topology of polymer chains. *Nature* **1975**, *258*, 398–402. [CrossRef] [PubMed]
26. Plesa, C.; Verschueren, D.; Pud, S.; van der Torre, J.; Ruitenberg, J.W.; Witteveen, M.J.; Jonsson, M.P.; Grosberg, A.Y.; Rabin, Y.; Dekker, C. Direct observation of DNA knots using solid state nanopore. *Nat. Nanotechnol.* **2016**, *11*, 1093–1097. [CrossRef] [PubMed]

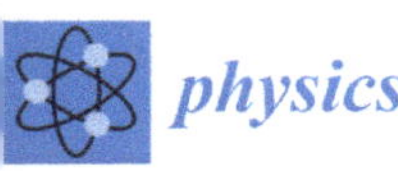

Article

Instability of Traveling Pulses in Nonlinear Diffusion-Type Problems and Method to Obtain Bottom-Part Spectrum of Schrödinger Equation with Complicated Potential

Michael I. Tribelsky [1,2]

[1] Faculty of Physics, M. V. Lomonosov Moscow State University, 119991 Moscow, Russia; mitribel@gmail.com
[2] National Research Nuclear University MEPhI, Moscow Engineering Physics Institute, 115409 Moscow, Russia

Abstract: The instability of traveling pulses in nonlinear diffusion problems is inspected on the example of Gunn domains in semiconductors. Mathematically, the problem is reduced to the calculation of the "energy" of the ground state in the Schrödinger equation with a complicated potential. A general method to obtain the bottom-part spectrum of such equations based on the approximation of the potential by square wells is proposed and applied. Possible generalization of the approach to other types of nonlinear diffusion equations is discussed.

Keywords: nonlinear diffusion; traveling waves; stability; Goldstone modes; Schrödinger equation; spectrum of low-exited states

Citation: Tribelsky, M.I. Instability of Traveling Pulses in Nonlinear Diffusion-Type Problems and Method to Obtain Bottom-Part Spectrum of Schrödinger Equation with Complicated Potential. *Physics* **2021**, *3*, 715–727. https://doi.org/10.3390/physics3030043

Received: 16 July 2021
Accepted: 9 August 2021
Published: 30 August 2021

Publisher's Note: MDPI stays neutral with regard to jurisdictional claims in published maps and institutional affiliations.

1. Historical Remarks

When the Editors kindly offered me to submit a paper to this Special Issue dedicated to my fifty years in physics, I began to think about a possible topic of the paper. Finally, I decided that the best is to generalize the results of my very first paper [1], which formally was published exactly fifty years ago. I said "formally" because *actually* this paper has never been published. Perhaps, its story is so remarkable that it is worth telling it here.

The point is that though I graduated from the Lomonosov Moscow State University (MSU)—the one where now I head a laboratory—I did not enter this university in the usual, standard manner. It so happened that the university I entered was the Belorussian State University (BSU) in Minsk. Now, Minsk is the capital of independent state Belarus, while, at that time, Minsk and Moscow both belonged to a single state: the Soviet Union. In Minsk, I met my first scientific adviser Mikhail Aleksandrovich El'yashevich [2].

Then, upon completing my first two university years in Minsk, I moved to Moscow. Thus, I became a student of MSU due to my transfer from BSU. Just one letter difference in the names meant the drastic difference in the ranks. Though BSU was quite a good university, MSU was (and is) the Number One.

Doing paperwork related to the transfer, I asked El'yashevich for a reference letter to one of his collaborators in Moscow. I then obtained a letter to his former Ph.D. student Sergei Ivanovich Anisimov [3], who became my next scientific adviser.

It was 1969. At that time, I could not even imagine how lucky I was. Anisimov was employed by the Landau Institute for Theoretical Physics. The Institute was created just five years ago to collect "under a single roof" the first generation of Lev Davidovich Landau's disciples [4]. By the time I am talking about, all of them had become first magnitude stars in the scientific sky.

Thus, suddenly and almost by chance, I became embedded in the scientific atmosphere representing the very top of theoretical physics in the USSR, and I would say in the entire world too. Moreover, I had even more good luck, though, I did not know it yet: In the very same year of my transfer to MSU, a prominent theoretical physicist Il'ya Mikhailovich Lifshits [5] succeeded the late Landau's position of the Head of the Theoretical Physics

Department at the Kapitza Institute [6]. To this end, he moved to Moscow from Khar'kov (a big Ukrainian city), where he resided before. In addition to this position, Il'ya Mikhailovich got a professorship at the Chair for Quantum Theory, the Faculty of Physics, MSU. By that time, another employee of the Landau Institute and a disciple of Il'ya Mikhailovich, namely, Mark Yakovlevich Azbel [7], already shared his position at the Landau Institute with a professorial position of this Chair. The second disciple of Il'ya Mikhailovich, who came from Khar'kov to Moscow and became a Professor of the same Chair, was Moisei Isaakovich Kaganov. Among other scientific accomplishments of this group was the galvanomagnetic theory of electrons with an arbitrary dispersion law. The theory describes the effects of both electric and magnetic fields acting together on free electrons in metals. In this theory, electrons are regarded as quasi-classical particles, but, instead of the conventional dependence of the energy on (quasi)momentum $\varepsilon(p) = p^2/(2m)$, this dependence may be arbitrary. The theory was a breakthrough in quantum solid-state physics, and was named after its creators—the LAK theory (Lifshitz, Azbel, Kaganov).

There is an interesting story related to this abbreviation. When another one of Landau's disciples, Alexander Solomonovich Kompaneetz, known, in addition to his outstanding scientific results, for his sense of humor, leant about LAK theory, he said, "It is excellent that the authors did not employ the inverted order of them." The joke is that *kal* in Russian means excrements.

To complete my description of the Chair for Quantum Theory, I should add that it was headed by one of the most prominent experts in theoretical physics, a very respectable person with the highest moral standards, Academician of the Soviet Academy of Sciences, Mikhail Aleksandrovich Leontovich [8]. Alas, all of them have already passed away.

In 1969, I knew nothing about these people and the Chair, but Anisimov did know. Therefore, when I asked his advice about the specific Chair at the Faculty of Physics for my specialization, he immediately replied, "The one where I.M. Lifshitz is a Professor." I took his advice and applied for the specialization at this Chair. Once again, I was lucky — my application was approved, and in addition to the excellent external scientific environment at the Landau Institute, I benefited from that at the Chair for Quantum Theory.

Soon after my appearance at the Chair, I began to attend lectures on the quantum theory of metals given by Kaganov. Bearing in mind that Kaganov was a brilliant lecturer, it is easy to understand that I admired the beauty of the lectures and that of the theory as a whole. Thus, it is easy to understand that, when Anisimov asked me about the preferences for the topic of the future study, my reply was, "Something from quantum solid-state physics".

It is worth mentioning that, at the time, I did not have any idea about the specific subfield, where the accomplishments of Anisimov lay (namely, laser–matter interaction, physical hydrodynamics, shock waves, plasma physics, and the like). Fortunately, he was a physicist with broad interests and understood physics far beyond the frames of his own subfield. It was a typical feature of physicists from the Landau Institute originated by Landau himself: Broad knowledge helps to see cross-links between different, seemingly unrelated, problems. This, in turn, sometimes helps to obtain very beautiful and unexpected results.

Then, according to my desire, Anisimov posed me a problem from quantum solid-state physics. It was related to the Gunn effect in semiconductors [9]. At that time, the effect was a fascinating, challenging topic, and, up to now, it still attracts a great deal of attention from researchers [10–15].

Naturally, now the understanding of the effect is more profound, and its mathematical description is much more elaborated than it was 50 years ago; see, e.g., Ref. [16]. However, since the goal of this paper is to generalize the methods and results discussed in Ref. [1], making them applicable to a broad class of related problems, rather than to inspect specific peculiarities of the Gunn effect itself; in what follows, I stick to the old model of the effect [17] employed by Knight and Peterson [18] and then, in my paper [1].

Very briefly, the essence of the phenomenon is as follows. In a strong enough electric field, **E**, applied to a semiconductor, the conductivity of the sample depends on **E**, and the current–voltage curve becomes nonlinear. In some cases, calculations based on the assumption of the spatially homogeneous distribution of the current density, **j**, and **E** along and across the sample give rise to very unusual behavior of this dependence so that, in a certain area of the **E** values, an increase in **E** results in a *decrease* in **j**.

In what follows, only one-dimensional cases will be considered, so I can replace **j**(**E**) $\to j(E)$. Then, by definition, the conductivity $\sigma = j/E$. Let us define the differential conductivity as $\sigma_{\mathrm{d}} = \mathrm{d}j/\mathrm{d}E$. Thus, the area mentioned above is characterized with a negative differential conductivity. Here, I will not discuss the microscopic mechanisms explaining the negativeness of σ_d; a detailed description may be found, e.g., in Ref. [19]. Further increase in E makes σ_d positive again so that the overall shape of the current–voltage curve resembles letter "N", see Figure 1.

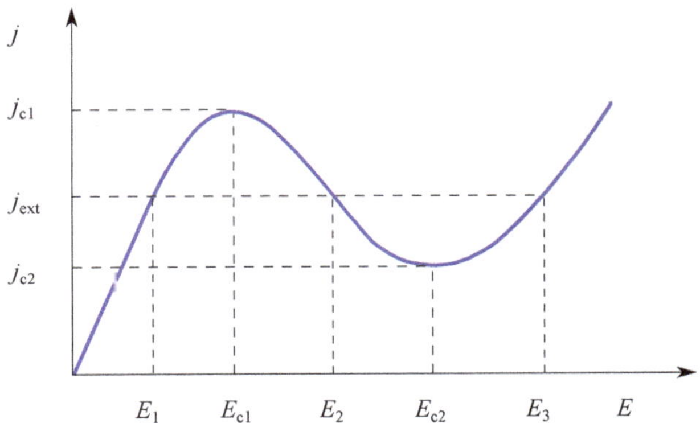

Figure 1. A letter-*N*-shape current–voltage characteristics obtained under the assumption that electric field $E = const$ along and across the sample: the differential conductivity, σ_{d} is negative at $E_{c1} < E < E_{c2}$.

It occurs that the assumption about the spatially uniform distribution of j and E in the regions with $\sigma_d < 0$ is erroneous. This distribution is unstable against small spatially-inhomogeneous perturbations and, eventually, is destroyed owing to their growth. In certain cases, the instability ends up forming a strong field domain bounded by the corresponding layers of charge density. The domain drifts along the sample with a constant speed until it hits the sample edge (anode). The domain disintegrates there, a new one emerges at the opposite side of the sample, and the process repeats. As a result, oscillations with the period L/v are generated. Here, L stands for the sample length and v is the drift speed of the domain. This is the Gunn effect [9]. It is successfully used in Gunn diodes to generate microwave oscillations [19].

Let us consider an idealized case of a single traveling strong-field Gunn domain drifting with a constant velocity along an infinitely-large sample. The "strong-field domain" means that the field outside it equals E_1 (see Figure 1), while inside the domain, it is greater than that. Then, if the voltage applied to the sample is constant, the single domain with a fixed shape is stable, while a configuration with several domains is not. However, if an external source fixes the current in the sample, even the single domain becomes unstable. The instability affects the faces of the domain, which begin to move in opposite directions with respect to the center of the traveling domain. If they move to each other, the domain contracts and, eventually, collapses. If the faces move in the opposite direction, the domain expands and transforms into two traveling layers [19].

Linear analysis of this secondary instability of a single traveling domain at a fixed current in the circuit was performed by Knight and Peterson [18]. Mathematically, the stability problem was reduced to the calculation of a gap between the ground and the first excited states in the one-dimensional Schrödinger equation with a complicated potential (see below). To this end, Knight and Peterson employed the Wentzel-Kramers-Brillouin (WKB) approximation. However, this approximation is accurate for highly excited states when the characteristic spatial scale of the wave function oscillations is small relative to the one for the variations of the potential. This is not the case for the ground state. Hence the accuracy of the results obtained in Ref. [18] through the WKB method, at least, was questionable. The problem, posed for me by Anisimov, was to check the results of Knight and Peterson employing for the calculations an approximation different from WKB.

If I faced this problem now, quite probably, I would have used the Ritz method supplemented by the orthogonality condition of the wave functions of the ground and excited states [20]. However, at that time, I was much more ignorant than I am now. Therefore, instead of taking a simple, known way (perhaps, at that time, it was neither simple nor known for me), I decided to go on my own one. Specifically, I decided no less than to find a new method to obtain approximate solutions to the Schrödinger equation opposite to the WKB-method, which could be suitable for the ground and low-excited states. Furthermore, I succeeded in doing that! So, maybe, ignorance is not always bad.

The main idea of the developed approach is somewhat unusual for quantum mechanics, where approximations conventionally are targeted to a wave function, while the potential is given and fixed. However, if one has ground and low-excited states in a complicated potential, the potential has a sharply varying profile relative to that of the wave functions, and the latter is not very sensitive to the fine details of the former. If so, why does one not try to approximate *the potential*, with some simple shapes, say, with square wells? Then, the Schrödinger equation becomes either exactly solvable or readily treated by perturbation methods.

The most challenging task was to set the first step in this way. The rest was just a matter of not so complicated calculations. I quickly did them and presented the results to Anisimov. "Very well," he said, "the problem is solved. Write a paper. One more point to be made. Il'ya Mikhailovich Lifshitz has organized a periodic scientific seminar at your Faculty. It takes place every second and fourth Thursday of a month from September to June. I recommend you to contact Il'ya Mikhailovich and ask him to put your talk about this study in the seminar program.".

Up to now, I remember how difficult it was for me (a fourth-year undergraduate student) to approach such a famous scientist as Il'ya Mikhailovich was and request a talk at his seminar. Finally, I gathered up all my courage and did it. "Excellent," replied Il'ya Mikhailovich, "Please contact the seminar's secretary, Mr. Rzhevsky, and ask him to find the nearest free spot in the program. Will 45 minutes be enough for you?".

To give a 45-min talk in front of an audience of top-rank experts, including a dozen of world-class scientists! My knees turned to jelly, but there was no way to retreat.

It is remarkable that I can vividly remember any moment before and after my talk, but nothing of the talk itself. However, it seems that I stood this test. Moreover, the talk at this seminar was a milestone for my relationship with Il'ya Mikhailovich. Since then, every one of my new results was discussed with him, either through a talk at the seminar or in a private manner at his office at the Kapitza Institute. Later on, I became a close associate and coauthor of Il'ya Mikhailovich [21]. We even had a joint Ph.D. student. Our close contact lasted until his unexpected premature death of a heart attack in 1982.

However, all this will be later on. At the time I am talking about, I could not imagine even a small part of that.

Thus, the first task (the talk) was complete, but the second remained: I had to write a paper. It was my very first paper, and it took a lot of my time and efforts to do that. Finally, an extended manuscript (in Russian) was submitted to *Fizika i Tekhnika Poluprovodnikov (Physics and Technology of Semiconductors)*.

It was the beginning of the bad luck for this paper. At that time, a new publication option was introduced. For some papers (especially lengthy ones), only abstracts were published. The papers themselves were *deposited* in specially assigned institutions. If an abstract of such a paper drew somebody's attention, and he/she was interested in the complete text, a copy of the one could be posted to him/her upon a request. Maybe it was an attempt to reduce the printed size of scientific journals and to solve, at least partly, the eternal Soviet problem of paper deficit. Anyway, my paper was accepted under this condition. Its abstract was published, see Figure 2, while the full text was deposited at *Research and Development (R&D) Institute Electronics*.

Since then, many events have occurred. The country named the Soviet Union does not exist anymore. Regarding *R&D Electronics,* I am afraid it has shared the destiny of the country. Now, I am residing within walking distance from the building where *R&D Electronics* used to be. It is a shopping mall there. Then, it is quite probable that the full text of my paper has ended up in a nearby scrap-heap, and the abstract reproduced in Figure 2 is the only remaining piece of the paper.

Вып. 10 ДЭ—416 от 30 июня 1971 г.

ОБ ИНКРЕМЕНТЕ НЕУСТОЙЧИВОСТИ ГАННОВСКИХ ДОМЕНОВ В РЕЖИМЕ ПОСТОЯННОГО ТОКА

М. И. Трибельский

Исследуется развитие неустойчивости в широких ганновских доменах (ширина вершины много больше ширины фронтов) в режиме стационарного внешнего тока. В основу расчетов положена феноменологическая модель, в которой полный ток складывается из тока проводимости и диффузионного тока. Неустойчивость заключается в том, что фронты домена начинают смещаться в противоположных направлениях. Вычислен инкремент нарастания неустойчивости. Коэффициент диффузии предполагался независящим от поля. Задача математически свелась к решению уравнения типа одномерного уравнения Шредингера с некоторым сложным потенциалом. Показано, что результаты слабо зависят от вида этого потенциала. Поэтому он был заменен потенциалом в виде двух прямоугольных ям, что позволило выполнить расчеты в конечном виде. Задача решена в двух предельных случаях: симметричного и сильно несимметричного доменов. В обоих случаях вычислены дополнительное падение напряжения на домене, а также ширина фронтов и вершины домена.

Существенно, что в конечный результат не входят какие-либо интегральные характеристики. Он зависит только от значения функций $j\,(E)$ и $dj/dE \equiv \sigma_d$ в некоторых характерных точках. Развитый метод без существенных изменений может быть применен к исследованию характера неустойчивости волн в виде двух и более доменов.

Поступило в Редакцию
13 марта 1971 г.

Figure 2. The only ever published piece of paper [1] (in Russian).

The English translation reads:

Vol. 10 DE-416 dated 30 June 1971
On the increment of the instability of the Gunn domains in the direct current regime
M. I. Tribel'skii

The growth of instability of wide Gunn domains (the width of the top is much larger than the widths of the faces) at the stationary external current regime is inspected. The basis of calculations is the phenomenological model, in which the total current is composed of the conductivity current and the diffusion one. Instability affects the domain faces so that they begin to shift in opposite directions. The instability increment is calculated. The diffusion coefficient is supposed to be independent of the field. Mathematically the stability problem is reduced to a one-dimensional Schrödinger equation with a certain complicated potential. It is shown that the results are weakly dependent on details of this potential. Therefore, the potential is approximated by two square potential wells (separated by a barrier), which made it possible to obtain an explicit expression for the increment. The problem is solved in two limiting cases, namely symmetric and highly asymmetric domains. In both cases, additional drops of voltage on the domain are calculated, as well as the width of the faces and the top of the domain.

It is essential that the final result does not include any integral characteristics of the problem. It depends only on the value of the functions $j(E)$ and $\mathrm{d}j/\mathrm{d}E \equiv \sigma_d$ at certain characteristic points. A developed method without significant changes may be extended to the study of the instability of waves in the form of two or more domains.

Received 13 March 1971.

The next attempt to publish these results I made after the defense of my Master Science Thesis. An appointed referee of the thesis was another disciple of Landau, Igor Ekhiel'evich Dzyaloshinskii [22]. After reading the thesis, he said that the results of this level should be available to the international community, and I should publish them abroad in English (when the first draft of the present paper was ready, I learned the sad news: Igor Ekhiel'evich Dzyaloshinskii passed away on 14 July 2021).

To publish abroad, ... it was easier to say than to do. Not to mention poor English, which I had at that time, sending a scientific paper abroad from the Soviet Union was not simple at all. The authors themselves were not eligible to do that. A manuscript had to be sent through specially authorized personnel. The personnel decided whether or not the paper could be submitted abroad, and, if the decision was affirmative, they took care of the submission.

Moreover, prior to the acceptance of the manuscript by the personnel, the authors had to do plenty of paperwork. On top of that, it took 2–3 months on average for mail to be delivered to the addressee. Up to now, I wonder why this was so much. Even if horses delivered the mail; it would not have taken such a long time!

I discussed the matter with Anisimov, and we decided to submit the paper to the East-German journal *Physica Status Solidi* published in English. There were two reasons for this choice. First, sending a paper to an Eastern bloc country required less paperwork, and chances to get permission for the submission were higher than that in the case of a Western journal. Second, the requirement for the English quality in this journal was not as strict as those in the West. The latter was important since my English was far from being perfect.

Thus, I wrote in English an elaborated version of Ref. [1] including some new results, did all the required paperwork, gave the bunch of documents to the "authorized personnel," and... lost control over the submission. Half a year elapsed, but I had not heard anything from the Editors. Then, I sent a postcard to *Physica Status Solidi* asking for the status of my paper. A reply came surprisingly fast—in just four months. However, it was pretty unexpected. The Editors informed me that they had never received my manuscript.

By that time, on the one hand, I had already published a paper [23], where the secondary instability of the Gunn domain was inspected just employing the Ritz method. On the other hand, I got a job and, owing to that, was forced to abandon my study in solid-state physics and focus on an entirely different topic.

Eventually, the results discussed in Ref. [1] have remained unpublished. Now, fifty years later, I try to realize the advice of Dzyaloshinskii and make these results available to the international community. Perhaps fifty years is a too long period to complete a task, but "that is not lost that comes at last!".

At the end of these, perhaps lengthy, remarks, I have to say that the results discussed below are not exactly the same as those in Ref. [1]. First, it is not good to publish the same results twice, even if the fifty years lie between the two publications. Second, I could not do this, even if I wanted to—the original manuscript is lost, and I do not remember all details. Last but not least: now I am a bit more experienced and educated than I was fifty years ago. Therefore, I extracted from this old problem the essential points and generalized them. These points are as follows: (i) the conclusion about the instability of traveling pulses in a broad class of nonlinear diffusion-type problems and (ii) a new method to obtain the bottom-part spectrum of the Schrödinger equation with a complicated potential. A discussion of these two issues is given below.

2. Problem Formulation

Thus, the problem is to find the instability increment for a single traveling Gunn domain at a fixed current in the circuit. According to what has been said above, the current–voltage characteristic of the semiconductor sample in question has the shape schematically shown in Figure 1. Regarding the external current, j_{ext}, let us suppose that it satisfies the restrictions $j_{c1} < j_{ext} < j_{c2}$, so that the equation $j(E) = j_{ext}$ always has three roots, $E_{1,2,3}$.

It is important to stress that the curve, shown in Figure 1, is not the *actual* current–voltage characteristic of the sample. As mentioned above, it would have been the one provided E is a constant along and across the sample. Obviously, this is not the case for the traveling domain, when E is coordinate- and time-dependent. Therefore, only the stable branches of the presented curve with $\sigma_d > 0$ coincide with the actual current–voltage characteristic. In contrast, the whole curve in Figure 1 should be regarded as the field dependence of the normalized average electron drift velocity [17].

It is convenient to normalize the electric field over E_2 and $j(E)$ over j_{ext} introducing the dimensionless quantities $\mathcal{E} \equiv E/E_2$ and $u(\mathcal{E}) \equiv j(E)/j_{ext}$. Then, under certain assumptions, in the traveling coordinate frame connected with the domain, the normalized electric field in the sample is described by the following equation [19]:

$$\mathcal{D}\mathcal{E}_{\xi\xi} + \alpha[s - u(\mathcal{E})]\mathcal{E}_{\xi} + [1 - u(\mathcal{E})] = \mathcal{E}_{\eta}(\xi, \eta), \tag{1}$$

where the subscripts indicate the corresponding derivatives. Equation (1) is written in dimensionless variables, whose detailed definition is not important for the subsequent analysis (it may be found in Ref. [19]). Note only that $\mathcal{D}$, s, ξ, and η stand for the diffusion coefficient, the domain velocity in the laboratory coordinate frame, traveling coordinate, and time, respectively; $\alpha = const > 0$ is the ratio of two characteristic spatial scales of the problem at $E = E_2$.

Let us suppose that $\mathcal{D} = const$. This assumption simplifies calculations, but it is not crucial for the analysis. A more general case, when $\mathcal{D} = \mathcal{D}(\mathcal{E})$, was inspected by Knight and Peterson [18].

It is important to stress that, if the dependence $u(\mathcal{E})$ is not related to the specific shape of $j(E)$, shown in Figure 1, Equation (1) is nothing but a nonlinear diffusion equation of quite a general type describing a wide diversity of problems. Accordingly, the results discussed below may be applied to a much broader class of problems, provided these problems have traveling solitary-wave-type solutions.

For a steady-state traveling wave, the right-hand side of Equation (1) vanishes, and the equation transforms into an ordinary differential equation. For the problem in question, a simple analysis reveals that its phase plane $(\mathcal{E}, \mathcal{E}_{\xi})$ has three singular points situated at the $\mathcal{E}$ axis at $\mathcal{E} = \mathcal{E}_{1,2,3}$ corresponding to $E = E_{1,2,3}$ in Figure 1. Note that, by definition, $\mathcal{E}_2 \equiv 1$ since $\mathcal{E} = E/E_2$. In the phase plane, a single traveling domain is described by a homoclinic path beginning in the saddle $(\mathcal{E}_1, 0)$, making a loop around the unstable focus $(\mathcal{E}_2, 0)$ and ending up in the same saddle $(\mathcal{E}_1, 0)$, see Figure 3.

It is possible to show that such a solution of Equation (1) exists at $s = 1$ solely [18]. Since this is the only case I am interested in, s below is always supposed to be equal to unity. Then, the homoclinic path may be found explicitly [18]; however, I do not need this expression for the subsequent inspection. Let us just designate the steady-state solution of Equation (1) as $\mathcal{E}_0(\xi)$. The goal of this paper is to analyze the stability of this solution against small time-dependent perturbations $\delta\mathcal{E}(\xi, \eta)$.

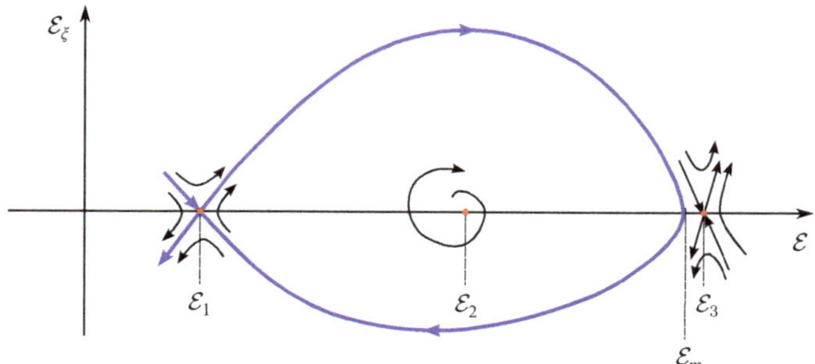

Figure 3. Phase plane $(\mathcal{E}, \mathcal{E}_\xi)$ (schematically). Three singular points are marked with red. The blue curve designates the homoclinic path corresponding to a single traveling domain. If the point $(\mathcal{E}_m, 0)$ merges with $(\mathcal{E}_3, 0)$, the homoclinic path is split into two independent heteroclinic ones (the upper and lower parts of the homoclinic path, respectively). See text for details.

3. Stability Analysis

The stability analysis performed in Ref. [18] generalizes a brilliant approach by Zeldovich and Barenblatt for the inspection of the stability of a slow combustion front [24] The main idea is as follows. Let us suppose that $\delta\mathcal{E}(\xi, \eta) = \mathcal{E}^{(1)}(\xi)\exp(-\lambda\eta)$, where λ is an eigenvalue of the stability problem. If there is a negative λ in the problem's spectrum, it means instability.

Substituting $\mathcal{E}(\xi, \eta) = \mathcal{E}^{(0)}(\xi) + \mathcal{E}^{(1)}(\xi)\exp(-\lambda\eta)$ in Equation (1) and linearizing the result in small $\mathcal{E}^{(1)}$, one arrives at the eigenvalue problem:

$$D\mathcal{E}^{(1)}_{\xi\xi} + \alpha[1 - u(\mathcal{E}^{(0)}(\xi))]\mathcal{E}^{(1)}_\xi - u_{\mathcal{E}}(\mathcal{E}^{(0)}(\xi))\mathcal{E}^{(1)} = -\lambda\mathcal{E}^{(1)}, \tag{2}$$

supplemented with the boundary conditions $\mathcal{E}^{(1)} \to 0$ at $\xi \to \pm\infty$. Then, introducing a new function $\psi(\xi)$ connected with $\mathcal{E}^{(1)}(\xi)$ by the relation,

$$\psi(\xi) = \exp\left(\frac{\alpha}{2\mathcal{D}}\int(1 - u(\mathcal{E}^{(0)}(\xi))d\xi\right)\mathcal{E}^{(1)}(\xi) \equiv F(\xi)\mathcal{E}^{(1)}(\xi), \tag{3}$$

one reduces Equation (2) to the standard Schrödinger equation:

$$\hat{H}\psi = \Lambda\psi, \tag{4}$$

$$\hat{H} = -\frac{d^2}{d\xi^2} + V(\xi), \tag{5}$$

$$V(\xi) = \left(\frac{\alpha(1 - u(\mathcal{E}^{(0)}(\xi))}{2\mathcal{D}}\right)^2 + \left(1 - \frac{\alpha\mathcal{E}^{(0)}_\xi(\xi)}{2}\right)\frac{u_{\mathcal{E}}(\mathcal{E}^{(0)}(\xi))}{\mathcal{D}}, \tag{6}$$

where $\Lambda \equiv \lambda/\mathcal{D}$. Let us remark that there is a misprint in the expression for $V(\xi)$ in Ref. [19] corrected in Equation (6).

Note that, since the homoclinic path begins and ends up at the same singular point $(\mathcal{E}_1, 0)$ and $u(\mathcal{E}_1) = 1$, the considered steady-state traveling domain solution satisfies the condition $u(\mathcal{E}^{(0)}(\xi)) \to 1$ at $\xi \to \pm\infty$. Therefore, as it follows from Equation (3), $\psi(\xi)$ and $\mathcal{E}^{(1)}(\xi)$, both have the same asymptotic behavior at $\xi \to \pm\infty$. This is important since it means that none of the solutions of the Schrödinger equation generate "false" solutions of the initial stability problem, which may not satisfy the boundary conditions $\mathcal{E}^{(1)}(\xi) \to 0$ at $\xi \to \pm\infty$.

Now, the most essential part of the stability analysis begins. If $\mathcal{E}^{(0)}(\xi)$ is a solution of the steady-state version of Equation (2), then, owing to the translational invariance of the problem $\mathcal{E}^{(0)}(\xi+\xi_0)$, where ξ_0 is *any* constant, also is its solution, i.e., being substituted in the left-hand side of Equation (2), $\mathcal{E}^{(0)}(\xi+\xi_0)$, turns it to zero *identically*.

Let us consider the limit $\xi_0 \to 0$. In this case, $\mathcal{E}^{(0)}(\xi+\xi_0) \approx \mathcal{E}^{(0)}(\xi) + \mathcal{E}^{(0)}_\xi(\xi)\xi_0$, and $\mathcal{E}^{(0)}_\xi(\xi)\xi_0$ here may be regarded as an infinitesimal perturbation to $\mathcal{E}^{(0)}(\xi)$. The perturbation transforms the steady-state solution into another steady-state solution. This means that such a perturbation is *neutrally-stable* and should not evolve in time. In other words, it means that $\mathcal{E}^{(0)}_\xi(\xi)$ is an eigenfunction of the stability problem with zero eigenvalue.

Note that we obtain this result based on the translational invariance solely, without the employment of a specific form of the differential operator in Equation (2). These neutrally-stable modes generated by a transformation of a continuous group of symmetry are called *Goldstone* modes. Since 2008, when the implementation of such a mode in strong-interaction physics (do you remember my remark about interconnections of different fields in physics?) resulted in the Nobel Prize being awarded to Prof. Yoichiro Nambu, they have also been called *Nambu–Goldstone* modes.

It is interesting to note that, twenty-five years after the publication of Ref. [1], I returned to the inspection of the role of Goldstone modes in stability problems. This study resulted in the discovery of a new type of chaos at the onset analogous to the second-order phase transitions in statistical physics, where the mean amplitudes of the turbulent modes played the role of the order parameter [25–27].

However, I have departed from the stability analysis of the Gunn domain. It is high time to be back. Actually, not so much remains to be done. Collecting together all mentioned above, one can conclude that

$$\psi(\xi) = F(\xi)\mathcal{E}^{(0)}_\xi(\xi) \tag{7}$$

is the eigenfunction of the Schrödinger equation, Equations (4)–(6) with zero eigenvalue.

Recall now the oscillation theorem [20]. The theorem states that, in a one-dimensional Schrödinger equation, the nth wave function of a discrete spectrum should vanish n times. Then, it is not a complicated task to show that the integral in the exponent in Equation (3) always remains finite, i.e., $F(\xi)$ never vanishes. Thus, all zeros of $\psi(\xi)$, if any, coincide with those of $\mathcal{E}^{(0)}_\xi(\xi)$. Finally, since for the traveling domain the profile $\mathcal{E}^{(0)}(\xi)$ has a single maximum (the homoclinic path in the phase plane $(\mathcal{E}, \mathcal{E}_\xi)$ crosses the $\mathcal{E}$-axis at $\mathcal{E} = \mathcal{E}_\mathrm{m}$, situated in between $\mathcal{E}_2$ and $\mathcal{E}_3$, see Figure 3), the product $F(\xi)\mathcal{E}^{(0)}_\xi(\xi)$ has a single zero at the value of ξ corresponding to the maximal field achieved in the domain, $\mathcal{E}_\mathrm{m}$. It means that the wave function (3) is the one for the first excited state. The "energy" of the ground state should be lower than those for excited states. Since the first excited state has zero "energy" this gives rise to the conclusion that the ground state has negative "energy", i.e., the spectrum of Equations (4)–(6) has a single negative eigenvalue. In other words, the solution $\mathcal{E}^{(0)}(\xi)$ is *unstable* with the instability increment equal to the modulus of λ, corresponding to the ground state of the Schrödinger equation.

Though the problem in question has several parameters, the actual control parameter, which relatively easily may be varied in an experiment, is the current in the circuit, j_ext. Varying j_ext, one can change the values of $E_{1,2,3}$ and hence the shape of the traveling domain.

At a certain value of $j_\mathrm{ext} = j_0$, the maximal field in the domain, $\mathcal{E}_\mathrm{m}$ merges with $\mathcal{E}_3$, and the homoclinic path in the plane $(\mathcal{E}, \mathcal{E}_\xi)$ is split into two independent heteroclinic paths connecting the singular points $(\mathcal{E}_1, 0)$ and $(\mathcal{E}_3, 0)$. One of these paths lies entirely in the upper semi-plane. For this solution, $\mathcal{E}^{(0)}_\xi$ is always positive at any finite ξ. The other lies entirely in the lower semi-plane and, for it, $\mathcal{E}^{(0)}_\xi < 0$ at any finite ξ, see Figure 3. Each of these solutions corresponds to a traveling charge layer transferring the sample from one steady-state to another steady-state.

It is important that, since, for the layers, $\mathcal{E}_{\xi}^{(0)}$ does not vanish at any finite ξ, the corresponding wave function given by Equation (7) is the one of the *ground* state of the Schrödinger equation. This means that, in contrast to the traveling domain, the traveling layers are *stable* [18,19].

4. Spectrum of Schrödinger Equation

Thus, to get the value of the instability increment and, hence, the characteristic time for the traveling domain decomposition, one has to obtain the energy level for the ground state in the Schrödinger equation with the complicated potential given by Equation (6). As it has been mentioned above, the main idea employed in Ref. [1] to fulfill this task is to approximate the actual smooth potential by a superposition of square potential wells. The parameters of the wells are selected so that the approximated potential keeps all main features of the initial smooth profile. One of the approximation parameters remains free. It is fixed by the condition that, for the first excited state, $\Lambda = 0$.

This procedure reduces the solution of the complicated initial problem to finding the roots of a set of transcendental equations. In the worst case, the latter may be readily done numerically. To illustrate this rather general approach, a simple case of a broad traveling domain is discussed below.

The domain becomes broad when j_{ext} approaches j_0. In the phase plane, it corresponds to the shift of the regular point of the path $(\mathcal{E}_{\mathrm{m}}, 0)$ toward the saddle $(\mathcal{E}_3, 0)$. In this case, the domain approximately may be presented as a nonlinear superposition of two layers with opposite charges, and the potential (6) has a shape of two wells separated by a barrier.

Each well is associated with the corresponding layer. The width of the barrier equals the distance between the layers and is large in the case under consideration. Then, the tunneling through the barrier is exponentially weak. If the tunneling were suppressed entirely, each well would have corresponded to the potential generated by a single layer and, in accordance with the mentioned above, had the ground state with $\Lambda = 0$. Thus, it is clear that the level with $\Lambda < 0$ for the domain occurs due to the finite tunneling resulting in the splitting of the ground states in the two wells with the same value of $\Lambda = 0$. Therefore, instead of the employment of a rather cumbersome general procedure of the solution of an entire problem with the finite tunneling, let us, first, neglect the tunneling and consider the solutions of the Schrödinger equation in each square well separately. Then, the finite tunneling is taken into account with the help of perturbation theory.

The values of the parameters of the square wells approximating the smooth profile of the potential may be obtained by inspection of Equation (6). For example, bearing in mind that at $\xi \to \pm\infty$, the solution $\mathcal{E}^{(0)}(\xi)$ describing the domain satisfies the conditions $\mathcal{E}^{(0)} \to \mathcal{E}_1$; $\mathcal{E}_{\xi}^{(0)} \to 0$ and that, by definition, $u(\mathcal{E}_{1,2,3}) = 1$, one immediately obtains that both outer walls of the wells have a height equal to $u_{\mathcal{E}}(\mathcal{E}_1)/\mathcal{D}$. Similarly, the barrier height is $u_{\mathcal{E}}(\mathcal{E}_{\mathrm{m}})/\mathcal{D} \approx u_{\mathcal{E}}(\mathcal{E}_3)/\mathcal{D}$. The widths of the wells and the barrier are estimated based on the exact solution describing the domain path in the phase plane obtained in Ref. [18]. The last remaining parameters are the depths of the wells. They are fixed by the conditions that the ground state in each well has $\Lambda = 0$.

I do not present here these simple but cumbersome calculations. Just note that the problem is rather robust against errors in the approximation of $V(\xi)$. The robustness is related to the smallness of the split of the ground levels in the wells due to the tunneling and the fact that, at the employed approach, the important condition $\Lambda = 0$ for the ground state in each separate well holds automatically.

Let us suppose that the wave functions, $|1,2\rangle$, of the ground state for each well are known and that these wave functions satisfy the equations $\hat{H}_{1,2}|1,2\rangle = 0$. Here, $\hat{H}_{1,2}$ designates the Hamiltonians, whose potentials, $U_{1,2}$ are the corresponding single-well potentials. Let us look for the wave function of the complete problem with the two-well potential in a form of a linear superposition of $|1,2\rangle$:

$$|\psi\rangle = c_1|1\rangle + c_2|2\rangle, \tag{8}$$

where $c_{1,2}$ are constants, which should be defined in the course of calculations. Then, since $|\psi\rangle$ is an eigenfunction of the complete Hamiltonian $\hat{H}$,

$$c_1\hat{H}|1\rangle + c_2\hat{H}|2\rangle = \Lambda(c_1|1\rangle + c_2|2\rangle). \tag{9}$$

Making scalar products with $\langle 1,2|$ and taking into account the normalization conditions $\langle 1|1\rangle = \langle 2|2\rangle = 1$, one arrives from Equation (9) to the following equations for $c_{1,2}$:

$$(H_{11} - \Lambda)c_1 + (H_{12} - \Lambda\langle 1|2\rangle)c_2 = 0, \tag{10}$$

$$(H_{21} - \Lambda\langle 2|1\rangle)c_1 + (H_{22} - \Lambda)c_2 = 0, \tag{11}$$

where H_{11}, H_{12}, H_{21} and H_{22} stand for the corresponding matrix elements. Note that the wave functions $|1\rangle$ and $|2\rangle$ are *not orthogonal* since they are the eigenfunctions of the *different* Hamiltonians, namely, $\hat{H}_1$ and $\hat{H}_2 \neq \hat{H}_1$.

The solvability condition requires vanishing of the determinant of Equations (10) and (11). This results in a quadratic equation for Λ. The difference, $\Delta = |\Lambda_1 - \Lambda_2|$, between the two roots of this equation approximately equals the desired instability increment.

This result has completed the instability analysis. However, the exact expression for Δ is rather cumbersome. Therefore, it is worth simplifying this result, employing the smallness of certain parameters in Equations (10) and (11). To this end, I have to estimate the matrix elements and the overlap integrals $\langle 1,2|2,1\rangle$.

To calculate the matrix elements, it is convenient to single out from the full double-well square potential, $V_{\rm DWS}(\xi)$, the part corresponding to a single well, i.e., to suppose that $V_{\rm DWS}(\xi) = U_{1,2} - V_{1,2}$, see Figure 4, where for the first well (left side of Figure 4), $V_1 = 0$ at $\xi < \xi_3$, while, at $\xi > \xi_3$, the sum of $V_{\rm DWS}$ and V_1 equals the height of the barrier. Similarly, for the second well (right), the sum $V_{\rm DWS} + V_2$ equals the height of the barrier at $\xi < \xi_2$, while, at $\xi > \xi_2$, the potential $V_2 = 0$. Thus, $U_{1,2}$ are the single-well potentials, and the corresponding Hamiltonians acting on their wave functions produce zero. Then, the leading terms in the matrix elements are estimated as follows:

$$H_{11} = -\langle 1|V_1|1\rangle \sim H_{22} = -\langle 1|V_2|1\rangle \sim \exp[-2d_{\rm b}\sqrt{u_{\mathcal{E}}(\mathcal{E}_3)/\mathcal{D}}], \tag{12}$$

where $d_{\rm b} = \xi_3 - \xi_2$ is the barrier width. The same estimate is true for H_{22} (for the sake of simplicity, the widths of the corresponding wells, $d_{1,2}\sqrt{u_{\mathcal{E}}(\mathcal{E}_3)/\mathcal{D}}$, are supposed to be not small). In the same manner, one obtains

$$H_{12} \sim H_{21} \sim \langle 2|1\rangle \sim \langle 1|2\rangle \sim \exp[-d_b\sqrt{u_{\mathcal{E}}(\mathcal{E}_3)/\mathcal{D}}], \tag{13}$$

Then, in the leading approximation,

$$\Delta \approx H_{12} + H_{21} \sim \exp[-d_b\sqrt{u_{\mathcal{E}}(\mathcal{E}_3)/\mathcal{D}}]. \tag{14}$$

Thus, as it could be expected, the value of the instability increment for the broad domain is exponentially small indeed.

Finally, note that, for each single square well, the normalized wave functions $|1,2\rangle$ may be readily obtained in the explicit form, including the normalization constant. Then, it is just a matter of more or less routine calculations to improve the accuracy of Equation (14), taking into account the prefactors and dropped higher-order exponentially small terms.

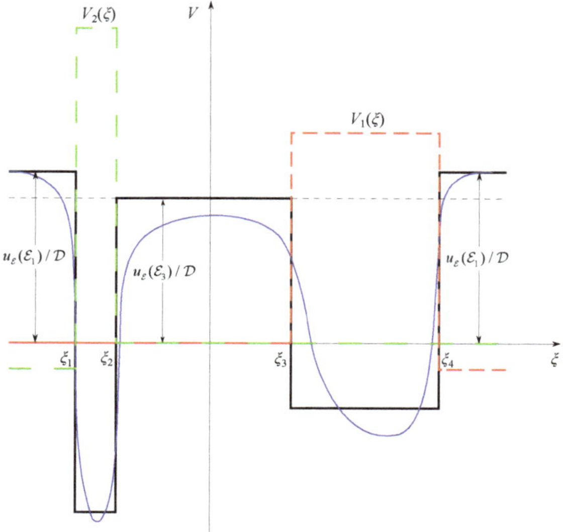

Figure 4. Schematically: The actual double-well potential $V(\xi)$ (smooth blue line). The approximation of $V(\xi)$ by the double-well square potential, $V_{\mathrm{DWS}}(\xi)$, is shown in black; $\xi_{1,2,3,4}$ designate the coordinates of the walls of the wells. $U_{1,2}(\xi) = V_{\mathrm{DWS}}(\xi) + V_{1,2}(\xi)$ are the potentials of the single-well approximation, when tunneling is neglected.

5. Conclusions

Summarizing and generalizing the discussed above, one arrives at the following conclusions:

- The analysis of the linear stability of traveling wave solutions in a wide class of nonlinear diffusion problems is reduced to inspection of a bottom part of the spectrum of the associated Schrödinger equation, whose potential is generated by the profile of the analyzed solution.
- The translational invariance transformation generates in the stability spectrum a neutrally-stable (Goldstone) mode.
- The qualitative answer to the question about the stability of the solution is readily obtained based on the oscillation theorem—if the Goldstone mode does not have any nodes, the solution is stable. Otherwise, it is unstable.
- To quantitatively characterize the instability (if any), the "energy" level of the ground state of the Schrödinger equation should be obtained.
- A powerful tool to make the problem of a bottom part of the Schrödinger equation spectrum tractable is to approximate the potential by square wells.

These conclusions are rather general. They are valid far beyond the frameworks of the Gunn effect and, hopefully, may help to analyze the stability of traveling waves in a broad class of nonlinear diffusion problems.

The developed approach to find an approximate solution and spectrum of the Schrödinger equation with a complicated potential valid for ground and low-excited states may be regarded as a complement to the known Wentzel-Kramers-Brillouin (WKB) method, which is good in the opposite case of high-excited states. A common disadvantage of both approaches is the approximation error, which is difficult to improve and even to control.

Funding: This work is financed by the Moscow Engineering Physics Institute Academic Excellence Project (in agreement with the Ministry of Education and Science of the Russian Federation of 27 August 2013, Project No. 02.a03.21.0005).

Acknowledgments: I am very grateful to the entire Editorial staff of *Physics* for this Special Issue dedicated to my humble person. My special thanks to my friends and colleagues Boris Malomed, Andrey Miroshnichenko, and Fernando Moreno, who volunteered to be Guest Editors of the Special Issue and completed this difficult task in an excellent manner. Thank you very much indeed!

Conflicts of Interest: The author declares no conflict of interest.

References

1. Tribel'skii, M. On the increment of the instability of the Gunn domains in the direct current regime. *Fiz. Tekhnika Poluprovodn.* **1971**, *5*, 2033. (In Russian)
2. Aleksandrov, E.B.; Alferov, Z.I.l; Apanasevich, P.A.; Bonch-Bruevich, A.M.; Borisevich, N.A.; Burakov, V.S.; Gol'danskii, V.I.; Voitovich, A.P.; Gribov, L.A.; Martynenko, O.M.; et al. In memory of Mikhail Aleksandrovich El'yashevich. *Phys.-Uspekhi* **1996**, *39*, 853–854. [CrossRef]
3. Dzyaloshinskii, I.E.; Zakharov, V.E.; Inogamov, N.A.; Kolokolov, I.V.; Kuznetsov, E.A.; Lebedev, V.V.; Novikov, S.P.; Sagdeev, R.Z.; Starobinskii, A.A.; Stishov, S.M.; et al. In memory of Sergei Ivanovich Anisimov. *Phys.-Uspekhi* **2019**, *62*, 1282–1283. [CrossRef]
4. Khalatnikov, I.M. History of the Landau Institute. Available online: https://www.itp.ac.ru/en/about/history/ (accessed on 15 August 2021).
5. Grosberg, A.Y.; Kaganov, M.I.; Tribel'skii, M.I.; Khokhlov, A.R. Il'ya Mikhailovich Lifshits (on the 80th anniversary of his birth). *Phys.-Uspekhi* **1997**, *40*, 225–226. [CrossRef]
6. P.L. Kapitza Institute for Physical Problems of Russian Academy of Sciences. Available online: https://www.kapitza.ras.ru/ (accessed on 15 August 2021).
7. Halperin, B.; Langer, J.; Mints, R. Mark Yakovlevich Azbel. *Phys. Today* **2020**, *73*, 67. [CrossRef]
8. Aleksandrov, A.P.; Velikhov, E.P.; Ginzburg, V.L.; Zel'dovich, Y.B.; Kadomtsev, B.B.; Lifshitz, E.M.; Prokhorov, A.M.; Rytov, S.M.; Sagdeev, R.Z.; Shafranov, V.D. Mikhail Aleksandrovich Leontovich (Obituary). *Phys.-Uspekhi* **1981**, *24*, 730–731. [CrossRef]
9. Gunn, J.B. Microwave oscillations of current in III–V semiconductors. *Solid State Commun.* **1963**, *1*, 88–91. [CrossRef]
10. Solymar, L.; Walsh, D.; Syms, R.R.A. *Principles of Semiconductor Devices*; Oxford University Press: Oxford, UK, 2004.
11. Hsu, H.W.; Dominguez, M.J.; Sih, V. Gunn threshold voltage characterization in GaAs devices with wedge-shaped tapering. *J. Appl. Phys.* **2020**, *128*, 074502. [CrossRef]
12. Hajo, A.S.; Yilmazoglu, O.; Dadgar, A.; Küppers, F.; Kusserow, T. Reliable GaN-based THz Gunn diodes with side-contact and field-plate technologies. *IEEE Access* **2020**, *8*, 84116–84122. [CrossRef]
13. Semyonov, E.V.; Malakhovskij, O.Y. Short-pulse properties of the Gunn diode. *IEEE Trans. Electron. Devices* **2020**, *67*, 2100–2105. [CrossRef]
14. Novoa-López, J.; Paz-Martínez, G.; Sánchez-Martín, H.; Lechaux, Y.; Íñiguez-de-la Torre, I.; González, T.; Mateos, J. Temperature behavior of Gunn oscillations in planar InGaAs diodes. *IEEE Electron. Device Lett.* **2021**, *42*, 1136–1139. [CrossRef]
15. Liang, F.; Wang, H.; Pan, J.; Li, J.; Xu, K.; Luo, Y. Voltage-driven mutual phase locking of planar nano-oscillators. *Int. J. Mod. Phys. B* **2021**, *35*, 2150072. [CrossRef]
16. Bonilla, L.L.; Teitsworth, S.W. *Nonlinear Wave Methods for Charge Transport*; John Wiley & Sons: Hoboken, NJ, USA, 2009; Chapter 6.
17. Kroemer, H. Theory of the Gunn effect. *Proc. IEEE* **1964**, *52*, 1736. [CrossRef]
18. Knight, B.; Peterson, G. Theory of the Gunn effect. *Phys. Rev.* **1967**, *155*, 393–404. [CrossRef]
19. Volkov, A.F.; Kogan, S.M. Physical phenomena in semiconductors with negative differential conductivity. *Phys.-Uspekhi* **1969**, *11*, 881–903. [CrossRef]
20. Landau, L.D.; Lifshitz, E.M. *Quantum Mechanics: Non-Relativistic Theory*; Elsevier: Amsterdam, The Netherlands, 2013.
21. Lifshits, I.M.; Rzhevskii, V.V.; Tribel'skii, M.I. Nonlinear acoustic effects in metals near the electron topological transition point. *Sov. Phys. JETP* **1981**, *54*, 810–817. Available online: http://www.jetp.ras.ru/cgi-bin/dn/e_054_04_0810.pdf (accessed on 18 August 2021).
22. Andreev, A.F.; Brazovskii, S.A.; Volovik, G.E.; Kats, E.I.; Kondratenko, P.S.; Kreines, N.M; Mineev, V.P.; Novikov, S.P.; Obukhov, S.P.; Pitaevskii, L.P.; et al. Igor Ekhiel'evich Dzyaloshinskii (on his 90th birthday). *Phys.-Uspekhi* **2021**, *64*, 214–215. [CrossRef]
23. Tribel'skii, M.I. Gunn domain instability increment. *Sov. Phys. Solid Sate* **1974**, *15*, 2448.
24. Zeldovich, Y.B.; Barenblatt, G.I. Theory of flame propagation. *Combust. Flame* **1959**, *3*, 61–74. [CrossRef]
25. Tribelsky, M.I.; Tsuboi, K. New scenario for transition to turbulence? *Phys. Rev. Lett.* **1996**, *76*, 1631–1634. [CrossRef] [PubMed]
26. Hidaka, Y.; Huh, J.H.; Hayashi, K.I.; Kai, S.; Tribelsky, M.I. Soft-mode turbulence in electrohydrodynamic convection of a homeotropically aligned nematic layer. *Phys. Rev. E* **1997**, *56*, R6256–R6259. [CrossRef]
27. Tribel'skii, M.I. Short-wavelength instability and transition to chaos in distributed systems with additional symmetry. *Phys.-Uspekhi* **1997**, *40*, 159–180. [CrossRef]

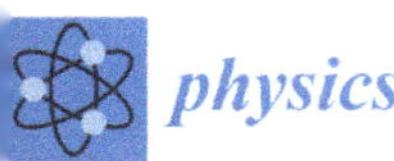 physics

Communication

Enhanced Chiral Mie Scattering by a Dielectric Sphere within a Superchiral Light Field

Haifeng Hu and Qiwen Zhan *

School of Optical-Electrical and Computer Engineering, University of Shanghai for Science and Technology, Shanghai 200093, China; hfhu@usst.edu.cn
* Correspondence: qwzhan@usst.edu.cn

Abstract: A superchiral field, which can generate a larger chiral signal than circularly polarized light, is a promising mechanism to improve the capability to characterize chiral objects. In this paper, Mie scattering by a chiral sphere is analyzed based on the T-matrix method. The chiral signal by circularly polarized light can be obviously enhanced due to the Mie resonances. By employing superchiral light illumination, the chiral signal is further enhanced by 46.8% at the resonance frequency. The distribution of the light field inside the sphere is calculated to explain the enhancement mechanism. The study shows that a dielectric sphere can be used as an excellent platform to study the chiroptical effects at the nanoscale.

Keywords: Mie scattering; superchirality; circular dichroism; T-matrix

Citation: Hu, H.; Zhan, Q. Enhanced Chiral Mie Scattering by a Dielectric Sphere within a Superchiral Light Field. *Physics* **2021**, 3, 747–756. https://doi.org/10.3390/physics3030046

Received: 30 June 2021
Accepted: 23 August 2021
Published: 2 September 2021

Publisher's Note: MDPI stays neutral with regard to jurisdictional claims in published maps and institutional affiliations.

1. Introduction

Chirality describes the asymmetric feature of a three-dimensional object, whose mirror image is different from itself [1]. Objects with chirality are quite common in nature. In the human body, most of the important molecules have chirality, such as DNA, enzyme, and protein. As a wave phenomenon, the light field also has chirality [2]. A well-known example is circularly polarized light (CPL). Left circularly polarized (LCP) and right circularly polarized (RCP) light beams are a pair of enantiomers that have opposite spins. The interaction between the light field and chiral objects has attracted great interest from researchers [3]. The chiral properties of the objects can be evaluated by the different responses under illumination of LCP and RCP lights. This chiral signal is useful to distinguish and measure the chirality of a sample. In natural materials, the chiral signal is very weak (10^{-3}–10^{-6}) [1]. The main reason is the mismatch between the size of the chiral molecules and the light wavelength. In previous studies, it has been shown that plasmonic structures can increase chiral signals by orders of magnitude owing to the strong light confinement near the metallic surface [4]. The plasmonic chiral structure is used to improve the sensing performance of enantiomers [5]. In a nonchiral structure, optical chirality can be largely increased [6]. It has been shown that the local chirality of the light field can be tailored near the plasmonic structure. The optical antenna theory can be used to design the chiral light [7]. Symmetric metal–dielectric–metal (MDM) metamaterial structures have also been proposed to enhance the chiral light–matter interaction [8]. However, metals have intrinsic loss, which fundamentally limits the practical performance of the designed plasmonic structures. Recently, dielectric nanoparticles with high refractive indices have also been used to manipulate the light field at the nanoscale [9]. The supercavity with high q-factor can be designed by subwavelength high-index particles [10]. The electric and magnetic resonances in silicon nanoparticles can be utilized to enhance the quantum efficiency of silicon nanoparticles [11]. The anapole mode in such kind of particles has been intensively studied recently [12]. The simplest case of an anapole is caused by the destructive interference between the electric dipole and the toroidal moment. In 2015, the radiationless anapole mode in visible was first observed [13]. By tuning the geometry of the scattering particle, an obvious dip can be found in the far-field scattering spectrum, which is caused by the excitation of the anapole mode. Moreover, by tailoring the incident field, the anapole mode can also be excited in a dielectric sphere [14]. The electric and magnetic resonances can be strongly excited inside the high-index particles with low loss which may further enable the chirality enhancement. Besides the enhancement mechanism arising from near-field structures, the enhancement of a chiral signal can also be realized by tailoring the light field. Traditionally, circularly polarized light is believed to have the highest chirality. However, Cohen proposed the superchiral field, which has larger chirality than circular polarized light [15,16]. Recently, it was demonstrated by us that a localized superchiral hotspot can be generated by tightly focusing a vectorial light beam with orbital angular momentum [17].

In this study, a method based on the T-matrix is employed to efficiently analyze the scattering process by a chiral sphere. It is demonstrated that two enhancement mechanisms can be employed to enhance the scattering circular dichroism (CD) signal of a chiral sphere simultaneously.

2. Materials and Methods

To consider the chiral property of the sphere in our theoretical model, the following constitutive relations are used [18]:

$$\mathbf{D} = \varepsilon_s \mathbf{E} - j\kappa_s \mathbf{H}, \tag{1}$$

$$\mathbf{B} = \mu_s \mathbf{H} + j\kappa_s \mathbf{E}. \tag{2}$$

In Equations (1) and (2), ε_s, μ_s, and κ_s are the permittivity, permeability, and chirality parameters of the material in the sphere. Following the procedure proposed by Bohren [19], the electromagnetic field in the chiral sphere ($\mathbf{E}_1$, $\mathbf{H}_1$) can be expressed by the linear transformation relation:

$$\begin{bmatrix} \mathbf{E}_1 \\ \mathbf{H}_1 \end{bmatrix} = \begin{bmatrix} j\sqrt{\varepsilon_s\mu_s} & \mu_s \\ \varepsilon_s & j\sqrt{\varepsilon_s\mu_s} \end{bmatrix} \begin{bmatrix} \mathbf{Q}_L \\ \mathbf{Q}_R \end{bmatrix}. \tag{3}$$

Because of the transformed fields, $\mathbf{Q}_L$ and $\mathbf{Q}_R$ should satisfy the wave equation, $\nabla^2\mathbf{Q}_{L/R} + k_{L/R}^2\mathbf{Q}_{L/R} = 0$. They can be expanded by the linear combination of vector spherical harmonics (VSHs):

$$\mathbf{Q}_L = \sum_{n=1}^{+\infty}\sum_{m=-n}^{+n} c_{mn}[Rg\mathbf{M}_{mn}(k_L) + Rg\mathbf{N}_{mn}(k_L)], \tag{4}$$

$$\mathbf{Q}_R = \sum_{n=1}^{+\infty}\sum_{m=-n}^{+n} d_{mn}[Rg\mathbf{M}_{mn}(k_R) - Rg\mathbf{N}_{mn}(k_R)], \tag{5}$$

where $k_{L/R} = \omega\sqrt{\varepsilon_s\mu_s} \mp \omega\kappa_s$, and m and n are both integers. In this study, $Rg\mathbf{M}_{mn}$ and $Rg\mathbf{N}_{mn}$ are regular VSHs whose values at origin are finite, and $\mathbf{M}_{mn}$ and $\mathbf{N}_{mn}$ are outgoing VSHs with singularity points at origin. The definitions of these VSH functions can be found in [20]. Outside the sphere, the incident field ($\mathbf{E}_{\text{inc}}$, $\mathbf{H}_{\text{inc}}$ and scattering field ($\mathbf{E}_s$, $\mathbf{H}_s$) can be expressed as the combinations of VSHs:

$$\mathbf{E}_{\text{inc}} = \sum_{n=1}^{+\infty}\sum_{m=-n}^{n} (u_{mn}Rg\mathbf{M}_{mn} + v_{mn}Rg\mathbf{N}_{mn}), \tag{6}$$

$$\mathbf{H}_{\text{inc}} = \frac{1}{iZ_0}\sum_{n=1}^{+\infty}\sum_{m=-n}^{n} (u_{mn}Rg\mathbf{N}_{mn} + v_{mn}Rg\mathbf{M}_{mn}), \tag{7}$$

$$\mathbf{E}_s = \sum_{n=1}^{+\infty}\sum_{m=-n}^{n} (a_{mn}\mathbf{M}_{mn} + b_{mn}\mathbf{N}_{mn}), \tag{8}$$

$$\mathbf{H}_s = \frac{1}{iZ_0}\sum_{n=1}^{+\infty}\sum_{m=-n}^{n}(a_{mn}\mathbf{N}_{mn} + b_{mn}\mathbf{M}_{mn}). \tag{9}$$

In Equations (6) and (7), the expansion coefficients u_{mn} and v_{mn} can be determined by the incident light field. The multipolar decomposition of various types of focused beams was considered in previous studies [21]. The expansion coefficients of u_{mn} and v_{mn} for the scattering field can be calculated by the T-matrix method. According to the boundary condition on the surface of the sphere (r = R_s), the T-matrix for a chiral sphere can be obtained from:

$$\begin{bmatrix} \mathbf{E}_s \\ \mathbf{H}_s \end{bmatrix} = \mathbf{T}\begin{bmatrix} \mathbf{E}_{\text{inc}} \\ \mathbf{H}_{\text{inc}} \end{bmatrix} = \left[\mathbf{T}^{(1)}\right]^{-1}\mathbf{T}^{(2)}\begin{bmatrix} \mathbf{E}_{\text{inc}} \\ \mathbf{H}_{\text{inc}} \end{bmatrix}. \tag{10}$$

In Equation (10), the elements of the matrices $\mathbf{T}^{(1)}$ and $\mathbf{T}^{(2)}$ are given below:

$$\mathbf{T}^{(1)}_{11,mn} = iZ_0\varepsilon_s\xi_n(k_0R_s)\psi'_n(k_RR_s) - i\sqrt{\varepsilon_s\mu_s}\psi_n(k_RR_s)\xi'_n(k_0R_s), \tag{11}$$

$$\mathbf{T}^{(1)}_{12} = iZ_0\varepsilon_s\psi_n(k_RR_s)\xi'_n(k_0R_s) - i\sqrt{\varepsilon_s\mu_s}\xi_n(k_0R_s)\psi'_n(k_RR_s), \tag{12}$$

$$\mathbf{T}^{(1)}_{21} = -Z_0\sqrt{\varepsilon_s\mu_s}\xi_n(k_0R_s)\psi'_n(k_LR_s) + \mu_s\psi_n(k_LR_s)\xi'_n(k_0R_s), \tag{13}$$

$$\mathbf{T}^{(1)}_{22} = Z_0\sqrt{\varepsilon_s\mu_s}\psi_n(k_LR_s)\xi'_n(k_0R_s) - \mu_s\xi_n(k_0R_s)\psi'_n(k_LR_s), \tag{14}$$

$$\mathbf{T}^{(2)}_{11} = -iZ_0\varepsilon_s\psi_n(k_0R_s)\psi'_n(k_RR_s) + i\sqrt{\varepsilon_s\mu_s}\psi_n(k_RR_s)\psi'_n(k_0R_s), \tag{15}$$

$$\mathbf{T}^{(2)}_{12} = -iZ_0\varepsilon_s\psi_n(k_RR_s)\psi'_n(k_0R_s) + i\sqrt{\varepsilon_s\mu_s}\psi_n(k_0R_s)\psi'_n(k_RR_s), \tag{16}$$

$$\mathbf{T}^{(2)}_{21} = Z_0\sqrt{\varepsilon_s\mu_s}\psi_n(k_0R_s)\psi'_n(k_LR_s) - \mu_s\psi_n(k_LR_s)\psi'_n(k_0R_s), \tag{17}$$

$$\mathbf{T}^{(2)}_{22} = -Z_0\sqrt{\varepsilon_s\mu_s}\psi_n(k_LR_s)\psi'_n(k_0R_s) + \mu_s\psi_n(k_0R_s)\psi'_n(k_LR_s). \tag{18}$$

In Equations (11)–(18), the Riccati–Bessel functions are defined as $\psi_n(\rho) = \rho j_n(\rho)$ and $\xi_n(\rho) = \rho h_n^{(1)}(\rho)$. Z_0 is the vacuum impedance. k_0 = ω/c is the wave vector in vacuum. According to the Mie theory, the scattering power can be calculated by:

$$W_{\text{scat}} = \frac{1}{2k_0^2}\sum_{n=1}^{+\infty}\sum_{m=-n}^{n}\left(|a_{mn}|^2 + |b_{mn}|^2\right). \tag{19}$$

The radius of the sphere is R_s = 75 nm, and the parameters of the chiral material are ε_s = 25, μ_s = 1, and κ_s = 0.01. The scattering spectrum under the illumination of linearly polarized light in free space (n_0 = 1) is shown in Figure 1a. In the frequency range of 200–700 THz, the scattering power is mainly caused by terms related to an electric dipole (ED), magnetic dipole (MD), and magnetic quadrupole (MQ) [22]. Their contributions to the scattering power can be analyzed by the following three terms from the total scattering power:

$$W_{\text{ED}} = \frac{1}{2k_0^2}\sum_m |a_{m1}|^2, \tag{20}$$

$$W_{\text{MD}} = \frac{1}{2k_0^2}\sum_m |b_{m1}|^2, \tag{21}$$

$$W_{\text{MQ}} = \frac{1}{2k_0^2}\sum_m |b_{m2}|^2. \tag{22}$$

In Figure 1b, the spectral curves for the three scattering terms are helpful to analyze the mechanisms of the Mie resonances. It can be clearly demonstrated that the three peaks of W_{scat} in Figure 1a are caused by the MD, ED, and MQ resonances at frequencies of 388.5, 538.5, and 563.5 THz, respectively. Linearly polarized light can be represented as the superposition of left- and right-hand circularly polarized light. The circularly polarized light is useful to explore the properties of materials [23]. To evaluate the chiral response of the sphere, the scattering CD parameter g_{scat} is calculated, which is defined as $g_{\text{scat}} = 2(W_{\text{scat}}^{+} - W_{\text{scat}}^{-})/(W_{\text{scat}}^{+} + W_{\text{scat}}^{-})$. In this definition, W_{scat}^{+} and W_{scat}^{-} are the scattering energies by the chiral sphere under the illuminations of LCP and RCP light. There are two regions in the spectrum in Figure 1c, where the scattering CD single is enhanced. One region is near the MD resonance wavelength, and the other is near the ED resonance wavelength. However, the signs of CD values in the two ranges are opposite. The inversion of the optical chiral response can be attributed to the main absorption mechanism, which is converted from MD to ED when the light frequency is increased. In the region where ED absorption is dominant, there exists a deep valley at a frequency of 563.5 THz, which is exactly the MQ resonance frequency.

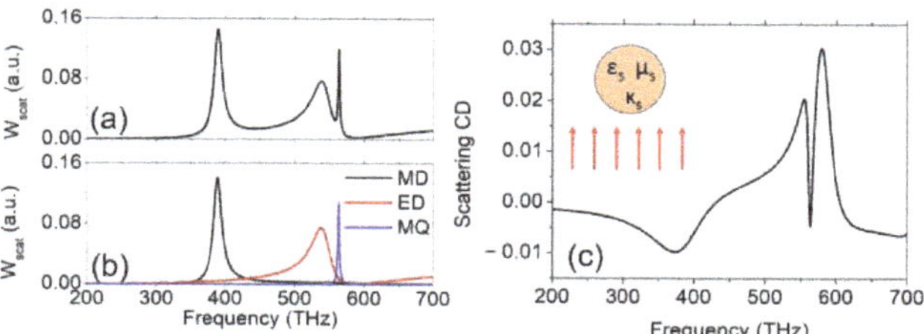

Figure 1. (**a**) The scattering spectrum of a chiral sphere with a radius of 75 nm under the illumination of linearly polarized light. The optical parameters of the sphere are $\varepsilon_s = 25$ (permittivity), $\mu_s = 1$ (permeability), and $\kappa_s = 0.01$ (chirality). The sphere is in free space (refractive index $n_0 = 1$). (**b**) The contributions to the total scattering energies from three different mechanisms (magnetic dipole, electric dipole, and magnetic quadrupole). (**c**) The circular dichroism (CD) spectrum of g_{scat} under the illumination of left circularly polarized (LCP) and right circularly polarized (RCP) light.

As discussed above, the light field can be tailored to gain larger chirality than CPL. To evaluate the chirality of the light field, the parameter of the CD enhancement factor is defined as g/g_{CPL}, where g and g_{CPL} are the CD values of a chiral dipole excited by the tailored light and CPL light. When $g/g_{\text{CPL}} > 1$, the tailored light can produce a larger CD signal than CPL light. This kind of light is called superchiral field. Recently, it was reported by us a highly localized superchiral hotspot by tightly focusing a radially polarized beam with orbital angular momentum near the dielectric interface [17]. In this study, the capability of the superchiral field to increase the scattering CD signal can be verified, when the particle is at the Mie resonance. It has been proved that when the incident angle of a focused beam is slightly smaller than the critical angle of totally internal reflection, the chirality of the light field (g/g_{CPL}) can be largely enhanced. Therefore, the configuration in Figure 2 is considered. The chiral sphere is on the substrate with a refractive index of n_1, which is larger than n_0. To analyze the interaction between the chiral sphere on the substrate and the superchiral field, the T-matrix in Equation (10) should be modified to incorporate the substrate into the theoretical model [22,24]. The scattering process is illustrated in Figure 2. The light propagates from the substrate (n_1) to the air (n_0), and the chiral sphere is illuminated by the transmitted light ($\mathbf{E}_{\text{inc}}$, $\mathbf{H}_{\text{inc}}$). The scattering field is expressed as $[\mathbf{E}_s, \mathbf{H}_s]^{\text{T}} = \mathbf{T}[\mathbf{E}_{\text{inc}}, \mathbf{H}_{\text{inc}}]^{\text{T}}$, where the T-matrix is given in Equations (11)–(18). As shown by the red arrow in Figure 2, a part of the scattering field can be reflected by the interface, and then the reflected field can also be scattered by the sphere again. One can see that there is a round trip of light between the sphere and the interface. Therefore, the multiple reflections between the sphere and the interface should be considered to build the accurate scattering model. In this study, the matrix of $\mathbf{L}_{\text{R}}$ is the reflection matrix, and the reflected field of the scattering field can be expressed as $\mathbf{L}_{\text{R}}\,[\mathbf{E}_s, \mathbf{H}_s]^{\text{T}}$. The final scattering field of the multiscattering process can be calculated by Equation (23) [24].

$$\begin{bmatrix} \mathbf{E}_s \\ \mathbf{H}_s \end{bmatrix} = \sum_{\alpha=0}^{+\infty} (\mathbf{T}\mathbf{L}_{\text{R}})^{\alpha}\mathbf{T}\begin{bmatrix} \mathbf{E}_{\text{inc}} \\ \mathbf{H}_{\text{inc}} \end{bmatrix} = [\mathbf{I} - \mathbf{T}\mathbf{L}_{\text{R}}]^{-1}\mathbf{T}\begin{bmatrix} \mathbf{E}_{\text{inc}} \\ \mathbf{H}_{\text{inc}} \end{bmatrix}. \tag{23}$$

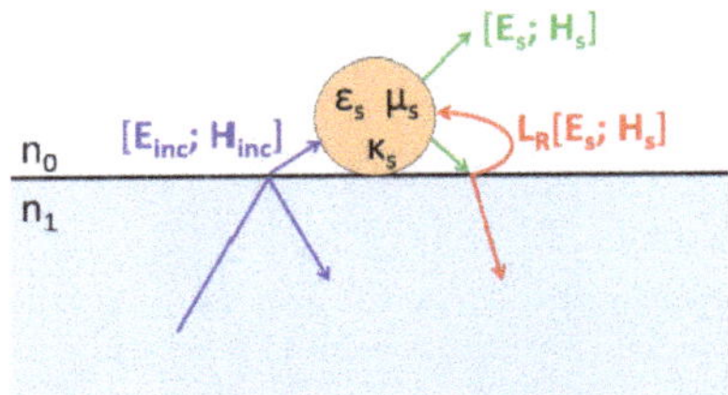

Figure 2. A schematic of the theoretical model to analyze light scattering by a chiral sphere on the substrate. The blue and green arrows represent the path of the incident light ($\mathbf{E}_{\text{inc}}$ and $\mathbf{H}_{\text{inc}}$) and scattering light ($\mathbf{E}_s$ and $\mathbf{H}_s$). The red arrows represent the light reflected and transmitted at the interface. $\mathbf{L}_\text{R}$ represent the reflection matrix.

According to Equation (23), the T-matrix for the sphere in free space should be modified as the effective one, which is $\mathbf{T}_{\text{eff}} = [\mathbf{I} - \mathbf{T}\mathbf{L}_\text{R}]^{-1}\mathbf{T}$. **I** represents the unit matrix. The elements of $\mathbf{L}_\text{R}$ are the reflection coefficients of VSHs by the interface. To calculate these coefficients, the VSH is first expanded by the plane waves. The directions of the plane waves are represented by $\left(\widetilde{\theta}, \widetilde{\varphi}\right)$, where $\widetilde{\theta}$ is the polar angle and $\widetilde{\varphi}$ is the azimuthal angle. The reflection coefficient of each plane wave can be calculated independently by the Fresnel equation. The reflected field of the VSH can be obtained by recombining all the reflected plane wave components. After these operations, the reflected field of each VSH field scattered by the sphere can be calculated. To calculate the elements of $\mathbf{L}_\text{R}$, the reflected fields have to be expended by the VSH basis, which has been used to express the incident field of $[\mathbf{E}_{\text{inc}}, \mathbf{H}_{\text{inc}}]^\text{T}$ in Equations (6) and (7). The expansions of the reflected VSHs, $\mathbf{M}_{mn,r}$ and $\mathbf{N}_{mn,r}$ can be expressed as:

$$\mathbf{M}_{mn,r} = \sum_{\nu=1}^{+\infty} \sum_{\mu=-\nu}^{\nu} \left[A_{\mu\nu,mn} Rg\mathbf{M}_{\mu\nu} + B_{\mu\nu,mn} Rg\mathbf{N}_{\mu\nu}\right], \tag{24}$$

$$\mathbf{N}_{mn,r} = \sum_{\nu=1}^{+\infty} \sum_{\mu=-\nu}^{\nu} \left[C_{\mu\nu,mn} Rg\mathbf{M}_{\mu\nu} + D_{\mu\nu,mn} Rg\mathbf{N}_{\mu\nu}\right]. \tag{25}$$

In Equations (24) and (25), the expansion coefficients can be determined by the following integrals with the integration path of $\widetilde{\theta}$ being $[\pi/2 + i\infty, \pi]$:

$$A_{\mu\nu,mn} = \delta_{\mu m} \gamma_{mn,\nu} \int_C \left(r_\theta m^2 \pi_n^m \pi_\nu^m - r_f \tau_n^m \tau_\nu^m\right) e^{2ik_0 \cos\widetilde{\theta} R_s} \sin\widetilde{\theta} d\widetilde{\theta}, \tag{26}$$

$$B_{\mu\nu,mn} = -m\delta_{\mu m} \gamma_{mn,\nu} \int_C \left(r_\theta \pi_n^m \tau_\nu^m - r_f \tau_n^m \pi_\nu^m\right) e^{2ik_0 \cos\widetilde{\theta} R_s} \sin\widetilde{\theta} d\widetilde{\theta}, \tag{27}$$

$$C_{\mu\nu,mn} = m\delta_{\mu m} \gamma_{mn,\nu} \int_C \left(r_\theta \tau_n^m \pi_\nu^m - r_f \pi_n^m \tau_\nu^m\right) e^{2ik_0 \cos\widetilde{\theta} R_s} \sin\widetilde{\theta} d\widetilde{\theta}, \tag{28}$$

$$D_{\mu\nu,mn} = -\delta_{\mu m} \gamma_{mn,\nu} \int_C e^{2ik_0 \cos\widetilde{\theta} R_s} \left(r_\theta \tau_n^m \tau_\nu^m - r_f m^2 \pi_n^m \pi_\nu^m\right) \sin\widetilde{\theta} d\widetilde{\theta}. \tag{29}$$

In Equations (26)–(29), r_θ and r_ϕ are the Fresnel reflection coefficients for p-polarized and s-polarized light, τ_n^m and π_n^m are the angle-dependent functions in the Mie theory for convenience [20]. The expression of $\gamma_{mn,\nu}$ is in Equations (26)–(29), which is only related to the order numbers (n, m and ν):

$$\gamma_{mn,\nu} = (-1)^{\nu+m} i^{\nu-n} \sqrt{\frac{(2n+1)(n-m)!}{n(n+1)(n+m)!}} \sqrt{\frac{(2\nu+1)(\nu-m)!}{\nu(\nu+1)(\nu+m)!}}. \tag{30}$$

After obtaining the coefficients in Equations (26)–(29), the reflected matrix $\mathbf{L}_\mathrm{R}$ can be expressed as:

$$\mathbf{L}_\mathrm{R} = \begin{bmatrix} \mathbf{A} & \mathbf{C} \\ \mathbf{B} & \mathbf{D} \end{bmatrix}. \tag{31}$$

Based on the theoretical model, light scattering by the sphere in the superchiral field can be analyzed efficiently. To further simplify the calculation, let us consider the vectorial Bessel beam as the incident light, whose field in a cylindrical coordinate can be expressed as [25]:

$$\mathbf{E}_\mathrm{inc} = e^{im_0\varphi + ik_1 z\cos\widetilde{\theta}} \begin{pmatrix} i\left[J_{m_0+1}(k_1 r\sin\widetilde{\theta}) - J_{m_0-1}(k_1 r\sin\widetilde{\theta})\right]\cos\widetilde{\theta}\mathbf{n}_r \\ \left[J_{m_0+1}(k_1 r\sin\widetilde{\theta}) + J_{m_0-1}(k_1 r\sin\widetilde{\theta})\right]\cos\widetilde{\theta}\mathbf{n}_\varphi \\ -2J_{m_0}(k_1 r\sin\widetilde{\theta})\sin\widetilde{\theta}\mathbf{n}_z \end{pmatrix}, \tag{32}$$

$$\mathbf{H}_\mathrm{inc} = \frac{1}{Z_1} e^{im_0\varphi + ik_1 z\cos\widetilde{\theta}} \begin{pmatrix} -\left[J_{m_0+1}(k_1 r\sin\widetilde{\theta}) + J_{m_0-1}(k_1 r\sin\widetilde{\theta})\right]\mathbf{n}_r \\ i\left[J_{m_0+1}(k_1 r\sin\widetilde{\theta}) - J_{m_0-1}(k_1 r\sin\widetilde{\theta})\right]\mathbf{n}_\varphi \\ 0 \end{pmatrix}. \tag{33}$$

In Equations (32) and (33), m_0 is the topological charge of the incident field. $k_1 = n_1 k_0$ is the wave vector in the substrate. The refractive indexes of the substrate and air are $n_1 = 1.518$ and $n_0 = 1$. As shown in Figure 3a, this vectorial light beam can be regarded as the superposition of the p-polarized plane waves, whose wave vector **k** lives in a cone with the polar angle of $\widetilde{\theta}$ and the azimuthal angle $\widetilde{\varphi} \in [0, 2\pi]$ in k-space. The chirality enhancement factor at the focus point ($r = 0$) is calculated when the polar angle is changed from zero to the critical angle of the total internal reflection, as shown in Figure 3b. When $\widetilde{\theta}$ is slightly smaller than the critical angle of $\widetilde{\theta}_c$, the hotspot region with high optical chirality can be realized near the focus point. The sign of optical chirality is the same with the topological charge m_0. Therefore, the scattering energies of W^+_scat and W^-_scat in the definition of g_scat are calculated when, $m_0 = 1$ and, $m_0 = -1$ in Equations (32) and (33).

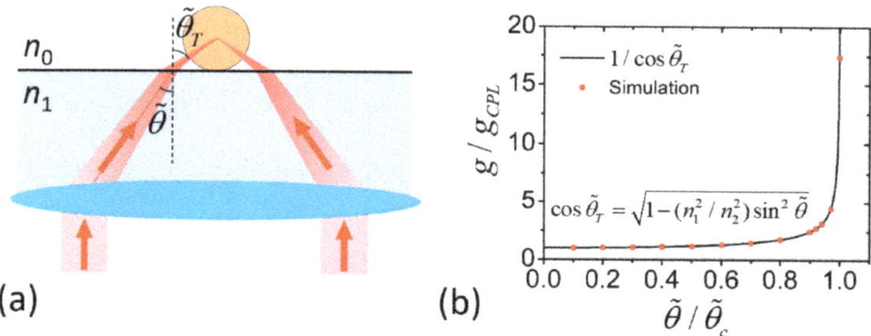

Figure 3. (**a**) The model to analyze light scattering by the sphere in the superchiral field. The incident field is the vectorial Bessel beam with orbital angular momentum of Equations (32) and (33), with the topological charge, $m_0 = \pm 1$. The refractive indexes of the substrate and air are $n_1 = 1.518$ and $n_0 = 1$. The incident angle is $\widetilde{\theta}$, and the refraction angle is $\widetilde{\theta}_\mathrm{T}$ (**b**) The enhancement factor of the chiral signal from a dipole at the focus ($r = 0$). g and g_CPL are the CD factors under the superchiral light and circularly polarized light (CPL) light. $\widetilde{\theta}_c$ represents the critical angle of total reflection.

In Figure 4a, the scattering CD spectrum for the sphere in the superchiral field with $\widetilde{\theta} = 0.98\theta_c$ is calculated as shown by the black curve. For comparison, the red curve represents the scattering CD spectrum for the same sphere in free space when illuminated by CPL light. As discussed above, the chiral response can be largely increased at the Mie resonance wavelength compared with that in the Rayleigh region, where the light wavelength is much larger than the sphere diameter. The peak values of the two CD spectra are 0.0445 and 0.0303 at frequencies of 546 and 579 THz, respectively. It means that, by introducing the superchiral field, the peak value of the scattering CD can be enhanced by 46.8%. The relationship between this enhancement factor and the incident angle has been calculated, as shown by the black curve in Figure 4b. The red dash line represents the level of the peak CD value for CPL light. When the incident angle approaches the critical angle, the CD enhancement is increased at first, and reaches the maximum value at $\widetilde{\theta} = 0.98\widetilde{\theta}_c$. Then the enhancement factor drops dramatically, which is quite different from the prediction by the chiral dipole model in Figure 3b. To explain the difference, one should notice that the CD value in Figure 3b is for a chiral dipole, which is an individual point. However, the curve in Figure 4b is for the chiral sphere with a diameter of R_s = 75 nm. As it has been reported previously [17], the area of the superchiral hotspot is decreased when $\widetilde{\theta}$ is close to $\widetilde{\theta}_c$, which may make it mismatch the size of the sphere. When $\widetilde{\theta} = 0.98\widetilde{\theta}_c$, the superchiral hotspot collapses to a point with infinite optical chirality. Around the singularity point, the light field becomes achiral. Therefore, there exists an optimized incident angle for a chiral sphere with finite size.

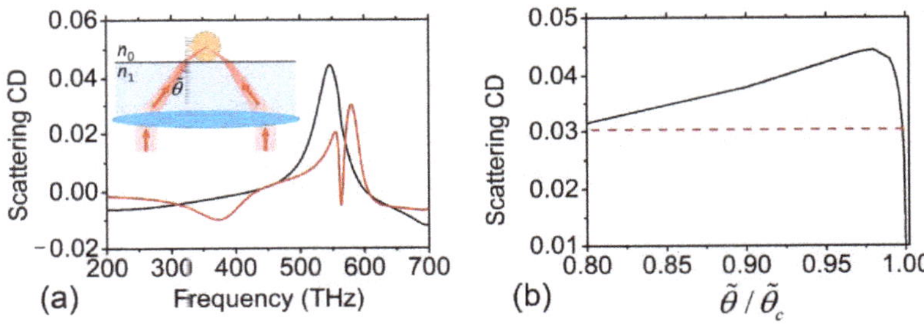

Figure 4. (**a**) The black curve represents the scattering CD spectrum of a chiral sphere on the substrate under the illumination of the superchiral field with $\widetilde{\theta} = 0.98\widetilde{\theta}_c$. The red curve is the scattering CD spectrum of the chiral sphere in free space excited by CPL. (**b**) The peak value of the scattering CD spectra with a different value of $\widetilde{\theta}$. The red dash line represents the peak value of the scattering CD spectra under the illumination of CPL, when the sphere is in free space.

3. Discussion

To explain the mechanism of the peak value of the scattering CD for the superchiral field in Figure 4a, the electrical and magnetic fields near the center of the chiral sphere are calculated. In Figure 5a–d, the distributions of the incident field with m_0 = 1 are plotted. The frequency of the incident light is 546 THz. For the electric field, $\mathbf{E_z}$ is dominant compared with the parallel component in Figure 5b around the focus point. Because the incident beam is entirely p-polarized, the field of $\mathbf{H_z}$ is zero.

By employing the theoretical model above, the scattering field by the chiral sphere is calculated in Figure 5e–h. The $\mathbf{E}_{||}$ and $\mathbf{H}_{||}$ of the light field are still the major components, as shown in Figure 5e,g. Because the scattering fields of $\mathbf{E_z}$ and $\mathbf{H_z}$ both have the helical phase term, the field intensity of $\mathbf{E_z}$ and $\mathbf{H_z}$ on the z axis is zero in Figure 5f,h. At the frequency of the Mie resonance, the light field is well confined within the sphere. From Figure 5e,g, it can be clearly seen that the resonance along the z direction is formed, and the spatial intensity distribution of $\mathbf{E}_{||}$ and $\mathbf{H}_{||}$ is well separated inside the sphere. The regions of the enhanced magnetic field are both in the upper and lower parts of the sphere. In these regions, the large magnetic field is helpful to increase the chiral response through the coupling between the electric and magnetic fields inside the chiral medium. It should be noted that, under the illumination of the superchiral field, the dip of the scattering CD curve caused by the interaction between ED and MQ disappears. It means that the chiral responses for different orders of the Mie resonances are not independent. In recent years, the scattering by a dielectric sphere has also been employed to enhance the circular dichroism of chiral materials [26,27]. The interaction between different chiral mechanisms should be further studied, which may lead to deeper understanding of the chiroptical effects and the development of novel technologies to distinguish and separate chiral molecules and nanoparticles.

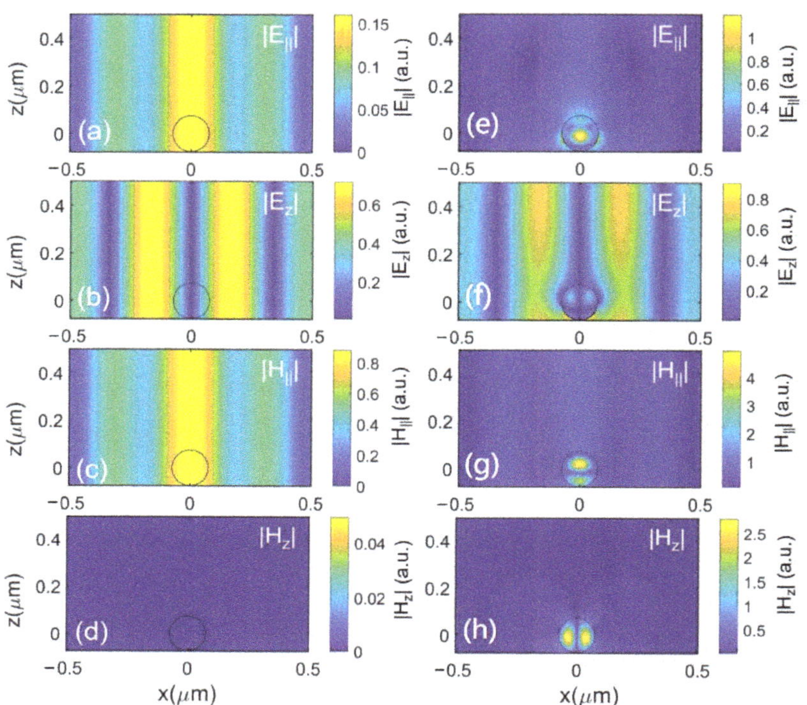

Figure 5. The light field of the vectorial Bessel beam above the air–dielectric interface. The topological charge is $m_0 = 1$. (**a–d**) The distributions of $\mathbf{E}_{||}$, $\mathbf{E_z}$, $\mathbf{H}_{||}$, and $\mathbf{H_z}$ without the chiral sphere. $\mathbf{E}_{||}$ and $\mathbf{E_z}$ ($\mathbf{H}_{||}$ and $\mathbf{H_z}$) represent the electric (magnetic) field components parallel and vertical to the interface. (**e–h**) The distributions of $\mathbf{E}_{||}$, $\mathbf{E_z}$, $\mathbf{H}_{||}$, and $\mathbf{H_z}$ scattering by the chiral sphere. The incident light frequency is 546 THz.

4. Conclusions

In summary, the theoretical method for analyzing light scattering by a chiral sphere in a vectorial light field is developed based on the T-matrix method. Compared with a pure numerical simulation (e.g., finite element method and finite-difference time-domain), this analytical method is more efficient for analyzing the mechanisms of chiroptical phenomena. Especially for a sphere with a subwavelength scale, the scattering field can be calculated accurately by employing a smaller number of vector spherical harmonics. It has been shown that the chiral response by a high-index nanosphere can be enhanced at the frequency of the Mie resonance. Moreover, the superchiral field can be employed to further increase the scattering circular dichroism value when the nanosphere is under resonance condition. Therefore, to obtain the highest chiral signal, both the enhancement mechanisms arising from the near-field nanostructure and the superchiral incident light field can be employed and optimized simultaneously. Further exploration of the chiral interaction between the artificially designed nanostructure (e.g., [28]) and the superchiral field is valuable to the field of optical activity for both fundamental and practical considerations.

Author Contributions: Conceptualization, H.H. and Q.Z.; methodology, H.H.; investigation, H.H.; writing—original draft preparation, H.H. and Q.Z.; writing—review and editing, Q.Z.; supervision, Q.Z. All authors have read and agreed to the published version of the manuscript.

Funding: This study is supported by the National Natural Science Foundation of China under Grant Nos. 62075132 and 92050202.

Acknowledgments: The authors acknowledge the helpful discussion with Mikhail Tribelsky on resonant light scattering by nano-objects during Mikhail Tribelsky visit to Shanghai. It is our honor to contribute to the Special Issue of *Physics*, which is dedicated to the celebration of his 70th birthday and outstanding achievements.

Conflicts of Interest: The authors declare no conflict of interest.

References

1. Barron, L.D. *Molecular Light Scattering and Optical Activity*; Cambridge University Press: Cambridge, UK, 2004.
2. Bliokh, K.Y.; Bekshaev, A.Y.; Nori, F. Dual electromagnetism: Helicity, spin, momentum and angular momentum. *New J. Phys.* **2013**, *15*, 033026. [CrossRef]
3. Mun, J.; Kim, M.; Yang, Y.; Badloe, T.; Ni, J.; Chen, Y.; Qiu, C.W.; Rho, J. Electromagnetic chirality: From fundamentals to nontraditional chiroptical phenomena. *Light Sci. Appl.* **2020**, *9*, 139. [CrossRef]
4. Hentschel, M.; Schäferling, M.; Duan, X.; Giessen, H.; Liu, N. Chiral plasmonics. *Sci. Adv.* **2017**, *3*, e1602735. [CrossRef] [PubMed]
5. Zhao, Y.; Askarpour, A.N.; Sun, L.; Shi, J.; Li, X.; Alu, A. Chirality detection of enantiomers using twisted optical metamaterials. *Nat. Commun.* **2017**, *8*, 14180. [CrossRef] [PubMed]
6. Vazquez-Guardado, A.; Chanda, D. Superchiral Light Generation on Degenerate Achiral Surfaces. *Phys. Rev. Lett.* **2018**, *120*, 137601. [CrossRef] [PubMed]
7. Poulikakos, L.V.; Thureja, P.; Stollmann, A.; Leo, E.D.; Norris, D.J. Chiral Light Design and Detection Inspired by Optical Antenna Theory. *Nano Lett.* **2018**, *18*, 4633–4640. [CrossRef]
8. Rui, G.; Hu, H.; Singer, M.; Jen, Y.J.; Zhan, Q.; Gan, Q. Symmetric Meta-Absorber-Induced Superchirality. *Adv. Opt. Mater.* **2019**, *7*, 1901038. [CrossRef]
9. Kuznetsov, A.I.; Miroshnichenko, A.E.; Brongersma, M.L.; Kivshar, Y.S.; Luk'Yanchuk, B. Optically resonant dielectric nanostructures. *Science* **2016**, *354*, aag2472. [CrossRef]
10. Rybin, M.V.; Koshelev, K.L.; Sadrieva, Z.F.; Samusev, K.B.; Bogdanov, A.A.; Limonov, M.F.; Kivshar, Y.S. High-Q Supercavity Modes in Subwavelength Dielectric Resonators. *Phys. Rev. Lett.* **2017**, *119*, 243901. [CrossRef] [PubMed]
11. Zhang, C.; Xu, Y.; Liu, J.; Li, J.; Xiang, J.; Li, H.; Li, J.; Dai, Q.; Lan, S.; Miroshnichenko, A.E. Lighting up silicon nanoparticles with Mie resonances. *Nat. Commun.* **2018**, *9*, 2964. [CrossRef] [PubMed]
12. Svyakhovskiy, S.E.; Ternovski, V.V.; Tribelsky, M.I. Anapole: Its birth, life, and death. *Opt. Express* **2019**, *27*, 23894–23904. [CrossRef] [PubMed]
13. Miroshnichenko, A.E.; Evlyukhin, A.B.; Yu, Y.F.; Bakker, R.M.; Chipouline, A.; Kuznetsov, A.I.; Luk'yanchuk, B.; Chichkov, B.N.; Kivshar, Y.S. Nonradiating anapole modes in dielectric nanoparticles. *Nat. Commun.* **2015**, *6*, 8069. [CrossRef]
14. Wei, L.; Xi, Z.; Bhattacharya, N.; Urbach, H.P. Excitation of the radiationless anapole mode. *Optica* **2016**, *3*, 799. [CrossRef]
15. Tang, Y.; Cohen, A.E. Optical chirality and its interaction with matter. *Phys. Rev. Lett.* **2010**, *104*, 163901. [CrossRef] [PubMed]
16. Tang, Y.; Cohen, A.E. Enhanced enantioselectivity in excitation of chiral molecules by superchiral light. *Science* **2011**, *332*, 333–336. [CrossRef]
17. Hu, H.; Gan, Q.; Zhan, Q. Generation of a Nondiffracting Superchiral Optical Needle for Circular Dichroism Imaging of Sparse Subdiffraction Objects. *Phys. Rev. Lett.* **2019**, *122*, 223901. [CrossRef]
18. Lindell, I.V.; Sihvola, A.H.; Tretyakov, S.A.; Viitanen, A.J. *Electromagnetic Waves in Chiral and Bi-Isotropic Media*; Artech House: Boston, MA, USA; London, UK, 1994.
19. Bohren, C.F. Light scattering by an optically active sphere. *Chem. Phys. Lett.* **1974**, *29*, 458–462. [CrossRef]
20. Mishchenko, M.I.; Travis, L.D.; Lacis, A.A. *Scattering, Absorption and Emission of Light by Small Particles*; Cambridge University Press: Cambridge, UK, 2002.
21. Lamprianidis, A.G.; Miroshnichenko, A.E. Excitation of nonradiating magnetic anapole states with azimuthally polarized vector beams. *Beilstein J. Nanotechnol.* **2018**, *9*, 1478–1490. [CrossRef]
22. Yang, Y.; Bozhevolnyi, S.I. Nonradiationg anapole states in nanophotonics: Form fundamentals to applications. *Nanotechnology* **2019**, *30*, 204001. [CrossRef]
23. Hernández-Acosta, M.A.; Soto-Ruvalcaba, L.; Martínez-González, C.L.; Trejo-Valdez, M.; Torres-Torres, C. Optical phase-change in plasmonic nanoparticles by a two-wave mixing. *Phys. Scr.* **2019**, *94*, 125802. [CrossRef]
24. Bauer, T.; Orlov, S.; Peschel, U.; Banzer, P.; Leuchs, G. Nanointerferometric amplitude and phase reconstruction of tightly focused vector beams. *Nat. Photonics* **2014**, *8*, 23–27. [CrossRef]
25. Zhan, Q. Cylindrical vector beams: From mathematical concepts to applications. *Adv. Opt. Photonics* **2009**, *1*, 1–57. [CrossRef]

26. Vestler, D.; Ben-Moshe, A.; Markovich, G. Enhancement of Circular Dichroism of a Chiral Material by Dielectric Nanospheres. J. *Phys. Chem. C* **2019**, *123*, 5017–5022. [CrossRef]
27. Ho, C.S.; Garcia-Etxarri, A.; Zhao, Y.; Dionne, J. Enhancing Enantioselective Absorption Using Dielectric Nanospheres. *ACS Photonics* **2017**, *4*, 197–203. [CrossRef]
28. Fernandez-Corbaton, I.; Fruhnert, M.; Rockstuhl, C. Dual and Chiral Objects for Optical Activity in General Scattering Directions. *ACS Photonics* **2015**, *2*, 376–384. [CrossRef]

MDPI

Article

Description of Nonlinear Vortical Flows of Incompressible Fluid in Terms of a Quasi-Potential

Andrei Ermakov [1] **and Yury Stepanyants** [2,3,*]

[1] Centre for Ocean Energy Research, Maynooth University, Co. Kildare, W23 F2H6 Maynooth, Ireland; ErmakovAndreiM@gmail.com
[2] School of Sciences, University of Southern Queensland, West St., Toowoomba, QLD 4350, Australia
[3] Department of Applied Mathematics, Nizhny Novgorod State Technical University n.a. R.E. Alekseev, 24 Minin St., 603950 Nizhny Novgorod, Russia
* Correspondence: Yury.Stepanyants@usq.edu.au

Abstract: As it was shown earlier, a wide class of nonlinear 3-dimensional (3D) fluid flows of incompressible viscous fluid can be described by only one scalar function dubbed the quasi-potential. This class of fluid flows is characterized by a three-component velocity field having a two-component vorticity field. Both these fields may, in general, depend on all three spatial variables and time. In this paper, the governing equations for the quasi-potential are derived and simple illustrative examples of 3D flows in the Cartesian coordinates are presented. The generalisation of the developed approach to the fluid flows in the cylindrical and spherical coordinate frames represents a nontrivial problem that has not been solved yet. In this paper, this gap is filled and the concept of a quasi-potential to the cylindrical and spherical coordinate frames is further developed. A few illustrative examples are presented which can be of interest for practical applications.

Keywords: incompressible fluid; vortical flow; vector-potential; vorticity

Citation: Ermakov, A.; Stepanyants, Y. Description of Nonlinear Vortical Flows of Incompressible Fluid in Terms of a Quasi-Potential. *Physics* **2021**, *3*, 799–813. https://doi.org/10.3390/physics3040050

Received: 9 August 2021
Accepted: 13 September 2021
Published: 22 September 2021

Publisher's Note: MDPI stays neutral with regard to jurisdictional claims in published maps and institutional affiliations.

1. Introduction

A great success in the solution of fluid dynamic problems is associated with the reduction of a primitive set of hydrodynamic equations to only one equation for some scalar function, e.g., velocity potential or stream function [1–5]. It has been shown in [6,7] that the class of exactly solvable hydrodynamic problems can be widened by introducing one more scalar function dubbed the quasi-potential. The starting point for the introduction of the quasi-potential is the condition of incompressibility of a fluid: $\operatorname{div}\mathbf{U} = 0$, where $\mathbf{U}$ is the velocity field. This condition allows us to introduce a vector-potential $\mathbf{A}$ such that $\mathbf{U} = \operatorname{curl}\mathbf{A}$ automatically satisfies this equation. However, there is a gauge invariance in the choice of the vector-potential as it is defined up to the gradient of any scalar function $f(t,x,y,z)$ of spatial variables x, y and z because $\operatorname{curl}(\nabla f) \equiv 0$. Therefore, the addition of the gradient of an arbitrary function $f(t,x,y,z)$ to the vector-potential, $\mathbf{A} + \nabla f$, does not affect the velocity field $\mathbf{U}$. Due to the freedom of choice of an arbitrary function $f(t,x,y,z)$, one of the components of the vector-potential can be eliminated. Therefore, without loss of generality, the vector potential $\mathbf{A}$ can be chosen consisting of two components only. It does not matter which component is eliminated in the Cartesian rectilinear coordinates because vector differential operations are symmetrical with respect to all spatial variables. However, it is not the case in curvilinear coordinates, in particular, in cylindrical or spherical coordinates. In any case, an arbitrary 3-dimensional velocity field can be described, in general, by two-component vector-potential, i.e., by two scalar functions—the corresponding components of the vector-potential. If there is any additional link between these two components, then the description of a fluid flow can be done in terms of only one scalar function. The governing equation for the corresponding scalar function can be derived from the primitive Navier–Stokes equation. This approach has been exploited in [6,7] in Cartesian coordinates and illustrated by nontrivial examples of fluid flows.

In this paper, fluid flows in the cylindrical and spherical coordinates are considered and it is shown how the quasi-potential can be introduced when one of the components of the vector-potential **A** is eliminated and two other components are linked with each other. Illustrative examples, which demonstrate three-dimensional velocity field with only two components of vorticity, are provided.

2. Governing Equations and Quasi-Potential in Cylindrical Coordinates

2.1. Case 1—Vector-Potential Does Not Contain the Radial Component

Consider first the case when the vector-potential does not contain the radial component A_r which is eliminated by the appropriate choice of function $f(t,x,y,z)$. Then, it can be presented as $\mathbf{A} = (F_1/r)\mathbf{e}_\varphi - F_2\mathbf{e}_z$, where F_1 and F_2 are functions of time and all spatial variables. Here φ is the azimuthal angle, and $\mathbf{e}_\varphi$ and $\mathbf{e}_z$ are the unit vectors. The velocity and vorticity fields for such a vector-potential are:

$$\mathbf{U} = F_3\,\mathbf{e}_r + \frac{\partial F_2}{\partial r}\mathbf{e}_\varphi + \frac{1}{r}\frac{\partial F_1}{\partial r}\mathbf{e}_z, \tag{1}$$

$$\omega = \left(\frac{1}{r^2}\frac{\partial^2 F_1}{\partial r\partial\varphi} - \frac{\partial^2 F_2}{\partial r\partial z}\right)\mathbf{e}_r + \left[\frac{\partial F_3}{\partial z} - \frac{\partial}{\partial r}\left(\frac{1}{r}\frac{\partial F_1}{\partial r}\right)\right]\mathbf{e}_\varphi + \frac{1}{r}\left[\frac{\partial}{\partial r}\left(r\frac{\partial F_2}{\partial r}\right) - \frac{\partial F_3}{\partial\varphi}\right]\mathbf{e}_z, \tag{2}$$

respectively, where $F_3 = -\dfrac{1}{r}\left(\dfrac{\partial F_2}{\partial\varphi} + \dfrac{\partial F_1}{\partial z}\right)$.

Consider a particular class of fluid flows having only two components of the vorticity, the φ- and z-components. For such a class of fluid motion, one has the following differential link between the functions F_1 and F_2:

$$\frac{1}{r^2}\frac{\partial^2 F_1}{\partial r\partial\varphi} = \frac{\partial^2 F_2}{\partial r\partial z} \tag{3}$$

Substitute then the expressions for $\mathbf{U}$ and ω in the Navier–Stokes equation in the Helmholtz form (4) [1]:

$$\frac{\partial\,\omega}{\partial t} + \operatorname{curl}\left[\,\omega\times\mathbf{U}\right] = \nu\Delta\,\omega. \tag{4}$$

Bearing in mind that thanks to the condition (3), the r-component of the vorticity is zero, one obtains from Equation (4):

$$(\operatorname{curl}\left[\,\omega\times\mathbf{U}\right]\cdot\mathbf{e}_r) \equiv \frac{1}{r}\left[\frac{\partial F_3}{\partial\varphi}\frac{\partial}{\partial r}\left(\frac{1}{r}\frac{\partial F_1}{\partial r}\right) - \frac{\partial F_3}{\partial z}\frac{\partial}{\partial r}\left(r\frac{\partial F_2}{\partial r}\right)\right] = 0. \tag{5}$$

Let us introduce function $S(\mathbf{r},t)$ such that

$$\frac{\partial}{\partial r}\left(\frac{1}{r}\frac{\partial F_1}{\partial r}\right) = \frac{\partial S}{\partial z}; \qquad \frac{\partial}{\partial r}\left(r\frac{\partial F_2}{\partial r}\right) = \frac{\partial S}{\partial\varphi}. \tag{6}$$

Then, the left-hand side of Equation (5) reduces to the Jacobian of two functions F_3 and S with respect to the variables φ and z. As well known, if the Jacobian is zero then, the functions are related by an arbitrary function. In our case this means that $\mathrm{J}(F_3,S) = 0$, so that $F_3 = G(S,r,t)$, where G is an arbitrary function of S, r and time t. Using the definition of function F_3 (see after Equation (2)) and Equation (6), one obtains:

$$\Delta\Phi = -\frac{1}{r}\frac{\partial}{\partial r}\left\{r\left[G(\Phi'_r) - \Phi'_r\right]\right\}, \tag{7}$$

where $\Phi = \int S\,dr$ is the *quasi-potential*, and prime stands for differentiation with respect to the corresponding variable indicated by a subscript.

In terms of the quasi-potential, the velocity and vorticity fields can be presented, respectively, as:

$$\mathbf{U} = \nabla\Phi + \left[G(\Phi_r') - \Phi_r'\right]\mathbf{e}_r\,, \tag{8}$$

$$\omega = \frac{\partial}{\partial z}\left[G(\Phi_r') - \Phi_r'\right]\mathbf{e}_\varphi - \frac{1}{r}\frac{\partial}{\partial\varphi}\left[G(\Phi_r') - \Phi_r'\right]\mathbf{e}_z. \tag{9}$$

By substitution of these expressions into two other components (the φ- and z-components) of the Helmholtz Equation (4), after some manipulations (see Appendix A) the equations reduce:

$$\frac{\partial}{\partial t}\left[G(\Phi_r') - \Phi_r'\right] + \nabla\Phi\cdot\nabla\left[G(\Phi_r') - \Phi_r'\right] + \frac{1}{2}\frac{\partial}{\partial r}\left[G(\Phi_r') - \Phi_r'\right]^2$$
$$- \nu\Delta\left[G(\Phi_r') - \Phi_r'\right] = Q(r,t), \tag{10}$$

where $Q(r,t)$ is an arbitrary function of the arguments.

In the particular case of $G(\Phi_r) \equiv \Phi_r$, the quasi-potential Φ becomes the conventional hydrodynamic velocity potential. Equation (7) reduces to the Laplace equation, Equation (10) disappears, and vorticity vanishes. In general, the main equations to be solved simultaneously are Equations (7) and (10) with two arbitrary functions $G(\Phi_r)$ and $Q(r,t)$. As there is big freedom in the choice of these functions, one obtains a good perspective to construct exact solutions to hydrodynamic equations and accommodate them to practical needs.

2.2. Case 2—Vector-Potential Does Not Contain the Azimuthal Component

Consider now the case when the vector potential has only r- and z-components: $\mathbf{A} = -r\,F_1\,\mathbf{e}_r + r\,F_2\,\mathbf{e}_z$. In this case, the velocity and vorticity fields become:

$$\mathbf{U} = \frac{\partial F_2}{\partial\varphi}\mathbf{e}_r + \frac{F_3}{r}\mathbf{e}_\varphi + \frac{\partial F_1}{\partial\varphi}\mathbf{e}_z\,, \tag{11}$$

$$\omega = \frac{1}{r}\left(\frac{\partial^2 F_1}{\partial\varphi^2} - \frac{\partial F_3}{\partial z}\right)\mathbf{e}_r + \frac{\partial}{\partial\varphi}\left(\frac{\partial F_2}{\partial z} - \frac{\partial F_1}{\partial r}\right)\mathbf{e}_\varphi + \frac{1}{r}\left(\frac{\partial F_3}{\partial r} - \frac{\partial^2 F_2}{\partial\varphi^2}\right)\mathbf{e}_z, \tag{12}$$

respectively, where $F_3 = -r\left[r\dfrac{\partial F_1}{\partial z} + \dfrac{\partial(r\,F_2)}{\partial r}\right]$.

Let us consider such a class of fluid motion which does not contain the φ-component of vorticity. In this case, there must be a relationship between functions F_1 and F_2 such that $\partial F_1/\partial r = \partial F_2/\partial z$. Then, the φ-component of the Helmholtz Equation (4) reduces to:

$$\left(\frac{\partial}{\partial z}\frac{F_3}{r^2}\right)\frac{\partial^2 F_2}{\partial\varphi^2} - \left(\frac{\partial}{\partial r}\frac{F_3}{r^2}\right)\frac{\partial^2 F_1}{\partial\varphi^2} = 0. \tag{13}$$

The left-hand side of this equation reduces to the Jacobian on variables r and z as soon as such function $S(\mathbf{r},t)$ is introduced that

$$\frac{\partial^2 F_2}{\partial\varphi^2} = \frac{\partial S}{\partial r}, \quad \frac{\partial^2 F_1}{\partial\varphi^2} = \frac{\partial S}{\partial z}. \tag{14}$$

Then, from Equation (13) one obtains: $F_3 = r^2 G(S,\varphi,t)$, where G is an arbitrary function of S, φ and t. Recalling the definition of function F_3 in terms of F_1 and F_2 and introducing a quasi-potential Φ such that $S = \partial\Phi/\partial\varphi$, Equation (13) reduces to:

$$\Delta\Phi = -\frac{1}{r^2}\frac{\partial}{\partial\varphi}\left[r^2 G(\Phi_\varphi') - \Phi_\varphi'\right]. \tag{15}$$

In terms of quasi-potential the velocity and vorticity fields read, respectively:

$$\mathbf{U} = \nabla\Phi + \frac{1}{r}\left[r^2 G(\Phi'_\varphi) - \Phi'_\varphi\right]\mathbf{e}_\varphi, \tag{16}$$

$$\omega = \frac{1}{r}\left\{-\frac{\partial}{\partial z}\left[r^2 G(\Phi'_\varphi) - \Phi'_\varphi\right]\mathbf{e}_r + \frac{\partial}{\partial r}\left[r^2 G(\Phi'_\varphi) - \Phi'_\varphi\right]\mathbf{e}_z\right\}. \tag{17}$$

In the case of a perfect fluid ($\nu = 0$), two other components, the r- and z-components, of the Helmholtz Equation (4) after simple but long manipulations similar to those presented in Appendix A can be reduced to only one equation:

$$\frac{\partial}{\partial t}\left[r^2 G(\Phi'_\varphi) - \Phi'_\varphi\right] + \nabla\Phi\cdot\nabla\left[r^2 G(\Phi'_\varphi) - \Phi'_\varphi\right] + \frac{1}{2r^2}\frac{\partial}{\partial\varphi}\left[r^2 G(\Phi'_\varphi) - \Phi'_\varphi\right]^2 = Q(\varphi, t), \tag{18}$$

where $Q(\varphi, t)$ is an arbitrary function of the arguments.

In the particular case of $G(\Phi'_\varphi) \equiv \Phi'_\varphi/r^2$, the quasi-potential Φ becomes a conventional hydrodynamic velocity potential. Equation (15) reduces to the Laplace equation Equation (18) vanishes, the velocity field becomes potential, and the vorticity field vanishes In general, the main equations to be solved simultaneously for Φ are Equations (15) and (18) with given functions $Q(\varphi, t)$, and $G(\Phi'_\varphi)$. Despite the more complex character of these equations, the freedom of choice of arbitrary functions allows one to obtain again a good perspective to construct exact solutions to hydrodynamic equations and accommodate them to practical needs. Unfortunately, in the case of a viscous fluid ($\nu \neq 0$) the introduction of a quasi-potential does not help to simplify the basic Equation (4).

Example of a Vortical Flow Illustrating Case 2

To illustrate this case, let us assume that function $G(\Phi'_\varphi) \equiv \lambda^2\Phi'_\varphi/r^2$, where $\lambda \neq \pm 1$ is a constant. One can readily verify that the following quasi-potential,

$$\Phi = \left(\frac{\sqrt{r^2+z^2}+z}{r}\right)^\lambda \sin\varphi, \tag{19}$$

satisfies Equations (15) and (18). Then, the velocity and vorticity fields read, respectively:

$$\mathbf{U} = \left(\frac{\sqrt{r^2+z^2}+z}{r}\right)^\lambda \frac{\lambda}{r\sqrt{r^2+z^2}}\left[-z\sin\varphi\,\mathbf{e}_r + \lambda\sqrt{r^2+z^2}\cos\varphi\,\mathbf{e}_\varphi + r\sin\varphi\,\mathbf{e}_z\right], \tag{20}$$

$$\omega = \left(\frac{\sqrt{r^2+z^2}+z}{r}\right)^\lambda \frac{\lambda(1-\lambda^2)\cos\varphi}{r^2\sqrt{r^2+z^2}}(r\,\mathbf{e}_r + z\,\mathbf{e}_z). \tag{21}$$

This solution describes stationary vortical flow periodic in azimuthal coordinate φ and has two components of vorticity. When $\lambda = 0$, the velocity and vorticity fields vanish, whereas when $\lambda = 1$, the flow becomes potential with the zero vorticity.

In Cartesian coordinates, both the velocity and vorticity fields have all three components:

$$\mathbf{U} = \lambda\left(\frac{R+z}{\sqrt{x^2+y^2}}\right)^\lambda \frac{-xy(\lambda R+z)\,\mathbf{i} + (\lambda x^2 R - y^2 z)\,\mathbf{j} + y(x^2+y^2)\,\mathbf{k}}{R(x^2+y^2)^{3/2}}, \tag{22}$$

where $R = \sqrt{x^2 + y^2 + z^2}$;

$$\omega = \left(\frac{R + z}{\sqrt{x^2 + y^2}} \right)^{\lambda} \frac{\lambda(1 - \lambda^2)x}{R(x^2 + y^2)^{3/2}} (x\,\mathbf{i} + y\,\mathbf{j} + z\,\mathbf{k}). \tag{23}$$

Figure 1 illustrates the velocity and vorticity fields as per Equations (22) and (23). The modulus of the velocity field is:

$$|\mathbf{U}| = \lambda \left(\frac{R + z}{\sqrt{x^2 + y^2}} \right)^{\lambda} \frac{\sqrt{\lambda^2 x^2 + y^2}}{(x^2 + y^2)}. \tag{24}$$

Furthermore, the modulus of the vorticity field is:

$$|\omega| = \lambda \left(1 - \lambda^2\right) \left(\frac{R + z}{\sqrt{x^2 + y^2}} \right)^{\lambda} \frac{x}{(x^2 + y^2)^{3/2}}. \tag{25}$$

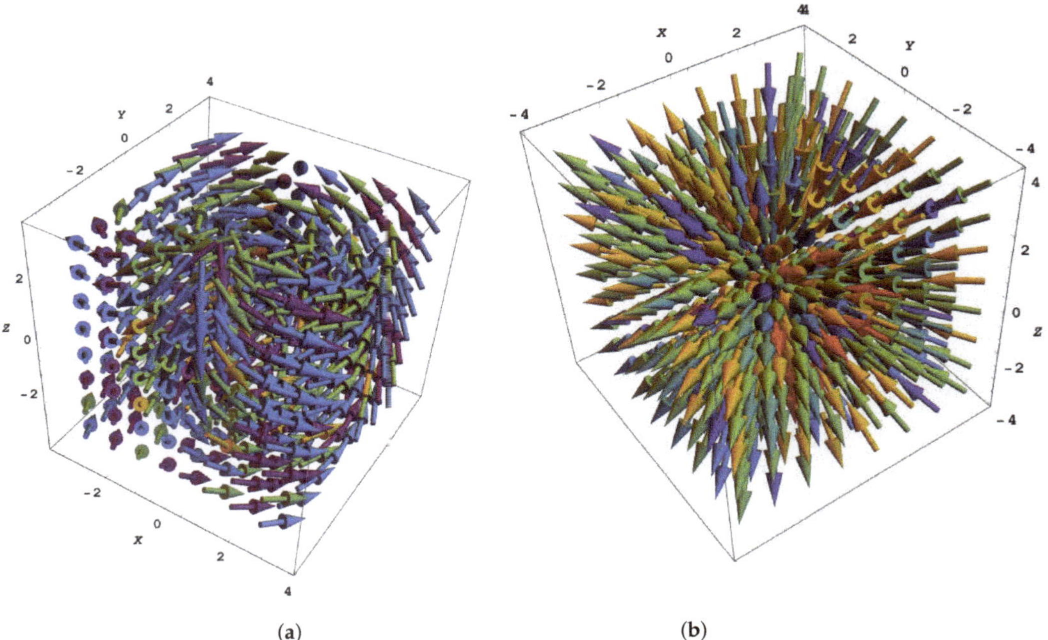

Figure 1. Fragments of (**a**) the velocity field as per Equation (22), and (**b**) the vorticity field as per Equation (23) for $\lambda = 2$.

2.3. Case 3—Vector-Potential Does Not Contain the Axial Component

Consider at last the case when the vector potential has only r- and φ-components and can be presented through the function F_1 and F_2 in the following way: $\mathbf{A} = (F_1/r)\,\mathbf{e}_r - F_2\,\mathbf{e}_\varphi$. In this case the velocity and vorticity fields become, respectively:

$$\mathbf{U} = \frac{\partial F_2}{\partial z}\,\mathbf{e}_r + \frac{1}{r}\frac{\partial F_1}{\partial z}\,\mathbf{e}_\varphi + F_3\,\mathbf{e}_z\,, \tag{26}$$

$$\omega = \frac{1}{r}\left(\frac{\partial F_3}{\partial \varphi} - \frac{\partial^2 F_1}{\partial z^2} \right)\mathbf{e}_r + \left(\frac{\partial^2 F_2}{\partial z^2} - \frac{\partial F_3}{\partial r} \right)\mathbf{e}_\varphi + \frac{1}{r}\frac{\partial}{\partial z}\left(\frac{\partial F_1}{\partial r} - \frac{\partial F_2}{\partial \varphi} \right)\mathbf{e}_z\,, \tag{27}$$

where $F_3 = -\frac{1}{r}\left(\frac{\partial(rF_2)}{\partial r} + \frac{1}{r}\frac{\partial F_1}{\partial \varphi}\right)$.

Assuming that the z-component of the vorticity is zero, one obtains the following relationship between the function F_1 and F_2:

$$\frac{\partial F_1}{\partial r} = \frac{\partial F_2}{\partial \varphi}. \tag{28}$$

By substitution the expressions for the velocity and vorticity in the Helmholtz Equation (4), one obtains:

$$(\mathrm{curl}\,[\boldsymbol{\omega} \times \mathbf{U}] \cdot \mathbf{e}_z) \equiv \frac{\partial F_3}{\partial r}\frac{\partial^2 F_1}{\partial z^2} - \frac{\partial F_3}{\partial \varphi}\frac{\partial^2 F_2}{\partial z^2} = 0. \tag{29}$$

Let us introduce function $S(\mathbf{r}, t)$ such that

$$\frac{\partial^2 F_1}{\partial z^2} = \frac{\partial S}{\partial \varphi}; \qquad \frac{\partial^2 F_2}{\partial z^2} = \frac{\partial S}{\partial r}. \tag{30}$$

Then, the left-hand side of Equation (5) reduces to the Jacobian of two functions F_3 and S with respect to the variables r and φ. Omitting the details which are similar to those presented in two previous section, one finds: $F_3 = G(S, z, t)$, where G is an arbitrary function of S, z and t. Using then the definition of function F_3 (see Equation (27)) and Equation (30), one obtains:

$$\Delta\Phi = -\frac{\partial}{\partial z}\left[G(\Phi_z') - \Phi_z'\right], \tag{31}$$

where the *quasi-potential* is $\Phi = \int S\,dz$.

In terms of the quasi-potential, the velocity and vorticity fields can be presented, respectively, as:

$$\mathbf{U} = \nabla\Phi + \left[G(\Phi_z') - \Phi_z'\right]\mathbf{e}_z\,, \tag{32}$$

$$\omega = \frac{1}{r}\frac{\partial}{\partial \varphi}\left[G(\Phi_z') - \Phi_z'\right]\mathbf{e}_r - \frac{\partial}{\partial r}\left[G(\Phi_z') - \Phi_z'\right]\mathbf{e}_\varphi. \tag{33}$$

By substitution of these expressions into two other components (the r- and φ-components) of the Helmholtz Equation (4) and neglecting viscosity ($\nu = 0$), after some manipulations, similar to those presented in the Appendix A, the equations reduce to only one equation:

$$\frac{\partial}{\partial t}\left[G(\Phi_z') - \Phi_z'\right] + \nabla\Phi \cdot \nabla\left[G(\Phi_z') - \Phi_z'\right] + \frac{1}{2}\frac{\partial}{\partial z}\left[G(\Phi_z') - \Phi_z'\right]^2 = Q(z, t), \tag{34}$$

where $Q(z, t)$ is an arbitrary function of the arguments.

In the particular case of $G(\Phi_z') \equiv \Phi_z'$, the quasi-potential Φ becomes the conventional hydrodynamic velocity potential. Equation (31) reduces to the Laplace equation, Equation (34) disappears, and vorticity vanishes. In general, the main equations to be solved simultaneously are Equations (31) and (34) with two arbitrary functions $G(\Phi_z')$ and $Q(z, t)$. There is again a big freedom in the choice of these functions, so that one can obtain a good perspective to construct exact solutions to hydrodynamic equations and accommodate them to practical needs. Unfortunately, in the case of a viscous fluid ($\nu \neq 0$), the introduction of a quasi-potential does not help to simplify the basic Equation (4).

Example of a Vortical Flow Illustrating Case 3

Consider now an example of the vortical flow for the third case when the vector potential has only r- and φ-components in cylindrical coordinates. Let us assume that

$G(\Phi_z) \equiv -\lambda^2\Phi_z$, where λ is a constant. One can readily verify that the following quasi-potential,

$$\Phi = \cos(\lambda r \sin\varphi)\cos z, \tag{35}$$

satisfies Equations (31) and (34). Then, the velocity and vorticity fields read, respcetively:

$$\mathbf{U} = -\lambda\left[\cos z\,\sin(r\lambda\sin\varphi)(\sin\varphi\,\mathbf{e}_r + \cos\varphi\,\mathbf{e}_\varphi) - \lambda\sin z\cos(r\lambda\sin\varphi)\,\mathbf{e}_z\right], \tag{36}$$

$$\omega = \lambda\left(1+\lambda^2\right)\sin z\,\sin(r\lambda\sin\varphi)(-\cos\varphi\,\mathbf{e}_r + \sin\varphi\,\mathbf{e}_\varphi). \tag{37}$$

This solution describes non-viscous stationary fluid flow. The vector fields $\mathbf{U}$ and ω are periodic in φ and z. In Cartesian coordinates, the velocity field becomes two-component:

$$\mathbf{U} = -\lambda(\sin\lambda y\cos z\,\mathbf{j} - \lambda\cos\lambda y\sin z\,\mathbf{k}), \tag{38}$$

whereas the vorticity field has only one x-component:

$$\omega = -\lambda\left(1+\lambda^2\right)\sin(\lambda y)\,\sin z\,\mathbf{i}. \tag{39}$$

Both these fields do not depend on x and therefore can be described by the conventional stream function $\psi = \sin\lambda y\sin z$, so that the velocity components are:

$$U_y = \frac{\partial\psi}{\partial z}, \quad U_z = -\frac{\partial\psi}{\partial y}.$$

For purely imaginary $\lambda = i$, the quasi-potential reduces to the conventional hydrodynamic potential, and the fluid field becomes potential with zero vorticity:

$$\mathbf{U} = \sinh y\cos z\,\mathbf{j} - \cosh\lambda y\sin z\,\mathbf{k}, \tag{40}$$

Thus, in general, this example describes a vortical flow double periodic in the (y, z) plane. Figure 2 illustrates the velocity and vorticity fields for $\lambda = 1$ when y and z vary in the range $[-4, 4]$.

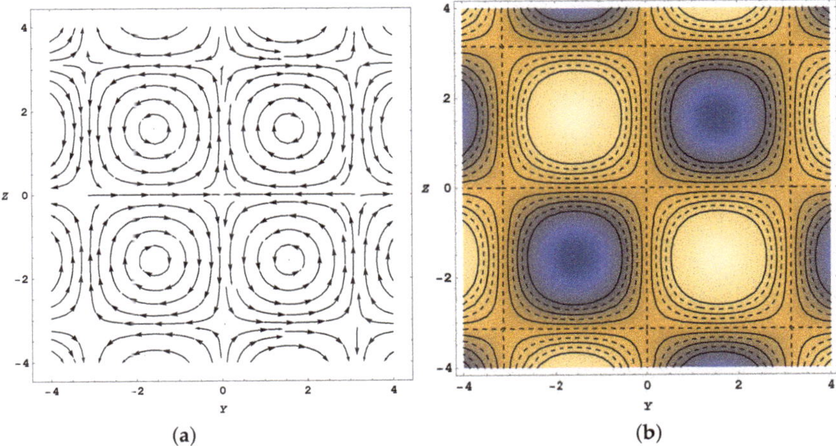

Figure 2. Fragments of (**a**) the velocity field as per Equation (38), and (**b**) the density of vorticity field as per Equation (39).

In this Section, only two simple examples have been presented to illustrate vortical flows of inviscid fluid flows. As one can see already from these two examples, the introduction of quasi-potentials allows us to construct exact solutions for fairly complex vortical flows in cylindrical coordinates. The transformation of a two-component vorticity field акщь cylindrical to Cartesian coordinates can lead to either one-component or three-component vorticity fields. There is no regular method to construct three-component vortical flows in Cartesian coordinates, whereas the developed theory allows us to find some exact solutions for such flows in terms of the quasi-potential.

3. Governing Equations and Quasi-Potential in Spherical Coordinates

The derivation of basic equations for the spherical case are similar but not the same to that presented in Section 2. Therefore, the derivation of the corresponding equations is described only briefly.

3.1. Case 1—Vector-Potential Does Not Contain the Radial Component

Consider the case when the vector-potential does not contain the component A_r and reads: $\mathbf{A} = -(F_2/r)\mathbf{e}_\varphi + (F_1/r)\mathbf{e}_\theta$ where θ is the polar angle, and F_1 and F_2 are functions of time and all spatial variables. The velocity and vorticity fields for such vector-potential are:

$$\mathbf{U} = F_3\mathbf{e}_r + \frac{1}{r}\frac{\partial F_1}{\partial r}\mathbf{e}_\varphi + \frac{1}{r}\frac{\partial F_2}{\partial r}\mathbf{e}_\theta, \tag{41}$$

$$\omega = \frac{1}{r^2\sin\theta}\left[\frac{\partial}{\partial\theta}\left(\frac{\partial F_1}{\partial r}\sin\theta\right) - \frac{\partial^2 F_2}{\partial r\partial\varphi}\right]\mathbf{e}_r$$
$$-\frac{1}{r\sin\theta}\left(\frac{\partial^2 F_1}{\partial r^2}\sin\theta - \frac{\partial F_3}{\partial\varphi}\right)\mathbf{e}_\varphi + \frac{1}{r}\left(\frac{\partial^2 F_2}{\partial r^2} - \frac{\partial F_3}{\partial\theta}\right)\mathbf{e}_\theta, \tag{42}$$

respectively, where $F_3 = -\dfrac{1}{r^2\sin\theta}\left[\dfrac{\partial F_1}{\partial\varphi} + \dfrac{\partial(F_2\sin\theta)}{\partial\theta}\right]$.

Consider again a particular class of fluid flows with the zero first component of the vorticity. Then, the relationship between functions F_1 and F_2:

$$\frac{\partial}{\partial\theta}(F_1\sin\theta) = \frac{\partial F_2}{\partial\varphi}. \tag{43}$$

Furthermore, from the r-component of the Helmholtz Equation (4) one obtains:

$$(\mathrm{curl}[\omega\times\mathbf{U}]\cdot\mathbf{e}_r) \equiv \sin\theta\frac{\partial^2 F_1}{\partial r^2}\frac{\partial F_3}{\partial\theta} - \frac{\partial^2 F_2}{\partial r^2}\frac{\partial F_3}{\partial\theta} = 0. \tag{44}$$

Let us introduce function S such that

$$\sin\theta\frac{\partial^2 F_1}{\partial r^2} = \frac{\partial S}{\partial\varphi}, \quad \frac{\partial^2 F_2}{\partial r^2} = \frac{\partial S}{\partial\theta}. \tag{45}$$

Then Equation (44) takes a form of the Jacobian of two functions, $\mathrm{J}(F_3, S) = 0$ with respect to the variables φ and θ. This gives $F_3 = G(S, r, t)$, where G is an arbitrary function of the arguments. Using the definition of function F_3 and Equation (45), one can reduce this relationship to the equation:

$$\Delta\int\Phi\,dr = \frac{1}{r^2}\frac{\partial}{\partial r}\left(r^2\Phi'_r\right) - G(\Phi'_r), \tag{46}$$

where the quasi-potential is defined by the equation $S = \partial\Phi/\partial r$, and Δ is the Laplacian in spherical coordinates. This equation can be rewritten in the form:

$$\begin{aligned}\Delta\Phi &= -\frac{1}{r^2}\frac{\partial^2}{\partial r^2}\left\{r^2\left[G(\Phi'_r) - \Phi'_r\right]\right\} + \frac{4}{r}\frac{\partial}{\partial r}\left[G(\Phi'_r) - \Phi'_r\right] \\ &- \frac{2}{r}\left(\frac{1}{r} + 1\right)\left[G(\Phi'_r) - \Phi'_r\right] + \frac{4}{r}\frac{\partial}{\partial r}(\Phi'_r - \Phi) + \frac{2}{r^2}(\Phi'_r - \Phi).\end{aligned} \tag{47}$$

The expressions for the velocity and vorticity fields in terms of the quasi-potential, are, respectively:

$$\mathbf{U} = \nabla\Phi + \left[G(\Phi_r') - \Phi_r'\right]\mathbf{e}_r, \tag{48}$$

$$\omega = \frac{1}{r}\left\{\frac{1}{\sin\theta}\frac{\partial}{\partial\varphi}\left[G(\Phi_r') - \Phi_r'\right]\mathbf{e}_\varphi - \frac{\partial}{\partial\theta}\left[G(\Phi_r') - \Phi_r'\right]\mathbf{e}_\theta\right\}. \tag{49}$$

Then, from the other two components of the Helmholtz Equation (4) neglecting the viscosity ($\nu = 0$), one obtains:

$$\frac{\partial}{\partial t}\left[G(\Phi_r') - \Phi_r'\right] + \nabla\Phi\cdot\nabla\frac{\partial}{\partial r}\left[G(\Phi_r') - \Phi_r'\right] + \frac{1}{2}\frac{\partial}{\partial r}\left[G(\Phi_r') - \Phi_r'\right]^2 = Q(r,t), \tag{50}$$

where $Q(r,t)$ is an arbitrary function of the arguments. Choosing $G(\Phi_r') \equiv \Phi_r'$, one obtains the expressions for the potential fluid flow with the zero vorticity and $\mathbf{U} = \nabla\Phi$.

The Illustrative Example

Let us construct the illustrative example of a stationary flow for this case with the following choice of functions $G(\Phi_r')$ and $Q(r,t)$:

$$\Phi(r,\varphi,\theta) = r^3\sin\varphi\sin\theta, \quad G(\Phi_r') = \frac{1}{6}\Phi_r', \quad Q(r,t) = -\frac{5}{2}r^3. \tag{51}$$

The velocity and vorticity fields with such choice of functions are, respectively:

$$\mathbf{U} = r^2\left(\frac{1}{2}\sin\varphi\sin\theta\,\mathbf{e}_r + \sin\varphi\cos\theta\,\mathbf{e}_\varphi + \cos\varphi\,\mathbf{e}_\theta\right), \tag{52}$$

$$\omega = \frac{5}{2}r\left(-\cos\varphi\,\mathbf{e}_\varphi + \sin\varphi\cos\theta\,\mathbf{e}_\theta\right). \tag{53}$$

These fields in the Cartesian coordinates look as follows:

$$\mathbf{U} = \frac{1}{2}\left[-xy\,\mathbf{e}_x + \left(2x^2 + y^2 + 2z^2\right)\mathbf{e}_y - yz\,\mathbf{e}_z\right]; \qquad \omega = \frac{5}{2}(-z\,\mathbf{e}_x + x\,\mathbf{e}_z). \tag{54}$$

Figure 3 illustrates the velocity field in the planes $y = -1$ and $y = 1$.

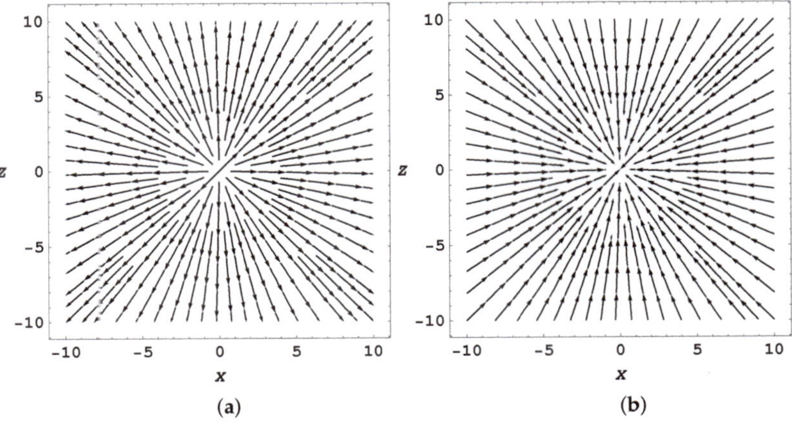

Figure 3. The velocity field in the planes $y = -1$ (**a**) and $y = 1$ (**b**) as per Equation (54).

Figure 4 illustrates the velocity field in the planes $x = 0$ and $x = \pm 5$, and Figure 5 illustrates the vorticity field in the (x,z)-plane.

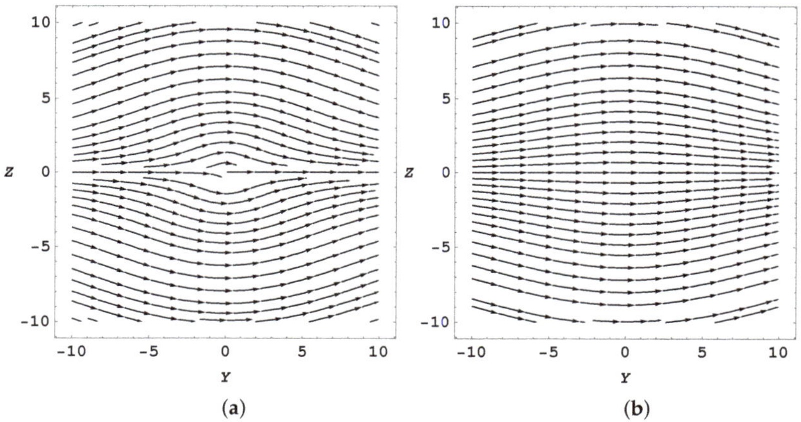

Figure 4. The velocity field in the planes $x = 0$ (**a**) and $x = \pm 5$ (**b**) as per Equation (54).

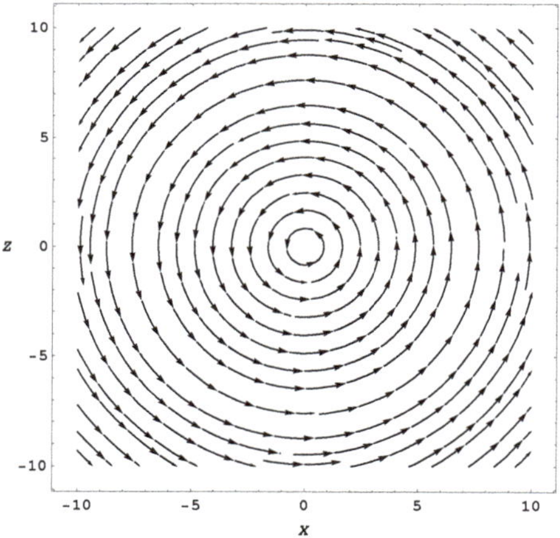

Figure 5. The vorticity field in the (x, z)-plane as per Equation (54).

3.2. Case 2—Vector-Potential Does Not Contain the Azimuthal Component

Consider now the case when the vector-potential does not contain the azimuthal component A_φ and reads: $\mathbf{A} = F_1 \sin\theta\, \mathbf{e}_r - F_2 r \sin\theta\, \mathbf{e}_\theta$, where F_1 and F_2 are functions of time and all spatial variables. The velocity and vorticity fields for such vector-potential read:

$$\mathbf{U} = \frac{\partial F_2}{\partial \varphi}\, \mathbf{e}_r + \frac{F_3}{r \sin\theta}\, \mathbf{e}_\varphi + \frac{1}{r}\frac{\partial F_1}{\partial \varphi}\, \mathbf{e}_\theta\,, \tag{55}$$

$$\omega = \frac{1}{r^2 \sin\theta}\left(\frac{\partial F_3}{\partial \theta} - \frac{\partial^2 F_1}{\partial \varphi^2}\right) \mathbf{e}_r + \frac{1}{r}\frac{\partial}{\partial \varphi}\left(\frac{\partial F_1}{\partial r} - \frac{\partial F_2}{\partial \theta}\right) \mathbf{e}_\varphi - \frac{1}{r \sin\theta}\left(\frac{\partial F_3}{\partial r} - \frac{\partial^2 F_2}{\partial \varphi^2}\right) \mathbf{e}_\theta, \tag{56}$$

respectively, where $F_3 = -\sin\theta \left[\frac{\partial}{\partial r}\left(F_2 r^2 \sin\theta\right) + \frac{\partial}{\partial \theta}(F_1 \sin\theta)\right]$.

Considering such a class of fluid flows which do not contain the φ-component of the vorticity, one obtains a relationship between the functions F_1 and F_2:

$$\frac{\partial F_1}{\partial r} = \frac{\partial F_2}{\partial \theta}. \tag{57}$$

Then, from the φ-component of the Helmholtz Equation (4) one obtains:

$$(\mathrm{curl}[\boldsymbol{\omega} \times \mathbf{U}] \cdot \mathbf{e}_\varphi)$$
$$= \frac{\partial}{\partial r}\left(\frac{F_3}{r^2 \sin^2\theta}\right)\left(\frac{\partial^2 F_1}{\partial \varphi^2} - \frac{\partial F_3}{\partial \theta}\right) - \frac{\partial}{\partial \theta}\left(\frac{F_3}{r^2 \sin^2\theta}\right)\left(\frac{\partial^2 F_2}{\partial \varphi^2} - \frac{\partial F_3}{\partial r}\right)$$
$$= \frac{\partial}{\partial r}\left(\frac{F_3}{r^2 \sin^2\theta}\right)\frac{\partial^2 F_1}{\partial \varphi^2} - \frac{\partial}{\partial \theta}\left(\frac{F_3}{r^2 \sin^2\theta}\right)\frac{\partial^2 F_2}{\partial \varphi^2} = 0. \tag{58}$$

Introducing function $S(\mathbf{r}, t)$ such that

$$\frac{\partial^2 F_1}{\partial \varphi^2} = \frac{\partial S}{\partial \theta}, \qquad \frac{\partial^2 F_2}{\partial \varphi^2} = \frac{\partial S}{\partial r}, \tag{59}$$

Equation (58) can be presented in the Jacobian form:

$$\mathrm{J}_{r,\theta}\left(\frac{F_3}{r^2 \sin^2\theta}, S\right) \equiv \frac{\partial}{\partial r}\left(\frac{F_3}{r^2 \sin^2\theta}\right)\frac{\partial S}{\partial \theta} - \frac{\partial}{\partial \theta}\left(\frac{F_3}{r^2 \sin^2\theta}\right)\frac{\partial S}{\partial r} = 0. \tag{60}$$

This implies that $F_3 = G(S, \varphi, t) r^2 \sin^2\theta$, where G is an arbitrary function of the arguments. Using the definition of function F_3 (see Equation (56)) and the relationship (57) between functions F_1 and F_2, one finally derives the static equation for the quasi-potential $\Phi = \int S\, d\varphi$:

$$\Delta\Phi = -\frac{\partial}{\partial \varphi}\left[G(\Phi'_\varphi) - \frac{\Phi'_\varphi}{r^2 \sin^2\theta}\right]. \tag{61}$$

Another equation, describing the time dependence of Φ, can be obtained from two other components (r- and θ-components) of the Helmholtz Equation (4). Omitting the details, which are similar to those shown in Appendix A, for the perfect fluid ($\nu = 0$), this equation reads:

$$\frac{\partial}{\partial t}\left[r^2 \sin^2\theta\, G(\Phi'_\varphi) - \Phi'_\varphi\right] + \nabla\Phi \cdot \nabla\left[r^2 \sin^2\theta\, G(\Phi'_\varphi) - \Phi'_\varphi\right]$$
$$+ \frac{1}{2r^2 \sin^2\theta}\frac{\partial}{\partial \varphi}\left[r^2 \sin^2\theta\, G(\Phi'_\varphi) - \Phi'_\varphi\right]^2 = Q(\varphi, t), \tag{62}$$

where $Q(\varphi, t)$ is an arbitrary function of the arguments.

In terms of quasi-potential the velocity and vorticity fields are, respectively:

$$\mathbf{U} = \nabla\Phi + \frac{1}{r \sin\theta}\left[r^2 \sin^2\theta\, G(\Phi'_\varphi) - \Phi'_\varphi\right], \tag{63}$$

$$\boldsymbol{\omega} = \frac{1}{r^2 \sin\theta}\frac{\partial}{\partial \theta}\left[r^2 \sin^2\theta\, G(\Phi'_\varphi) - \Phi'_\varphi\right]\mathbf{e}_r - \frac{1}{r \sin\theta}\frac{\partial}{\partial r}\left[r^2 \sin^2\theta\, G(\Phi'_\varphi) - \Phi'_\varphi\right]\mathbf{e}_\theta. \tag{64}$$

Choosing $G(\Phi'_r) \equiv \Phi'_r / r^2 \sin^2\theta$, one obtains the expressions for the potential fluid flow with the zero vorticity and $\mathbf{U} = \nabla\Phi$.

3.3. Case 3—Vector-Potential Does Not Contain the Polar Component

Consider, at last, the case when the vector-potential does not contain the polar component A_θ and reads: $\mathbf{A} = -F_1\,\mathbf{e}_r + F_2 / r \sin\theta\,\mathbf{e}_\varphi$, where F_1 and F_2 are functions of time and all spatial variables. The velocity and vorticity fields for such vector-potential are, respectively:

$$\mathbf{U} = \frac{1}{r}\left(\frac{1}{r \sin\theta}\frac{\partial F_2}{\partial \theta}\,\mathbf{e}_r + \frac{\partial F_1}{\partial \theta}\,\mathbf{e}_\varphi + F_3\,\mathbf{e}_\theta\right), \tag{65}$$

$$\boldsymbol{\omega} = \frac{1}{r \sin\theta}\left\{\frac{1}{r}\left[\frac{\partial}{\partial \theta}\left(\sin\theta \frac{\partial F_1}{\partial \theta}\right) - \frac{\partial F_3}{\partial \varphi}\right]\mathbf{e}_r + \sin\theta\left[\frac{\partial F_3}{\partial r} - \frac{\partial}{\partial \theta}\left(\frac{1}{r^2 \sin\theta}\frac{\partial F_2}{\partial \theta}\right)\right]\mathbf{e}_\varphi\right.$$

$$+\left[\frac{\partial}{\partial\varphi}\left(\frac{1}{r^2\sin\theta}\frac{\partial F_2}{\partial\theta}\right)-\frac{\partial}{\partial r}\left(\sin\theta\frac{\partial F_1}{\partial\theta}\right)\right]\mathbf{e}_\theta\Big\}. \tag{66}$$

where $F_3=-\frac{1}{\sin\theta}\left(\frac{\partial F_2}{\partial r}+\frac{\partial F_1}{\partial\varphi}\right)$.

Consider now a particular class of fluid flows with the zero polar component of the vorticity. Then, one obtains the relationship between functions F_1 and F_2:

$$\frac{1}{r^2\sin\theta}\frac{\partial^2 F_2}{\partial\theta\partial\varphi}=\sin\theta\,\frac{\partial^2 F_1}{\partial\theta\partial r}. \tag{67}$$

Then, from the θ-component of the Helmholtz Equation (4) one obtains:

$$\left(\frac{\partial}{\partial\varphi}\frac{F_3}{r^2}\right)\left[\frac{\partial}{\partial\theta}\left(\frac{1}{r^2\sin\theta}\frac{\partial F_2}{\partial\theta}\right)-\frac{\partial F_3}{\partial r}\right]-\left(\frac{\partial}{\partial r}\frac{F_3}{r^2}\right)\left[\frac{\partial}{\partial\theta}\left(\sin\theta\frac{\partial F_1}{\partial\theta}\right)-\frac{\partial F_3}{\partial\varphi}\right]=0. \tag{68}$$

Introducing function S such that

$$\frac{\partial}{\partial\theta}\left(\frac{1}{r^2\sin\theta}\frac{\partial F_2}{\partial\theta}\right)-\frac{\partial F_3}{\partial r}=\frac{\partial S}{\partial r},\qquad \frac{\partial}{\partial\theta}\left(\sin\theta\frac{\partial F_1}{\partial\theta}\right)-\frac{\partial F_3}{\partial\varphi}=\frac{\partial S}{\partial\varphi}, \tag{69}$$

Equation (68) transforms to the form of Jacobian:

$$\mathrm{J}_{\varphi,r}\left(\frac{F_3}{r^2},S\right)\equiv\frac{\partial}{\partial\varphi}\left(\frac{F_3}{r^2}\right)\frac{\partial S}{\partial r}-\frac{\partial}{\partial r}\left(\frac{F_3}{r^2}\right)\frac{\partial S}{\partial\varphi}=0. \tag{70}$$

This implies that $F_3=r^2G(S,\theta,t)$, where G is an arbitrary function of the arguments Using the definition of function F_3 (see Equation (66) and the relationship between functions F_1 and F_2 (67), one finally obtains the static equation for the function S:

$$\Delta\Phi=\frac{1}{r^2\sin\theta}\frac{\partial(S\sin\theta)}{\partial\theta}, \tag{71}$$

where quasi-potential is $\Phi=\int\left[S+r^2G(S)\right]d\theta$, and Δ is the Laplacian in spherical coordinates.

Another equation, describing time dependence of S, can be obtained from two other components (r- and φ-components) of the Helmholtz Equation (4) with $\nu=0$. Omitting the details, which are similar to those presented in Appendix A, the final expression is:

$$\frac{\partial S}{\partial t}+\nabla\Phi\cdot\nabla S-\frac{1}{2r^2}\frac{\partial S^2}{\partial\theta}=Q(\theta,t). \tag{72}$$

Then, the formulae for the velocity and vorticity fields are, respectively:

$$\mathbf{U}=\nabla\Phi-\frac{S}{r}\,\mathbf{e}_\theta, \tag{73}$$

$$\omega=\frac{1}{r^2\sin\theta}\frac{\partial S}{\partial\varphi}\mathbf{e}_r-\frac{1}{r}\frac{\partial S}{\partial r}\mathbf{e}_\varphi. \tag{74}$$

These formulae exhaust all possible versions of introduction of a quasi-potential in spherical coordinates. No an illustrative example is found for this case but let believe that this may be possible and, hopefully, an example will be found in the future, moreover, it can be of a practical interest.

4. Conclusions

Thus, in this paper it has been demonstrated that the introduction of a proper quasi-potential is possible in the cylindrical and spherical geometry albeit it is not a trivial generalisation of the quasi-potential theory developed in Refs. [6,7]. The quasi-potential approach helps one to construct exact solutions describing fairly complex class of three-dimensional velocity fields with two-component vorticity fields. After transformation from the curvilinear coordinates to the Cartesian coordinates, the velocity and vorticity fields can become fairly complex containing all three components of velocity and vorticity. In the particular cases, the quasi-potential theory reduces to the conventional potential theory or to the theory of a vortex flow based on the introduction of a stream-function. Our preliminary study shows that the quasi-potential approach can be used also in other curvilinear coordinates however, the basic equations and formulae for the velocity and vorticity fields become fairly complex and not so illustrative. Summarising the results obtained in this paper and in Refs. [6,7], one can see that the traditional potential theory describes fluid flows with zero vorticity; the stream-function approach allows us to describe flows with only one component of vorticity; and the introduction of a quasi-potential allows us to describe flows with two components of vorticity. It can be noted in conclusion that, unfortunately, new ideas in the description of classical hydrodynamics appear seldom. Three relatively recent publications [8–10] can be mentioned to illustrate that the development of such ideas leads to the discovery of new non-trivial exact solutions in classical fluid mechanics.

In Appendix B one of the authors of this paper (Y.S.) presents his brief reminiscent about contacts with M. Tribelsky to whose honour this issue is dedicated.

Author Contributions: Conceptualization, Y.S.; methodology, Y.S.; validation, A.E. and Y.S.; formal analysis, A.E.; investigation, A.E. and Y.S.; writing—original draft preparation, A.E.; writing—review and editing, Y.S.; visualization, A.E.; supervision, Y.S. All authors have read and agreed to the published version of the manuscript.

Funding: Y.S. acknowledges the funding of this study provided by the State task program in the sphere of scientific activity of the Ministry of Science and Higher Education of the Russian Federation (project No. FSWE-2020-0007) and the grant of President of the Russian Federation for the state support of Leading Scientific Schools of the Russian Federation (grant No. NSH- 2485.2020.5). The paper was completed when Y.S. was working under the SMRI visiting program at The University of Sydney in March–April 2021. Y.S. is thankful for the provided grant and hospitality of SMRI staff.

Conflicts of Interest: The authors declare no conflict of interests.

Appendix A

The φ-component of the Helmholtz Equation (4) after substitution the expressions for the velocity (8) and vorticity (9) fields gives:

$$\frac{\partial^2}{\partial t \partial z}\left[G(\Phi'_r, r) - \Phi'_r\right] - \frac{\partial}{\partial r}\left\{G(\Phi'_r, r)\frac{\partial}{\partial z}\left[G(\Phi'_r, r) - \Phi'_r\right]\right\} - \nu\Delta\frac{\partial}{\partial z}\left[G(\Phi'_r, r) - \Phi'_r\right]$$
$$- \frac{\partial}{\partial z}\left\{\frac{\partial\Phi'_r, r)}{\partial z}\frac{\partial}{\partial z}\left[G(\Phi'_r, r) - \Phi'_r\right] + \frac{1}{r^2}\frac{\partial\Phi'_r, r)}{\partial\varphi}\frac{\partial}{\partial\varphi}\left[G(\Phi'_r, r) - \Phi'_r\right]\right\} = 0.$$

This equation can be rewritten in the form:

$$\frac{\partial^2}{\partial t \partial z}\left[G(\Phi'_r, r) - \Phi'_r\right] + \frac{\partial}{\partial z}\left\{\nabla\Phi \cdot \nabla\left[G(\Phi'_r, r) - \Phi'_r\right]\right\} + \frac{1}{2}\frac{\partial^2}{\partial z \partial r}\left[G(\Phi'_r, r) - \Phi'_r\right]^2$$
$$- \nu\Delta\frac{\partial}{\partial z}\left[G(\Phi'_r, r) - \Phi'_r\right] = 0.$$

After integration over z this equation reduces to:

$$\frac{\partial}{\partial t}\left[G(\Phi'_r, r) - \Phi'_r\right] + \nabla\Phi \cdot \nabla\left[G(\Phi'_r, r) - \Phi'_r\right] + \frac{1}{2}\frac{\partial}{\partial r}\left[G(\Phi'_r, r) - \Phi'_r\right]^2$$
$$- \nu\Delta\left[G(\Phi'_r, r) - \Phi'_r\right] = Q_1(r, \varphi, t), \quad \text{(A1)}$$

where $Q_1(r, \varphi, t)$ is an arbitrary function of the arguments.

Then, the similar manipulations with the z-component of the Helmholtz Equation (4) lead to the equation:

$$\frac{\partial}{\partial t}\left[G(\Phi'_r, r) - \Phi'_r\right] + \nabla\Phi \cdot \nabla\left[G(\Phi'_r, r) - \Phi'_r\right] + \frac{1}{2}\frac{\partial}{\partial r}\left[G(\Phi'_r, r) - \Phi'_r\right]^2 - \nu\Delta\left[G(\Phi'_r, r) - \Phi'_r\right] = Q_2(r, z, t), \tag{A2}$$

The left-hand side of Equations (A1) and (A2) are identical; therefore, one can conclude that functions $Q_1(r, z, t)$ and $Q_2(r, z, t)$ can depend only on r and t and must be equal. Denoting them by $Q(r, t)$, one obtains Equation (10).

Appendix B

Here, one of the authors, Y.S., presents his reminiscent about friendship contacts with Michael Tribelsky (Misha). We met relatively late, in 2007, when we both worked as Visiting Professors at the Max Planck Institute for the Physics of Complex Systems, Dresden, Germany. We quickly found many commons between us, including age, education, interests, attitude to science and life. One of the attractive Misha's features is a great sense of humour and an optimistic attitude to life even when it was hard to him. We spent much time in discussions of physical and mathematical problems, as well as stories of science history.

Figure A1. Yury Stepanyants (**left**) and Mikhail Tribelsky (**right**) at the Circular Quay, Sydney, in 2008.

Our contacts continued when Misha visited me in Sydney the next year. I was attracted to them by persistence and perseverance in achieving goals (this is clearly seen from his autobiographical notes presented in this Special Issue). This pertains not only to physical problems but to all aspects of life. I observed how carefully Misha chose the best spots for photography of exotic trees, beautiful buildings, and other objects, taking photos in Sydney when we walked continuously discussing something or disputing on different subjects.

Misha obtained an excellent education in one of the world's most prestigious physical faculties, namely the one at Lomonosov Moscow State University. He had an opportunity to communicate and jointly work with brilliant world-renowned scientists such as his supervisor Prof. S. Anisimov and then, Profs. Ya. Zeldovich, I. Lifshits, V. Zakharov. He boldly tackled difficult physical problems and found their own, often unconventional solutions to them. I was delighted with his story about solving an important industrial problem for one of the Japanese companies. The problem was related to the calculation of the extinction coefficient in complicated multi-component media. At first, it was not even clear how to approach the task. Still, after persistent reflections, Misha found the key in a relatively short time which caused the admiration of company bosses.

I should also mention Misha's deep knowledge of computer design and the internal devices of a computer. I had a chance to be convinced that Misha's level in this field is the same as that (or even higher than that) of professional IT (information technology) experts. Misha generously shared his knowledge with me. More than once, he helped me resolve the difficult issues related to installing and fine-tuning computer programs. It is a pleasure to have such a kind and reliable friend as Misha Tribelsky. I wish him good health and many years of creative work in science.

References

1. Lamb, H. *Hydrodynamics*; Cambridge University Press: Cambridge, UK, 1932.
2. Milne-Thomson, L.M. *Theoretical Hydromechanics*; Macmillan Publishers: London, UK, 1960.
3. Kochin, N.E.; Kibel, I.A.; Roze, N.V. *Theoretical Hydromechanics*; Interscience Publishers: New York, NY, USA, 1964.
4. Batchelor, G.K. *An Introduction to Fluid Mechanics*; Cambridge University Press: Cambridge, UK, 1967.
5. Landau, L.D.; Lifshitz, E.M. *Fluid Mechanics*; Pergamon Press: Oxford, UK, 1993.
6. Stepanyants, Y.A.; Yakubovich, E.I. Scalar description of three-dimensional vortex flows of incompressible fluid. *Dokl. Phys.* **2011**, *56*, 130–133. [CrossRef]
7. Stepanyants, Y.A.; Yakubovich, E.I. The Bernoulli integral for a certain class of non-stationary viscous vortical flows of incompressible fluid. *Stud. Appl. Math.* **2015**, *135*, 295–309. [CrossRef]
8. Makinde, O.D.; Khan, W.A.; Chinyoka, T. New developments in fluid mechanics and its engineering applications. *Math. Probl. Eng.* **2013**, *2013*, 797390. [CrossRef]
9. Yakubovich, E.I.; Zenkovich, D.A. Matrix approach to Lagrangian fluid dynamics. *J. Fluid Mech.* **2001**, *443*, 167–196. [CrossRef]
10. Yakubovich, E.I.; Shrira, V.I. Non-steady columnar motions in rotating stratified Boussinesq fluids: Exact Lagrangian and Eulerian description. *J. Fluid Mech.* **2011**, *691*, 417–439. [CrossRef]

 *physics*

Article

Measuring α-FPUT Cores and Tails

Sergej Flach

Center for Theoretical Physics of Complex Systems, Institute for Basic Science (IBS), Daejeon 34126, Korea; sflach@ibs.re.kr

Abstract: Almost 70 years ago, the Fermi–Pasta–Ulam–Tsingou (FPUT) paradox was formulated in, observed in, and reported using normal modes of a nonlinear, one-dimensional, non-integrable string. Let us recap the paradox. One normal mode is excited, which drives three or four more normal modes in the core. Then, that is it for quite a long time. So why are many normal modes staying weakly excited in the tail? Furthermore, how many? A quantitative, analytical answer to the latter question is given here using resonances and secular avalanches A comparison with the previous numerical data is made and extremely good agreement is found.

Keywords: Fermi–Pasta–Ulam–Tsingou (FPUT) problem; normal modes; resonances; secular avalanche

Citation: Flach, S. Measuring α-FPUT Cores and Tails. *Physics* **2021**, 3, 879–887. https://doi.org/10.3390/physics3040054

Received: 25 August 2021
Accepted: 19 September 2021
Published: 30 September 2021

Publisher's Note: MDPI stays neutral with regard to jurisdictional claims in published maps and institutional affiliations.

1. Introduction

In 1955, Fermi, Pasta, Ulam and Tsingou published their celebrated report on the thermalization of weakly nonlinear strings [1], bringing forth a fundamental physical and mathematical problem of energy equipartition and ergodicity. The study was reportedly performed by Enrico Fermi, John Pasta, Stanislav Ulam, and Mary Tsingou, and the internal Los Alamos report was written and authored by Fermi, Pasta and Ulam [1]. A series of numerical simulations showed that energy, initially placed in a low-frequency normal mode of the linear problem with a frequency ω_q and a corresponding wave number q, stayed almost completely locked within a few neighbor low-frequency modes in the presence of nonlinear mode–mode interactions, instead of being distributed among all modes of the system. Moreover, the recurrence of energy to the originally excited mode was observed after a long simulation time. It has been known since as the Fermi–Pasta–Ulam–Tsingou (FPUT) problem, paradox, and discovery [2–5].

A number of studies have focused on the explanation of recurrences. Zabusky and Kruskal pioneered the pathway of integrable approximations and soliton counting in real space [6–8]. To connect to the limit of weak nonlinear dynamics, Ford and Jackson followed the path of resonances in normal mode space [9–11]. Tuck and Menzel (née Tsingou) studied in detail the fate of recurrences for longer times. To their surprise, they observed super-recurrences, i.e., beatings of the recurrence amplitudes [12]. Sholl and Henry searched for scaling relations from recurrence time computations [13]. Lin, Goedde, and Lichter arrived at more detailed scaling relations for the recurrence times, and in addition also produced intriguing numerical data for the dependence of the number of excited modes of the energy [14]. The framework of periodic orbits in dynamical systems was used to rigorously prove the existence of exact time-periodic orbits, coined q-breathers, which are nonlinearity-induced deformed normal-mode periodic orbits of the linear limit [15,16]. FPUT trajectories correspond to perturbed q-breather solutions. An advanced perturbation analysis in mode space which uses secular avalanches was derived by Ponno et al. [17], which arrived at an approximate estimate of the excited mode number in the FPUT experiment. Recently, Pace and Campbell arrived at an elegant theoretical quantitative explanation of super-recurrences [18]. What remains, then, is to quantitatively explain the numerical observations on the excited mode number by Lin et al. [14] which is what is done below.

2. The α-FPUT Chain

FPUT-studied models have cubic (α-FPUT) and quartic (β-FPUT) nonlinearities in the Hamiltonian potential energy, and the α-FPUT case is considered here. The Hamiltonian of the α-FPUT lattice for N particles is given by

$$H_\alpha(\mathbf{q},\mathbf{p}) = \sum_{n=1}^{N} \frac{p_n^2}{2} + \sum_{n=0}^{N} \frac{1}{2}(q_{n+1}-q_n)^2 + \frac{\alpha}{3}(q_{n+1}-q_n)^3. \tag{1}$$

Fixed boundary conditions, $q_0 = q_{N+1} = 0$ and $p_0 = p_{N+1} = 0$ are used, where $q_n(t)$ and $p_n(t)$ are canonical coordinates and momenta, respectively.

The normal-mode representation is introduced via a canonical Fourier transform,

$$\begin{bmatrix} q_n \\ p_n \end{bmatrix} = \sqrt{\frac{2}{N+1}} \sum_{k=1}^{N} \begin{bmatrix} Q_k \\ P_k \end{bmatrix} \sin\left(\frac{nk\pi}{N+1}\right), \tag{2}$$

which diagonalizes the harmonic oscillator Hamiltonian part. Rewriting Equation (1) in these normal-mode coordinates $(\mathbf{Q},\mathbf{P})$ yields:

$$H_\alpha(\mathbf{Q},\mathbf{P}) = \sum_{k=1}^{N} \frac{P_k^2 + \omega_k^2 Q_k^2}{2} + \frac{\alpha}{3} \sum_{k,j,l=1}^{N} A_{k,j,l} Q_k Q_j Q_l, \tag{3}$$

where the normal mode frequencies are

$$\omega_k = 2\sin\left(\frac{k\pi}{2(N+1)}\right), \tag{4}$$

and the normal mode energies are defined as

$$E_k = \frac{P_k^2 + \omega_k^2 Q_k^2}{2}. \tag{5}$$

Note that these normal mode energies are conserved quantities for $\alpha = 0$, but cease to be preserved for the nonlinear case. The coupling constants $A_{k,j,l}$ are given by [16]

$$A_{k,j,l} = \frac{\omega_k \omega_j \omega_l}{\sqrt{2(N+1)}} \sum_{\pm}\left(\delta_{k,\pm j \pm l} - \delta_{k \pm j \pm l, 2(N+1)}\right). \tag{6}$$

Here, the sums $\sum_\pm$ are overall combinations of plus and minus signs among the $\pm$ symbols, and $\delta_{j,l}$ is the Kronecker delta function.

One can rescale the normal-mode coordinate and momentum [19] pairs in Equation (3) by $(\mathbf{Q},\mathbf{P}) \to (\mathbf{Q}/\alpha, \mathbf{P}/\alpha)$. If E represents the total energy in the system, this leads to

$$H_{\alpha=1}(\mathbf{Q},\mathbf{P}) = \alpha^2 E, \tag{7}$$

which allows one to investigate results as functions of the combined parameter, $E\alpha^2$, rather than using the separate parameters E and α.

Let us show the evolution of the original α-FPUT trajectory for $\alpha = 0.25$, $N = 32$ and energy $E = 0.077$ placed initially into the mode with $k_0 = 1$. (All variables in this paper are considered dimensionless.) In Figure 1, the time dependence of the mode energies $E_k(t)$ is plotted for the first five modes of the data from [16].

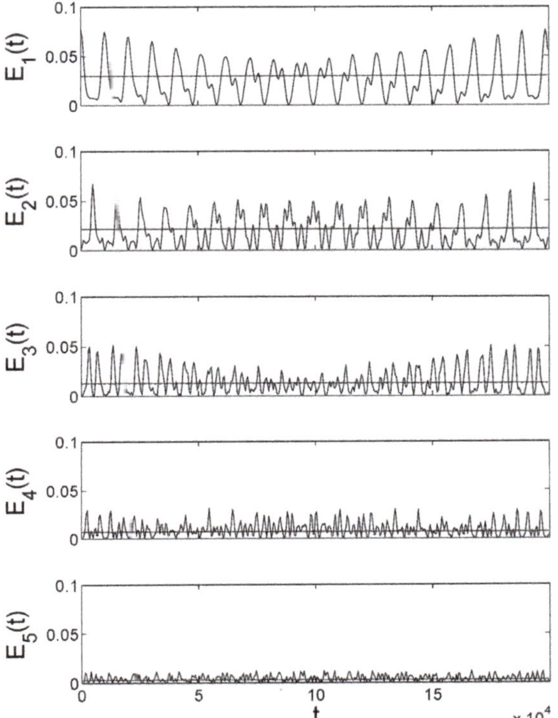

Figure 1. Evolution of the linear mode energies for the first five modes on a large timescale for the original Fermi–Pasta–Ulam-Tsingou (FPUT) trajectory for the parameter $\alpha = 0.25$, $theeneregyE = 0.077$, and the number of particles $N = 32$ [1] (oscillating curves). The almost-straight horizontal lines indicate the weak time dependence of the linear mode energies on corresponding exact time-periodic q-breather solutions from [16]. Figure is adapted from [16].

The period of the slowest ($k_0 = 1$) harmonic mode is $T_1 = 2\pi/\omega_1 \approx 66.02$. One observes slow processes of the redistribution of mode energies, with recurrence time amounting to $T_R \approx 10530$. One can also observe even slower modulations of recurrence amplitudes on time scales of the order of 10^5, which are the celebrated super-recurrences with $T_{SR} \approx 2 \times 10^6$ [12,18]. The localization in q-space is also well observed, with the maximum of E_5 being eight times smaller than that of E_1. The number of strongly participating modes can be therefore estimated to be around three or four. The almost straight horizontal lines indicate the weak time dependence of the linear mode energies on the corresponding exact time-periodic q-breather solutions from [16].

3. Mode Coupling Approach

The equations of motion for the normal mode amplitudes follow from Equations (1)–(6) and read:

$$\ddot{Q}_k + \omega_k^2 Q_k = -\alpha \sum_{l,m=1}^{N} A_{k,l,m} Q_l Q_m, \tag{8}$$

where the dots stay for time derivative.

The system of Equation (8) describes a network of oscillators with different eigenfrequencies. These oscillators interact with each other via nonlinear interaction terms. The interaction network is long-ranged in k-space. To be more specific, each normal-mode oscillator is interacting with a set of other doublets of oscillators. The total number of multiplets one normal-mode oscillator is connected to is proportional to N^2. The number N is the total number of oscillators (particles), or, more generally, the volume of the system. The coupling constants $A_{k,l,m}$ depend on the oscillator frequencies; see Equation (6). Still their values do not decay exponentially fast, with some growing distance between oscillators (after introducing a proper metric). Therefore, essentially all oscillators interact with all others. This is what is meant by "long range".

If there is long-range interaction in mode space, why do modes not quickly excite other modes and thermalize? The reason is that the interaction is nonlinear. Indeed, with linear interactions, exciting one mode will inevitably excite other modes in some proportion to the coupling coefficient amplitudes. However, with nonlinear couplings, things are more complicate as shown below. Actually, it is insightful to recall the seemingly simple problem of the periodic motion of one oscillator in an anharmonic potential,

$$\ddot{x} = -x - \alpha x^2 - \beta x^3 \,, \tag{9}$$

where x is the space coordinate.

The bounded motion at energy E yields a solution which is periodic with some period $T(E) = 2\pi/\Omega$ (with Ω denoting the frequency), and can be represented by a Fourier series

$$x(t) = \sum_k A_k e^{ik\Omega t} \,, \tag{10}$$

which leads to algebraic equations for the Fourier coefficients,

$$A_k = k^2\Omega^2 A_k - \alpha \sum_{k_1} A_{k_1} A_{k-k_1} - \beta \sum_{k_1,k_2} A_{k_1} A_{k_2} A_{k-k_1-k_2} \,. \tag{11}$$

Let us note that Equation (11) has similar properties as compared to Equation (8)—the coupling between the Fourier coefficients is nonlinear but long ranged. Yet, it is known that the bounded solutions (10) to (9) are analytic functions $x(t)$, and thus the Fourier series coefficients A_k converge exponentially fast with k [20].

3.1. Complex Mode Variables

Ponno et al. [17] attempted to obtain analytical expressions for the mode dynamics of the α-FPUT model at times shorter than, or at best of the order of the time of first recurrence. Following their approach, let us perform a change from real to complex variables:

$$u_k \equiv \frac{\omega_k Q_k + iP_k}{\sqrt{2E}} \,, \quad |u_k(t)|^2 = \frac{E_k(t)}{E} = e_k(t) \,, \tag{12}$$

The α-FPUT Hamiltonian now reads:

$$\mathcal{H}(u,u^*) = \underbrace{\sum_{k=1}^{N-1} |u_k|^2}_{\mathcal{H}_2} + \underbrace{\frac{\mu}{12} \sum_{k_1,k_2,k_3=1}^{N-1} \Delta_{k_1,k_2,k_3} \prod_{j=1}^{3} (u_{k_j} + u_{k_j}^*)}_{\mathcal{H}_3} \,, \tag{13}$$

where

$$\Delta_{k_1,k_2,k_3} \equiv \delta_{k_1+k_2,k_3} + \delta_{k_2+k_3,k_1} + \delta_{k_3+k_1,k_2} - \delta_{k_1+k_2+k_3,2N} \,, \tag{14}$$

and

$$\mu \equiv \alpha \sqrt{\frac{E}{N}} \,. \tag{15}$$

The quadratic, $\mathcal{H}_2$, and cubic, $\mathcal{H}_3$, parts of the Hamiltonian are indicated in Equation (13). The equations of motion then read:

$$\dot{u}_k = -i\omega_k \left[u_k + \frac{\mu}{4} \sum_{p,q=1}^{N-1} \Delta_{k,p,q}(u_p + u_p^*)(u_q + u_q^*) \right] . \quad (16)$$

The FPUT initial condition turns to

$$u_k(0) = \delta_{k,1} . \quad (17)$$

3.2. Resonances

With the FPUT initial condition of exciting mode $k = 1$, and a small value of μ in Equation (15), the first mode starts evolving in an almost periodic fashion with a frequency almost equal to ω_1. Assuming $u_k = e^{-i\omega_1 t}\delta_{k,1}$ as a solution to zero order in μ, and inserting this into the righ-hand side (r.h.s.) of Equation (16) leaves us with

$$\dot{u}_2 + i\omega_2 u_2 = -i\mu\omega_2 \cos^2 \omega_1 t \equiv -i\frac{\mu\omega_2}{2}(1 + \cos 2\omega_1 t) . \quad (18)$$

Thus, mode $k = 2$ is driven by a periodic force with frequency $2\omega_1$. This is as close to resonance as $\omega_2 - 2\omega_1$ is close to zero, and its smallness is to be compared with the drive amplitude $\sim\mu\omega_2$. Assuming $\pi/(2(N+1)) \ll 1$, which is correct for $N = 32$, let us expand the dispersion relation: $\omega_k = 2\sin\left(\frac{k\pi}{2(N+1)}\right) \approx 2\left(\frac{\pi k}{2(N+1)} - \frac{\pi^3 k^3}{48(N+1)^3}\right)$. Then:

$$\Delta_2 \equiv |\omega_2 - 2\omega_1| \approx \frac{\pi^3}{4(N+1)^3} . \quad (19)$$

Stripping Equation (18) off its nonresonant terms (whose contribution to the solution is reduced by a factor of $\Delta_2/\omega_{1,2} \sim 1/N^2$), one is left with

$$\dot{u}_2 + i\omega_2 u_2 \approx -i\frac{\mu\omega_2}{4} \exp(-2i\omega_1 t) . \quad (20)$$

The solution to Equation (20) reads:

$$u_2(t) = \frac{\mu\omega_2}{4\Delta_2} \mathrm{e}^{-i\omega_2 t} \left[\mathrm{e}^{i\Delta_2 t} - 1 \right] . \quad (21)$$

As long as $u_2(t) \ll 1$, the above approach is valid. The border of its validity is reached when the energy E takes the critical value E_2^{sa}, at which mode $k = 2$ is involved in a secular avalanche [17]:

$$\frac{\mu\omega_2}{4\Delta_2} = 1 \;\rightarrow\; E_2^{\mathrm{sa}} = \frac{\pi^4}{36\alpha^2 N^3} . \quad (22)$$

For energies $E \ll E_2^{\mathrm{sa}}$, the small frequency Δ_2 leads to a slow modulation in the r.h.s. of Equation (21), which results in a corresponding slow modulation of the energy stored in mode $k = 1$ due to energy conservation. Then, the corresponding zero order (or perturbative) recurrence time estimate is:

$$T_{\mathrm{R}}^{(0)} \equiv \frac{2\pi}{\Delta_2} = \frac{8}{\pi^2} N^3 \; , \; E \ll E_2^{\mathrm{sa}} . \quad (23)$$

This coincides with earlier results by Sholl and Henry [13] (see also Lin et al. [14]), and the relevant resonance was already worked out in Ford's paper [9].

Let us calculate some numbers. Figure 1 in Ref. [14] uses parameters $E = 2.2$, $\alpha = 0.1$, and $N = 32$. On one side, it follows $T_{\mathrm{R}}^{(0)}$ = 26,560, but due to the large energy, it also follows $E \gg E_2^{\mathrm{sa}} = 0.0083$, implying that the recurrence time concept is invalid since perturbation theory is inapplicable. Still, the measured $T_{\mathrm{R}} = 6400$ is orders of magnitude larger than the typical mode period, T_1, and only a factor of four smaller than the perturbation theory estimate. The original FPUT trajectory was investigated in Figure 1 of Ref. [16], with parameters $E = 0.077$, $\alpha = 0.25$ and $N = 32$, and shown here in Figure 1. Again, $E \gg E_2^{\mathrm{sa}} = 0.0013$. The measured recurrence time T_{R} = 10,500 is smaller than the perturbation result $T_{\mathrm{R}}^{(0)}$, but still orders of magnitude larger than the mode period T_1. Part of the FPUT surprise must have been that even for $E \gg E_2^{\mathrm{sa}}$, recurrence times still stayed large and reasonably close to their perturbation theory estimates, and a fast approach to equipartition was missing.

4. The Number of Excited Modes

For $E \gg E_2^{\mathrm{sa}}$, one concludes that the FPUT trajectory is resonant. Mode $k = 2$ will be resonantly pumped up by mode $k = 1$ until mode $k = 1$ is depleted. It is needless to state that the process continues into higher modes, showing a complex resonant avalanche, as studied in detail in Ref. [17]. Furthermore, this is what FPUT observed, since they evidently chose the proper parameters to ensure that the system is in the nonperturbative regime of a resonant avalanche, which one enters for energies $E \geq E_2^{\mathrm{sa}}$.

Why is the secular avalanche stopping and not continuing to flood all the modes? According to Ref. [17], this is simply because the modes in the mode packet can be separated into core modes and tail modes. Core modes are strongly and resonantly interacting with each other. Tail modes fail to be resonantly pumped as they are tuned out of resonance due to the nonlinear dispersion relation. The boundary-separating core and tail modes are functions of the energy. For $E \ll E_2^{\mathrm{sa}}$, all modes are tail modes except for the one core mode initially excited.

To see that the above approach is extended to higher orders of perturbation theory; see Ref. [17] for details. Mode $k = 2$ is driven by mode $k = 1$ through the resonant term u_1^2. Mode $k = 3$ is driven by the resonant term $u_1 u_2$, and so on. One arrives at

$$\dot{u}_k + i\omega_k u_k \approx -i\frac{\mu\omega_k}{4}\mathrm{e}^{-ik\omega_1 t} . \tag{24}$$

The relevant resonances are $\Delta_k = |k\omega_1 - \omega_k|$, and lead to

$$u_k(t) \approx \frac{\mu\omega_k}{4\Delta_k}\mathrm{e}^{-i\omega_k t}\left[\mathrm{e}^{i\Delta_k t} - 1\right] . \tag{25}$$

The critical energy E_k^{sa}, above which mode k becomes part of the core and the secular avalanche, then reads:

$$\frac{\mu\omega_k}{4\Delta_k} = 1 \rightarrow E_k^{\mathrm{sa}} = \frac{2\pi^4(k^2-1)^2}{24^2\alpha^2 N^3} . \tag{26}$$

Since $E_k^{\mathrm{sa}} \sim k^4$, it follows that for some reasonably small value of $k \equiv k_c$, the corresponding mode will be out of resonance:

$$k_c = \sqrt{1 + \frac{6}{\pi^2}\mu N^2} . \tag{27}$$

This agrees very well with the detailed derivations in Ref. [17], which culminate in a rough scaling estimate $k_c \approx \sqrt{\mu}N$ for large k_c and large N. At the same time, Equation (27) is accurate for small values of k_c, which is the case for, e.g., the original FPUT trajectory (see precise numbers just below). This can happen despite a large value of $N \gg 1$ since other relevant (small) parameters include the energy E and the coupling constant α, which make the product μN^2 small. The mode energies for $k > k_c$ are decreasing exponentially with increasing k, as observed numerically in Ref. [17,21], and as also derived for the mode energy profiles in q-breather solutions [15,16]. Therefore, the number of modes participating in an FPTU trajectory is simply given by k_c.

Let us calculate numbers again. Figure 1 in Ref. [14] uses parameters $E = 2.2$, $\alpha = 0.1$ and $N = 32$. It follows $k_c = 4.16$ in good agreement with the numerical observations. About four modes are involved in the resonant dynamics of the core, while all other modes stay out of resonance. The original FPUT trajectory, which was investigated in Figure 1 in Ref. [16] with parameters $E = 0.077$, $\alpha = 0.25$ and $N = 32$, yields $k_c = 2.9$, which is again in good agreement with numerical observations; see also Figure 1.

One is now in a position to quantitatively compare the central result obtained, Equation (27), with numerical results from Lin et al. [14]. The authors of that study measured the effective number,

$$n_{\rm eff} = {\rm e}^S \,,\; S = -\sum_k e_k \ln e_k, \tag{28}$$

which ranges from 1 to N as $S = 0$ for one mode excited, and $S = \ln N$ for equally distributed mode energies. According to the derivation made:

$$k_c = n_{\rm eff}. \tag{29}$$

In order to test the above equality, the data on $n_{\rm eff}$ versus μN^2 for $N = 3, 264, 128$ are extracted from Figure 4 of Ref. [14]. The result is plotted in Figure 2 along with the theoretical result for k_c in Equation (27). Very good agreement can be observed.

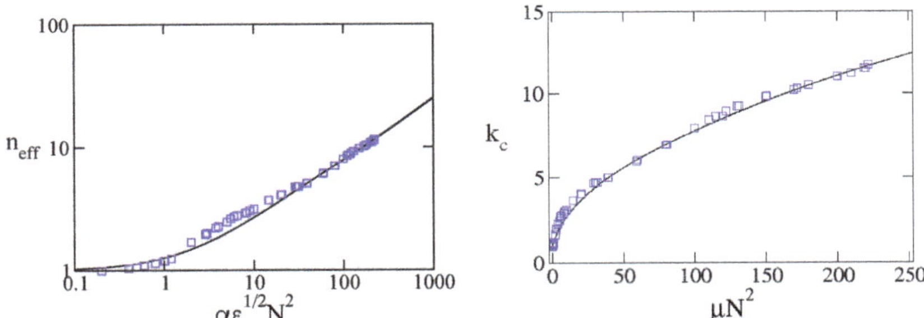

Figure 2. (**Left panel**) The effective number, $n_{\rm eff}$ versus $\alpha\varepsilon^{1/2}N^2 \equiv \mu N^2$ with $\varepsilon \equiv E/N$. The symbols represent the data read off from Figure 4b of Ref. [14], correspond to a set of different system sizes, $N = 3, 264, 128$, and demonstrate a single scaling curve existence. The black line represents the theoretical result (27). (**Right panel**) same as left panel but using linear scales. The square-root law of Equation (27) provides an extremely good fit .

One concludes that the FPUT trajectory will, for long times, excite a mode packet with k_c modes, which are the core modes of the packet. The remaining $(N - k_c)$ modes belong to the tail of the packet, and decay exponentially with increasing k. Let us further note that the theoretical result obtained here can be rewritten in the limit of large system size as:

$$\frac{k_c}{N} \approx \frac{\sqrt{6\alpha}}{\pi}\varepsilon^{1/4} \,,\; \varepsilon = \frac{E}{N} \,,\; N \gg 1\,. \tag{30}$$

Therefore, the packet size becomes system size-independent if expressed entirely through intensive quantities—wave numbers and energy densities.

5. Discussion

It is now understood that both FPUT recurrences and super-recurrences are part of the dynamics of so-called metastable states (mode packets) [22], which eventually relax into equipartition at some time T_2. These metastable states are formed at a time T_1, persisting over potentially huge time intervals, $\sim(T_2 - T_1)$. They are characterized by a localized distribution of energy in mode space. This distribution has a core which, e.g., in the case of the original FPUT test, contains a few low-frequency modes. The distribution has also a tail. Galgani and Scott observed that this tail is exponentially decaying [21]. The distribution appears to be almost stationary when using proper averaging times which are much shorter than T_m, though of course lots of dynamics is going on at various time scales, e.g., recurrences and super-recurrences. The core shows recurrence and super-recurrence but also various forms of chaos (at even larger time scales). The tails are characterized by decay structures in normal mode space, resonances, and slow incoherent heating. The decay structures can be exponential, thus leading to length scales in normal mode space or algebraic implying the absence of the latter. Resonances in the decay profiles show the driven nature of these tails, with the core being the driving source. Incoherent heating results from the same core driving, which at larger time scales may exhibit chaotic incoherent dynamics.

The core dynamics are quantified by their recurrence, T_{R}, and super-recurrence, T_{SR} times and the core size. While the recurrence and super-recurrence times were assessed in previous studies a quantitative calculation of the core size is provided here, which agrees very well with the measured data. One can therefore conclude that the regular core dynamics have been to some extent exhaustively studied. What remains for the core is to assess its chaotic incoherent dynamics. These dynamics are the reason for the slow heating of the tail modes, and will ultimately explain the time scale T_2 of a final reaching of equipartition, as studied numerically in detail in Ref. [23].

6. Conclusions

In this paper, a quantitative estimates for the size of the core and tail of a low-frequency normal-mode excitation in a Fermi–Pasta–Ulam–Tsingou (FPUT) chain is provided. How will these results be modified if high frequency modes are excited? Even more intriguing is the question of whether any of these results are carried over in some form for an FPUT system at thermal equilibrium—what are the details of the energy transfer between low and high frequency modes in such a situation? What will happen at higher lattice dimensions? What changes if the model supports optical normal modes with a finite frequency gap in the band structure? It seems there a plenty of questions to be addressed 70 years after the FPUT experiment.

7. HB2U

I had a great time working together with Mikhail Tribelsky back in Dresden at the Max Planck Institute for Physics of Complex Systems. We both shared a passion for puzzles and paradoxes, probably intuitively realizing that however remote the topic of the puzzle is, solving it will advance the mind in potentially unexpected areas of science and general understanding of the world. I dedicate this paper to Mikhail's 70th birthday. It solves some aspects of a paradox which is almost the same age, paradoxically.

Funding: This research received no external funding.

Data Availability Statement: The data presented in this study are available on request from the author.

Acknowledgments: I thank David Campbell and Carlo Danieli for helpful discussions. This work was supported by the Institute for Basic Science, Project Code (IBS-R024-D1).

Conflicts of Interest: The author declares no conflict of interest.

Abbreviations

The following abbreviations are used in this manuscript:

FPUT	Fermi, Pasta, Ulam, Tsingou
HB2U	Happy Birthday To You

References

1. Fermi, E.; Pasta, J.; Ulam, S. Studies of the nonlinear problems. I. LASL Report LA-1940; Los Alamos Scientific Laboratory: Los Alamos, CA, USA, 1955. doi: 10.2172/4376203. Reprinted in *The Collected Papers of Enrico Fermi*; Segré, E., Ed.; University of Chicago Press: Chicago, IL, USA, 1965; Volume II. [CrossRef]

2. Ford, J. The Fermi–Pasta-Ulam problem: Paradox turns discovery. *Phys. Rep.* **1992**, *213*, 271–310. [CrossRef]
3. Weissert, T.P. *The Genesis of Simulation in Dynamics: Pursuing the Fermi–Pasta-Ulam Problem*; Springer-Verlag: New York, NY, USA, 1997.
4. Campbell, D.K.; Rosenau, P.; Zaslavsky, G. Introduction: The Fermi-Pasta-Ulam problem—The first fifty years. *Chaos* **2005**, *15*, 015101. [CrossRef] [PubMed]
5. Gallavotti, G. *The Fermi–Pasta-Ulam Problem: A Status Report*; Springer: Berlin/Heidelberg, Germany, 2007.
6. Zabusky, N.J. Exact solution for the vibrations of a nonlinear continuous model string. *J. Math. Phys.* **1962**, *3*, 1028–1039. [CrossRef]
7. Kruskal, M.D.; Zabusky, N.J. Stroboscopic-perturbation procedure for treating a class of nonlinear wave equations. *J. Math. Phys.* **1964**, *5*, 231–244. [CrossRef]
8. Zabusky, N.J.; Kruskal, M.D. Interaction of "solitons" in a collisionless plasma and the recurrence of initial states. *Phys. Rev. Lett.* **1965**, *15*, 240–243. [CrossRef]
9. Ford, J. Equipartition of energy for nonlinear systems. *J. Math. Phys.* **1961**, *2*, 387–393. [CrossRef]
10. Jackson, E.A. Nonlinear coupled oscillators. I. Perturbation theory; Ergodic Problem. *J. Math. Phys.* **1963**, *4*, 551–558. [CrossRef]
11. Jackson, E.A. Nonlinear coupled oscillators. II. Comparison of theory with computer solutions. *J. Math. Phys.* **1963**, *4*, 686–700. [CrossRef]
12. Tuck, J.; Menzel, M. The superperiod of the nonlinear weighted string (FPU) problem. *Adv. Math.* **1972**, *9*, 399–407. [CrossRef]
13. Sholl, D.S.; Henry, B. Recurrence times in cubic and quartic Fermi–Pasta-Ulam chains: A shifted-frequency perturbation treatment. *Phys. Rev. A* **1991**, *44*, 6364. [CrossRef] [PubMed]
14. Lin, C.; Goedde, C.; Lichter, S. Scaling of the recurrence time in the cubic Fermi–Pasta-Ulam lattice. *Phys. Lett. A* **1997**, *229*, 367–374. [CrossRef]
15. Flach, S.; Ivanchenko, M.; Kanakov, O. q-Breathers and the Fermi–Pasta-Ulam problem. *Phys. Rev. Lett.* **2005**, *95*, 064102. [CrossRef] [PubMed]
16. Flach, S.; Ivanchenko, M.; Kanakov, O. q-Breathers in Fermi–Pasta-Ulam chains: Existence, localization, and stability. *Phys. Rev. E* **2006**, *73*, 036618. [CrossRef] [PubMed]
17. Ponno, A.; Christodoulidi, H.; Skokos, C.; Flach, S. The two-stage dynamics in the Fermi–Pasta-Ulam problem: From regular to diffusive behavior. *Chaos Interdiscip. J. Nonlinear Sci.* **2011**, *21*, 043127. [CrossRef] [PubMed]
18. Pace, S.D.; Campbell, D.K. Behavior and breakdown of higher-order Fermi-Pasta-Ulam-Tsingou recurrences. *Chaos Interdiscip. J. Nonlinear Sci.* **2019**, *29*, 023132. [CrossRef] [PubMed]
19. De Luca, J.; Lichtenberg, A.J.; Lieberman, M.A. Time scale to ergodicity in the Fermi–Pasta-Ulam system. *Chaos* **1995**, *5*, 283–297. [CrossRef] [PubMed]
20. Zygmund, A. *Trigonometric Series*; Cambridge University Press: Cambridge, UK, 1968.
21. Galgani, L.; Scott, A. Planck-like distributions in classical nonlinear mechanics. *Phys. Rev. Lett.* **1972**, *28*, 1173. [CrossRef]
22. Benettin, G.; Carati, A.; Galgani, L.; Giorgilli, A. The Fermi-Pasta-Ulam problem and the metastability perspective. In *The Fermi–Pasta-Ulam Problem*; Gallavotti, G., Ed.; Springer: Berlin/Heidelberg, Germany, 2007; pp. 151–189. [CrossRef]
23. Danieli, C.; Campbell, D.; Flach, S. Intermittent many-body dynamics at equilibrium. *Phys. Rev. E* **2017**, *95*, 060202. [CrossRef] [PubMed]

 physics

Article

Fluctuating Number of Energy Levels in Mixed-Type Lemon Billiards

Črt Lozej [1], Dragan Lukman [2,†] and Marko Robnik [2,*]

1 Faculty of Mathematics and Physics, University of Ljubljana, Jadranska 19, SI-1000 Ljubljana, Slovenia; clozej@gmail.com
2 CAMTP—Center for Applied Mathematics and Theoretical Physics, University of Maribor, Mladinska 3, SI-2000 Maribor, Slovenia; dragan.lukman@gmail.com
* Correspondence: Robnik@uni-mb.si
† Deceased.

Abstract: In this paper, the fluctuation properties of the number of energy levels (mode fluctuation) are studied in the mixed-type lemon billiards at high lying energies. The boundary of the lemon billiards is defined by the intersection of two circles of equal unit radius with the distance $2B$ between the centers, as introduced by Heller and Tomsovic. In this paper, the case of two billiards, defined by $B = 0.1953,\ 0.083$, is studied. It is shown that the fluctuation of the number of energy levels follows the Gaussian distribution quite accurately, even though the relative fraction of the chaotic part of the phase space is only 0.28 and 0.16, respectively. The theoretical description of spectral fluctuations in the Berry–Robnik picture is discussed. Also, the (golden mean) integrable rectangular billiard is studied and an almost Gaussian distribution is obtained, in contrast to theory expectations. However, the variance as a function of energy, E, behaves as $\sqrt{E}$, in agreement with the theoretical prediction by Steiner.

Keywords: nonlinear dynamics; quantum chaos; mixed-type systems; energy level statistics; billiards; lemon billiards

Citation: Lozej, Č.; Lukman, D.; Robnik, M. Fluctuating Number of Energy Levels in Mixed-Type Lemon Billiards. *Physics* **2021**, 3, 888–902. https://doi.org/10.3390/physics3040055

Received: 15 July 2021
Accepted: 16 August 2021
Published: 2 October 2021

Publisher's Note: MDPI stays neutral with regard to jurisdictional claims in published maps and institutional affiliations.

1. Introduction

The purpose of this paper is to analyze the energy spectra of two characteristic complex mixed-type lemon billiards within the scope of quantum chaos. The boundary of the lemon billiards is defined by the intersection of two circles of equal unit radius with the distance $2B$ between the centers, as introduced by Heller and Tomsovic in 1993 [1,2]. The present study represents a continuation of our recent paper [3] on the classical and quantum *ergodic* billiard ($B = 0.5$) with strong stickiness effects, from the family of lemon billiards, as well as on three simple mixed-type lemon billiards with only one regular region, surrounded by a uniform chaotic sea without stickiness regions, namely, with the shape parameters $B = 0.42,\ 0.55,\ 0.6$ [4].

In the present paper, two lemon billiards with the shape parameters $B = 0.1953,\ 0.083$ are studied. These lemon billiards are mixed-type billiards with several independent regular regions embedded in a chaotic sea with no significant stickiness regions, which serve as examples of systems with a complex divided phase space. These lemon billiards were selected by the criterion of the maximally complex chaotic component generated by a single chaotic orbit. The discovery of the present and past physically and dynamically different and interesting lemon billiards has only been made possible thanks to the recent extensive analysis of Lozej [2]. The entire family of classical lemon billiards for a dense set of about 4000 values of $B \in [0.01, 0.99975]$ (in steps of $dB = 0.00025$) was systematically analyzed as for the corresponding phase space structure and stickiness effects. It must be emphasized that although all the lemon billiards belong to the same family of billiards as for the mathematical definition, individually, the lemon billiards have quite different, in fact, very rich, dynamical properties, important in the classical and quantum context.

A general introduction to the subjects in quantum chaos, related to this study, can be found in [3]. Let us also mention the books by Stöckmann [5] and Haake [6] on general quantum chaos and the recent reviews [7,8] on stationary quantum chaos in generic (mixed type) systems.

The main purpose of the present paper is to analyze the two selected quantum lemon billiards of $B = 0.1953$ and 0.083, with the goal to study the energy spectra, while the structure of the Poincaré–Husimi functions in the phase space, the separation of the regular eigenstates and chaotic eigenstates, as well as the localization properties of the chaotic eigenstates and their statistics will be treated separately [9].

The main results are as follows. The energy spectra are calculated by the scaling method of Vergini–Saraceno [10] in two versions, one based on the plane waves and the other one based on the circular waves (Bessel functions for the radial part and trigonometric functions for the angular part). This is done for high-lying eigenstates with the wavenumber k (in the specific units) $k = 2880$ for up to 300,000 consecutive levels for each of the four symmetry classes (odd-odd, odd-even, even-odd, and even-even). The energy of the level at k is $E = k^2$. It turns out that about 0.1% of the levels are lost for technical reasons, which is a known and experienced fact in numerically calculating billiard spectra with the scaling method, while the accuracy of individual levels is better than 1% of the mean energy level spacing. The spectral statistics are found to be stable with respect to these losses. The cumulative (integrated) energy level density (spectral staircase function) $\mathcal{N}(k)$ is well described by the Weyl formula (with the leading area term and the perimeter term) $W(k)$ if there are no missing levels. Then the fluctuations of the actual staircase function $\mathcal{N}(k)$ around the Weyl function, the difference $R(k) = \mathcal{N}(k) - W(k)$, and the $R(k)$ distribution are studied. In the literature, $R(k)$ is called mode fluctuation [11–14]. Due to the lost levels, this difference has a drift to negative values and fluctuates around the mean value. In order to separate the drift and the fluctuations, the quantity $w(k) = R - m(R)$, where $m(R)$ is the local average of $R(k)$ over 100 consecutive levels, is investigated. Then, the distribution of $w(k)$ for about 300,000 levels of each symmetry class separately, starting at $k = k_0 = 2880$, within the interval approximately $k \in [2880, \approx 3700]$ is studied. One finds that the distribution of w follows a Gaussian distribution fairly well, which is a surprising result as the two billiards have the relative fraction of the chaotic phase space of only 0.28 and 0.16, respectively. For comparison, also the entirely regular, integrable, case of the maximally irrational rectangular billiard is investigated and it is surprisingly found an almost Gaussian distribution for w, but with the variance rising linearly with k, as generally predicted by the theory of Steiner [11–14], while the distribution itself in integrable systems is predicted to be nonuniversal, varying from case to case, which here is not confirmed. The validity of the theoretical predictions has also been checked in experiments with superconducting microwave billiards [15]. Finally, the level spacing distribution for the two lemon billiards is presented which provides good agreement with the Berry–Robnik–Brody distribution.

The paper is organized as follows. In Section 2, the definition are given and classical dynamical properties of the lemon billiards are examined. In Section 3, the analysis of the fluctuation quantity $w(k)$ is performed. In Section 4, the statistical analysis of the entire energy spectra is made by calculating the level spacing distribution, $P(S)$. In Section 5, the results are discussed and conclusions are presented.

2. The Lemon Billiards and Their Phase Portraits

The family of lemon billiards was introduced by Heller and Tomsovic in 1993 [1] and has been studied in a number of papers [16–20], most recently, in [2–4,21]. and in the recent studies [3,4] by us. The lemon billiard boundary is defined by the intersection of two circles of equal unit radius with the distance between the centers, $2B$, being less than the diameters and $B \in (0,1)$, and is given by the following implicit equations in Cartesian coordinates:

$$(x+B)^2 + y^2 = 1, \quad x > 0, \qquad (1)$$
$$(x-B)^2 + y^2 = 1, \quad x < 0.$$

As usual, the canonical variables are used to specify the location, s, and the momentum component, p, on the boundary at the collision point. Namely, the arc length, s, is counted in the mathematical positive sense (counterclockwise) from the point $(x,y) = (0, -\sqrt{1-B^2})$ as the origin, while p is equal to the sine of the reflection angle θ; thus $p = \sin\theta \in [-1,1]$, as $\theta \in [-\pi/2, \pi/2]$. The bounce map $(s,p) \Rightarrow (s',p')$ is area preserving as in all billiard systems [22]. Due to the two kinks, the Lazutkin invariant tori (related to the boundary glancing orbits) do not exist. The period-2 orbit connecting the centers of the two circular arcs at the positions $(1-B,0)$ and $(-1+B,0)$ is always stable (and therefore surrounded by a regular island) except for the case $B = 1/2$, where it is a marginally unstable periodic orbit, the case being ergodic and treated by us earlier [3].

The circumference of the entire billiard, $\mathcal{L}$, is given by:

$$\mathcal{L} = 4\arctan\sqrt{B^{-2}-1}. \qquad (2)$$

The area $\mathcal{A}$ of the billiard is:

$$\begin{aligned} \mathcal{A} &= 2\arctan\sqrt{B^{-2}-1} - 2B\sqrt{1-B^2} \\ &= \frac{\mathcal{L}}{2} - 2B\sqrt{1-B^2}. \end{aligned} \qquad (3)$$

The structure of the phase space is shown in Figure 1 for the lemon billiard of $B = 0.1953$, with $\mathcal{L} = 5.4969$. The relative fraction of the area of the chaotic component of the bounce map is $\chi_c = 0.3585$, while the relative fraction of the phase space volume of the same chaotic component is $\rho_2 = 0.2804$, the Berry–Robnik parameter. Three independent regular island chains are clearly visible, the largest one around the period-2 orbit, which is densely covered by the invariant tori, with no visible thin chaotic layers inside. Let us denote the largest island chain by L, the second largest one by M, and the smallest one by S. The relative phase space volume of all three regular regions taken together is $\rho_1 = 1 - \rho_2 = 0.7196$. The chaotic sea is quite uniform, with no significant stickiness regions, and is generated by a single chaotic orbit.

The structure of the phase space, as shown in Figure 2 for the lemon billiard of $B = 0.083$, is more complex. Here, $\mathcal{L} = 5.9508$. The relative fraction of the area of the chaotic component of the bounce map is $\chi_c = 0.2168$, while the relative fraction of the phase space volume of the same chaotic component is $\rho_2 = 0.1617$. Thus, the relative phase space volume fraction of the complementary regular regions is $\rho_1 = 1 - \rho_2 = 0.8383$. In this case also, the chaotic sea is rather uniform, with no significant stickiness regions, and is generated by a single chaotic orbit, creating a very complex structure, perhaps mimicking stickiness in some thin regions. Both billiards are clearly of the Kolmogorov-Arnold-Moser (KAM) type, generic systems; examples of stochastic transition were studied already in [23].

One can conclude that the two cases $B = 0.1953,\ 0.083$ are interesting to verify the Berry–Robnik picture of quantum billiards [24], including the possible quantum localization of chaotic eigenstates, leading to the Berry–Robnik–Brody level spacing distribution and the universal statistical properties of the localization measures [4,9], as there are no significant stickiness effects, based on the results of the analysis of the recurrence time statistics in [2], unlike in the ergodic case, $B = 0.5$, studied in [3].

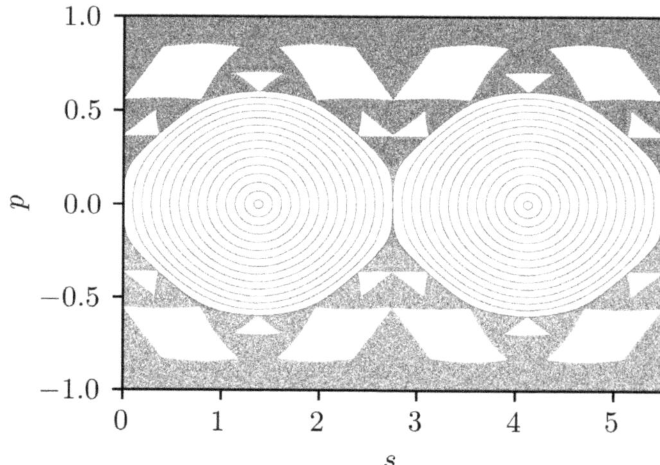

Figure 1. The phase portrait of the lemon billiard of $B = 0.1953$. The parameters are: the relative fraction of the area of the chaotic component of the bounce map, $\chi_c = 0.3585$, the relative fraction of the phase space volume, $\rho_2 = 0.2804$, and the complementary regular region, $\rho_1 = 1 - \rho_2 = 0.7196$. The abscissa is the location point, $s \in [0, 5.4969]$, and the ordinate is the momentim, $p \in [-1, 1]$, on the boundary at the collisions point. The chaotic component was created by a single chaotic orbit.

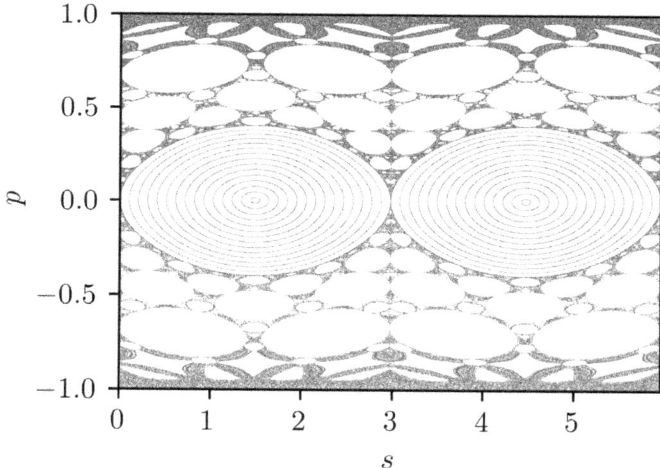

Figure 2. Same as Figure 1, but for $B = 0.083$ and $s \in [0, 5.9508]$. The parameters are: $\chi_c = 0.2168$, $\rho_2 = 0.1617$, and $\rho_1 = 1 - \rho_2 = 0.8383$.

3. The Energy Spectra and the Fluctuation of the Number of Energy Levels

Let us now turn to the quantum billiard $\mathcal{B}$ described by the stationary Schrödinger equation in the chosen units ($\hbar^2/2m = 1$) given by the Helmholtz equation:

$$\Delta\psi + k^2\psi = 0 \tag{4}$$

with the Dirichlet boundary conditions $\psi|_{\partial\mathcal{B}} = 0$, and the energy is $E = k^2$. Here $\hbar$ is the reduced Planck constant and m is the mass of the particle.

The mean number of energy levels $W(E)$ below $E = k^2$ is determined quite accurately, especially at large energies, asymptotically exact, by the celebrated Weyl formula (with perimeter corrections) using the Dirichlet boundary conditions, namely:

$$W(E) = \frac{\mathcal{A}\,E}{4\pi} - \frac{\mathcal{L}\,\sqrt{E}}{4\pi} + \text{c.c.}, \tag{5}$$

where "c.c." stays for small constants, determined by the corners and the curvature of the billiard boundary, which differentially play no role. Thus, the mean density of levels $d(E) = dW/dE$ is equal to:

$$d(E) = \frac{\mathcal{A}}{4\pi} - \frac{\mathcal{L}}{8\pi\sqrt{E}}. \tag{6}$$

The numerical method used here to solve the Helmholtz equation is based on the Heller's plane wave decomposition method and the Vergini–Saraceno scaling method [10,25]. Both versions of the Vergini–Saraceno method are implemented, namely, the one, based on plane waves, and the other one, based on the circular waves, and the same results are obtained within an accuracy of 0.1% of the mean level spacing. As mentioned, typically at most about 0.1% of the levels are lost. The numerical accuracy was checked by the convergence test, by varying the method's parameters (the number of basis waves and the number of nodes on the boundary).

The billiard considered has two reflection symmetries; thus, the eigenstates have four symmetry classes: odd-odd, odd-even, even-odd, and even-even. For the purpose of analyzing the spectral statistics and the wave functions, let us consider only the quarter billiard. In this case, the Weyl formula for the four symmetry classes reads:

$$\bar{W}(E) = \frac{\mathcal{A}\,E}{16\pi} - \frac{\bar{\mathcal{L}}\,\sqrt{E}}{4\pi} + \text{c.c.}, \tag{7}$$

where $\bar{\mathcal{L}}$ is defined for each of the above-defined symmetry classes as follows:

$$\begin{aligned} \bar{\mathcal{L}}_{\text{oo}} &= \frac{\mathcal{L}}{4} + a + b, \\ \bar{\mathcal{L}}_{\text{oe}} &= \frac{\mathcal{L}}{4} + a - b, \\ \bar{\mathcal{L}}_{\text{eo}} &= \frac{\mathcal{L}}{4} - a + b, \\ \bar{\mathcal{L}}_{\text{ee}} &= \frac{\mathcal{L}}{4} - a - b, \end{aligned} \tag{8}$$

where $a = \sqrt{1 - B^2}$ and $b = 1 - B$. Note that summing up the four contributions in Equation (7) with Equation (8), one obtains the Weyl formula for the entire spectrum Equation (5). For each symmetry class for each of the two billiards, about 300,000 levels were calculated and the difference, $R(k) = \mathcal{N}(k) - \bar{W}(k)$, was studied between the staircase function, $\mathcal{N}(k)$, and the Weyl function, $\bar{W}(k)$.

Figure 3 shows the results for the $B = 0.1953$ billiard. It is clear that $R(k)$ decreases linearly with k due to the numerical loss of the levels. Typically, about 0.1% of the levels are missing. In order to study the fluctuation properties, one needs to subtract the local mean value, $m(R(k))$, obtained by averaging over 100 consecutive levels, and then analyze the resulting fluctuation quantity, $w(k) = R(k) - m(R(k))$. The agreement with the Gaussian distribution is extremely good which is surprising, as the billiard is predominantly regular, with only 28% of the chaotic component. Steiner's theory [11–14] predicts a Gaussian distribution only for entirely chaotic (ergodic) systems, while the distribution of the fluctuation quantity, w, in integrable systems is expected to be nonuniversal. In the mixed-type case, the distribution of w should be in between, which here is not the case. Moreover, even for the rectangle as an example of the integrable systems, an (almost) Gaussian distribution is found (see below).

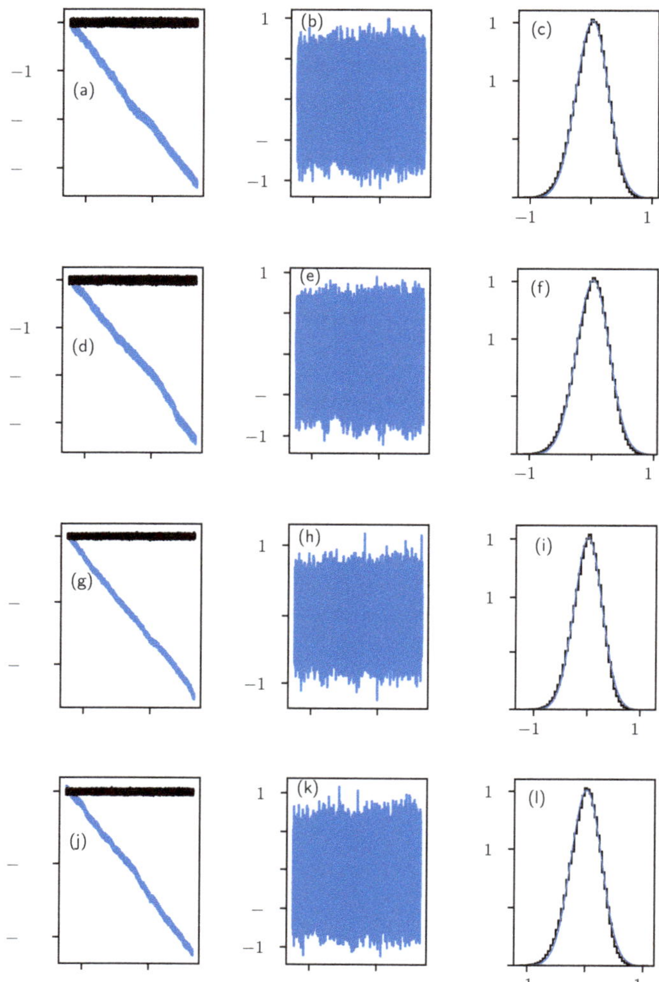

Figure 3. Results for the $B = 0.1953$ billiard: **left column (a,d,g,j)**: the difference, $R(k) = \mathcal{N}(k) - \bar{W}(k)$, between the staircase function, $\mathcal{N}(k)$, around the Weyl function and the Weyl function as a function of the wavenumber k (in the mean linearly decreasing line); **middle column (b,e,h,k)**: the fluctuation quantity, $w(k) = R(k) - m(R(k))$ as afunction of k, where $m(R(k))$ is the local average of $R(k)$ over 100 consecutive levels; **right column (c,f,i,l)**: the distribution of w. The rows top to bottom refer to the symmetry classes odd-odd, odd-even, even-odd, and even-even, respectively. In each parity case, there are about 300,000 levels. The agreement with the Gaussian distribution is extremely good. The small shift of the maximum around $w \approx 0.1$ is due to the imperfection of local averaging. The values of the mean, μ, and the standard deviation, σ are (top to bottom): (0.126, 2.624), (0.139, 2.636), (0.132, 2.635), (0.148, 2.641), respectively. Within the expected statistical error, the value of σ is the same for all parities.

The same analysis was performed for the case of the $B = 0.083$ billiard, and the results are shown in Figure 4, and lead to the same conclusions.

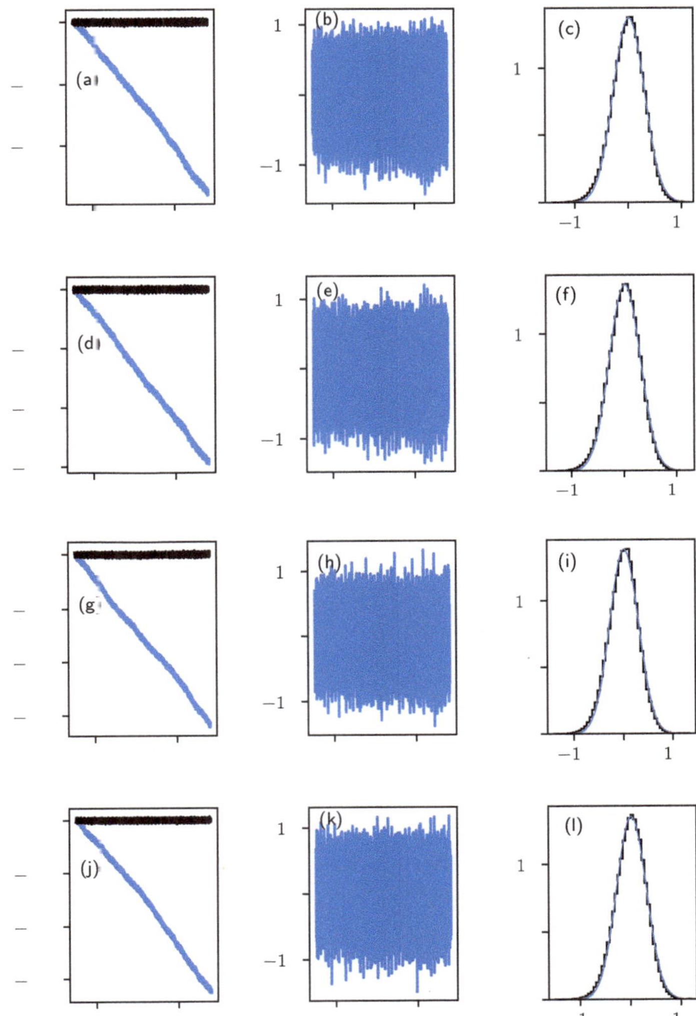

Figure 4. Results for the $B = 0.083$ billiard:**left column (a,d,g,j)**: the difference, $R(k) = \mathcal{N}(k) - \bar{W}(k)$, between the staircase function, $\mathcal{N}(k)$, around the Weyl function and the Weyl function as a function of the wavenumber k (in the mean linearly decreasing line); **middle column (b,e,h,k)**: the fluctuation quantity, $w(k) = R(k) - m(R(k))$ as afunction of k, where $m(R(k))$ is the local average of $R(k)$ over 100 consecutive levels; **right column (c,f,i,l)**: the distribution of w. The rows top to bottom refer to the symmetry classes odd-odd, odd-even, even-odd, and even-even, respectively. In each parity case, there are about 300,000 levels. The agreement with the Gaussian distribution is again extremely good. The values of the mean μ and standard deviation σ are: (0.148, 2.915), (0.137, 2.925), (0.154, 2.900), (0.146, 2.946) (top to bottom). Within the expected statistical error, the value of σ is the same for all parities.

The local averaging procedure over 100 consecutive levels is, of course, somewhat arbitrary. Therefore, the results were checked for 50, 200, 400 levels. In all cases, the distribution was found to represent Gaussian extremely well. The standard deviation σ at each level number keeps the same value for all parities, but increases slowly with the number of levels such that for 50, 100, 200, 400 for $B = 0.1953$, one gets $\sigma \approx$ 2.08, 2.63, 2.88, 2.92, respectively. For the $B = 0.083$ billiard, the corresponding values are: $\sigma \approx$ 2.22, 2.92, 3.50, 3.52.

In order to explore an integrable, entirely regular system, an example of the maximally irrational rectangle billiard was studied whose aspect ratio was taken the golden mean $g = (1+\sqrt{5})/2 \approx 1.61803$. The spectrum in this case is known analytically, and is exact: $E_{ln} = l^2/g^2 + n^2$, where l and n are two positive integers. Figure 5 shows the fluctuation quantity $w(k)$, which has no drift, because the spectrum is exact (no lost levels) over a wide range of k.

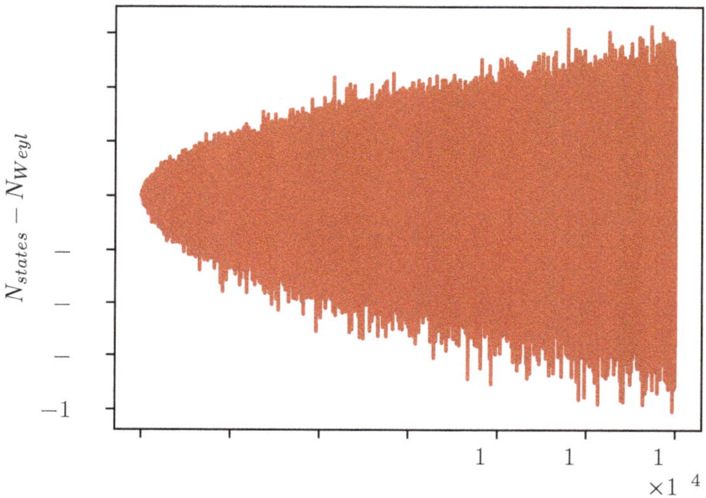

Figure 5. Results for the rectangle billiard, the fluctuation quantity $w(k) = R(k) = \mathcal{N}(k) - W(k)$ for a wide range of k.

The corresponding distributions at various wavenumber k intervals starting at k_0 are shown in Figure 6. Each histogram comprises about 100,000 objects. They are surprisingly close to a Gaussian distribution, contrary to the theoretical expectation, and thus, the distribution is just close to the case of ergodic chaotic systems. Therefore, the distribution of the number of energy levels (distribution of mode fluctuations) is not a good criterion to distinguish ergodic chaotic systems from the regular integrable systems. In Table 1, the data for the mean, μ, standard deviation, σ, skewness, and kurtosis are given.

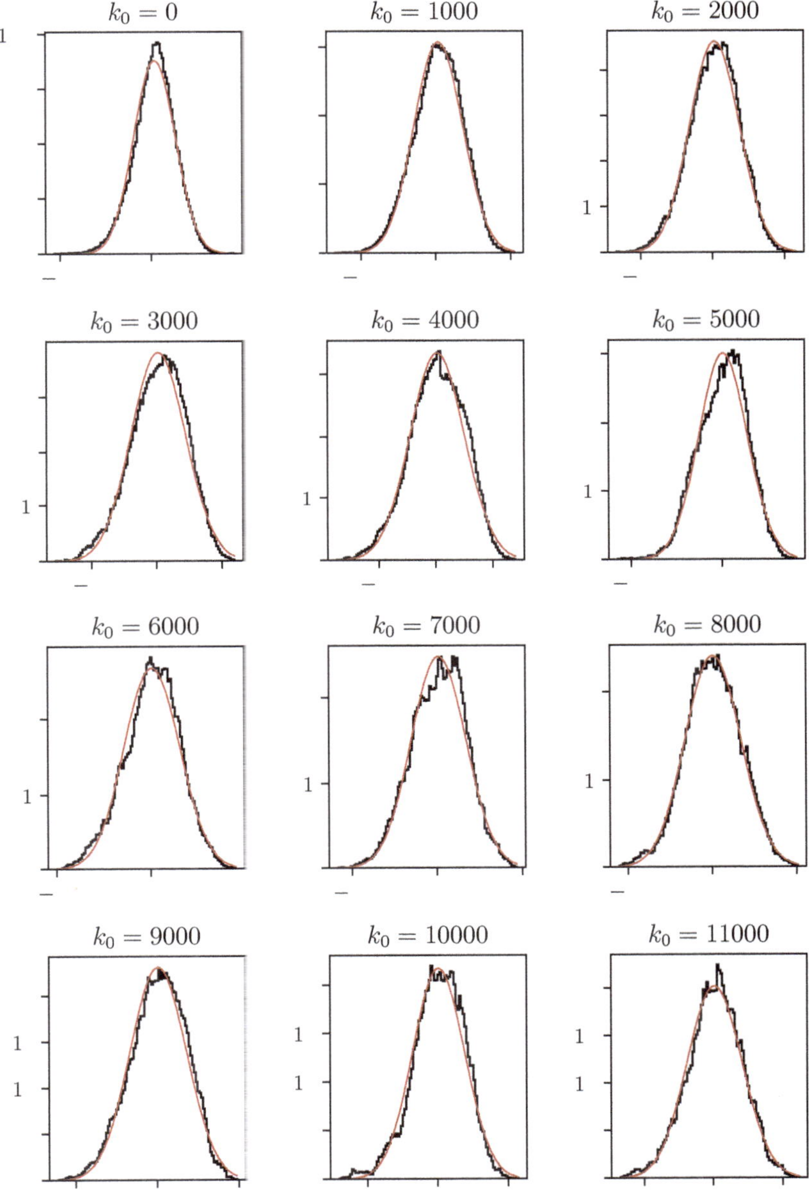

Figure 6. Distributions of w, $w(k) = R(k) = \mathcal{N}(k) - W(k)$, for the rectangle billiard in twelve intervals of k. In each histogram, there are about 100,000 objects, distributed in 100 bins. The distribution is surprisingly close to Gaussian; there is no significantly nonuniversal distribution for integrable systems. In Table 1, the values of the mean μ, standard deviation σ, skewness, and kurtosis are given. The fitting (red) curve is the Gaussian distribution with the same μ and σ as those obtained. However, the variance as a function of k is linear for all integrable systems; it is thus universal in the class of integrable systems, see Figure 7.

Table 1. Parameters of the distribution of Figure 6 for the wavenumber k-intervals starting at k_0: the mean μ, standard deviation σ, skewness, and kurtosis.

Parameters of the Distributions in Figure 6				
k_0	μ	σ	**Skewness**	**Kurtosis**
0	0.747	4.530	−0.309	0.345
1000	0.757	6.483	−0.234	−0.132
2000	0.764	8.587	−0.222	−0.121
3000	0.741	10.490	−0.306	−0.171
4000	0.795	11.593	−0.274	−0.190
5000	0.624	13.258	−0.285	−0.113
6000	0.727	14.795	−0.261	−0.088
7000	1.067	16.041	−0.204	−0.223
8000	0.590	16.532	−0.117	−0.054
9000	0.752	17.271	−0.274	−0.192
10,000	0.852	18.394	−0.433	0.318
11,000	1.051	19.772	−0.231	0.032

In Figure 7, the standard deviation of the distributions of w is plotted as a function of k, which clearly accurately follows the prediction by Steiner [11–14], namely the standard deviation σ rises as $\sqrt{k}$, shown also in the log-log plot, and the variance is a linear function of k. Therefore, the dependence of the variance on k is a good signature of chaos, unlike the distribution itself: in ergodic chaotic systems, the standard deviation behaves as $\sqrt{\log k}$.

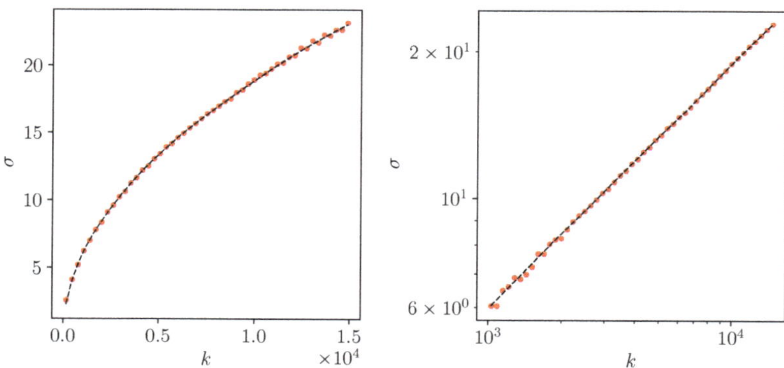

Figure 7. The standard deviation σ of the distribution of the fluctuation quantity $w(k) = R(k) = \mathcal{N}(k) - W(k)$ for a wide range of k, shown along with the fitting curve, $\sigma = a\,k^\gamma$, with $a = 0.187$ and $\gamma = 0.502$: in the linear scale (**left panel**) and in log-log scale **(right panel)**. The log-log plot clearly shows a power-law increase of k with the slope 0.5.

4. Level Spacing Distribution for the Entire Spectrum

One of the most important statistical measures of the (unfolded) energy spectra is the level spacing distribution, $P(S)$. For integrable systems, one gets Poissonian statistics and $P_P(S) = \exp(-S)$, while for classically ergodic (fully chaotic) systems the Gaussian orthogonal ensemble (GOE) of random matrix theory applies. The Wigner distribution (Wigner surmise) is 2-dimensional GOE distribution and is a very good approximation for the GOE level spacing distribution (infinite-dimensional),

$$P_W(S) = \frac{\pi S}{2} \exp\left(-\frac{\pi S^2}{4}\right). \tag{9}$$

There is a general useful relationship, namely, one using the gap probability, $\mathcal{E}(S)$, being the probability of having no level on an arbitrary interval of length S: the level spacing distribution $P(S)$ is, in general, equal to the second derivative of the gap probability, $P(S) = d^2\mathcal{E}(S)/dS^2$.

For Poisson statistics: $\mathcal{E}_\mathrm{P}(S) = \exp(-S)$, while for the Wigner distribution, one finds:

$$\mathcal{E}_\mathrm{W}(S) = 1 - \mathrm{erf}\left(\frac{\sqrt{\pi}S}{2}\right) = \mathrm{erfc}\left(\frac{\sqrt{\pi}S}{2}\right). \tag{10}$$

For mixed-type systems, there is typically one dominant chaotic component with the relative density of levels ρ_2 (equal to the relative fraction of the chaotic phase space volume), while the complement is typically a regular component of relative density, $\rho_1 = 1 - \rho_2$. If the regular and chaotic levels superimpose statistically independent of each other, then obviously, the gap probability factorizes:

$$\mathcal{E}(S) = \mathcal{E}_\mathrm{P}(\rho_1 S)\ \mathcal{E}_\mathrm{W}(\rho_2 S), \tag{11}$$

and therefore, in this case, the level spacing distribution is given by the Berry–Robnik (BR) formula [24]:

$$\begin{aligned} P_\mathrm{BR}(S) &= e^{-\rho_1 S}\exp\left(-\frac{\pi\rho_2^2 S^2}{4}\right)\left(2\rho_1\rho_2 + \frac{\pi\rho_2^3 S}{2}\right) \\ &+ e^{-\rho_1 S}\rho_1^2\mathrm{erfc}\left(\frac{\sqrt{\pi}\rho_2 S}{2}\right). \end{aligned} \tag{12}$$

The above statements are true provided the Heisenberg time is larger than any classical transport time in the system [8]. (The Heisenberg time is defined as $2\pi\hbar\ d(E)$, where $d(E)$ is the mean energy level density, also the reciprocal mean energy level spacing.) If this is not the case, the chaotic eigenstates can be quantum (or dynamically) localized, which implies localized chaotic Poincaré–Husimi functions in the phase space. The level spacing distribution for such localized chaotic eigenstates becomes (approximately) the known Brody distribution [26,27]:

$$P_\mathrm{B}(S) = cS^\beta\exp\left(-dS^{\beta+1}\right), \tag{13}$$

where by the normalization of the total probability and the first moment, one gets:

$$c = (\beta+1)d, \quad d = \left[\Gamma\left(\frac{\beta+2}{\beta+1}\right)\right]^{\beta+1}, \tag{14}$$

where $\Gamma(x)$ is the Gamma function. The Brody distribution interpolates the exponential and Wigner distribution as β goes from zero to one. The important feature of the Brody distribution is the fractional level repulsion effect, meaning the power law at small S, $P(S) \propto S^\beta$. The corresponding gap probability is:

$$\mathcal{E}_\mathrm{B}(S) = \frac{1}{\gamma(\beta+1)}Q\left(\frac{1}{\beta+1}, (\gamma S)^{\beta+1}\right), \tag{15}$$

where $\gamma = \Gamma\left(\frac{\beta+2}{\beta+1}\right)$, and $Q(a,x)$ is the incomplete Gamma function:

$$Q(a,x) = \int_x^\infty t^{a-1}e^{-t}dt. \tag{16}$$

Here, the only parameter is β, the level repulsion exponent in (13), which measures the degree of localization of the chaotic eigenstates: if the localization is maximally strong, the eigenstates practically do not overlap in the phase space (of the Wigner functions), and one finds $\beta = 0$ and a Poissonian distribution, while in the case of maximal extendedness (no localization), one finds $\beta = 1$ and the GOE statistics of levels applies. Thus, by replacing $\mathcal{E}_W(S)$ by $\mathcal{E}_B(S)$, the Berry–Robnik–Brody (BRB) distribution is obtained which generalizes the BR distribution (12) such that the localization effects in chaotic eigenstates are included [28]. Note that in the semiclassical limit, $\hbar \to 0$ or $k \to \infty$, the Heisenberg time becomes arbitrarily large (larger than any classical transport time), the localization effects of chaotic eigenstates disappear, and the BRB distribution becomes the BR distribution. However, the theoretical derivation of the Brody distribution for the localized chaotic states remains an important open problem. Furthermore, while the local behavior at small S, described by the power law $P(S) \propto S^\beta$, is certainly correct, the global feature of the Brody distribution is surely approximate, although empirically is well founded.

In Figure 8, the present study, the classical transport time of the billiards is very short; therefore, one expects $\beta \approx 1$, and the level spacing distribution is almost BR one (12). Thus, in the level statistics, one does not detect large localization effects. For the spectral unfolding procedure, the Weyl formula, Equation (5), is used, which at high energies is quite accurate.

Figure 8, shows the level spacing distributions $P(S)$ for the two billiards—one of $B = 0.1953$ (left column), and another one of $B = 0.083$ (right column)—for the spectral stretches each about 20,000 levels long, starting at $k_0 = 2880$, for each parity: even-even even-odd, odd-even and odd-odd and for all four parities together (about 80,000 levels), each of them along with the best fitting BRB distribution.

As one can see, in the case of $B = 0.1953$, Figure 8a,c,e,g,i, the β parameter is very close to one, while the parameter ρ_1 is close to the classical value, $\rho_1 = 1 - \rho_2 = 0.7196$, being the relative fraction of the volume of the regular part of the phase space (see Section 2). Thus, the system exhibits the Berry–Robnik picture with weak localization effects in the chaotic part of the energy spectrum. $P(S)$ is well fit by the BRB distribution. The picture is very similar for other lower values of k_0 (not shown).

For $B = 0.083$, one observes a substantial variation of both the β and ρ_1 parameters, the latter one being far from the classical value $\rho_1 = 0.8383$. This is certainly due to the complexity of the phase portrait shown in Figure 2, where many structural features are quantum mechanically not yet resolved, even at such high lying energies $E = k^2$, starting at $k_0 = 2880$. This will be studied in more detail in the forthcoming paper by us [9], where the Poincaré–Husimi functions in the phase space are analyzed in a similar way as in [4].

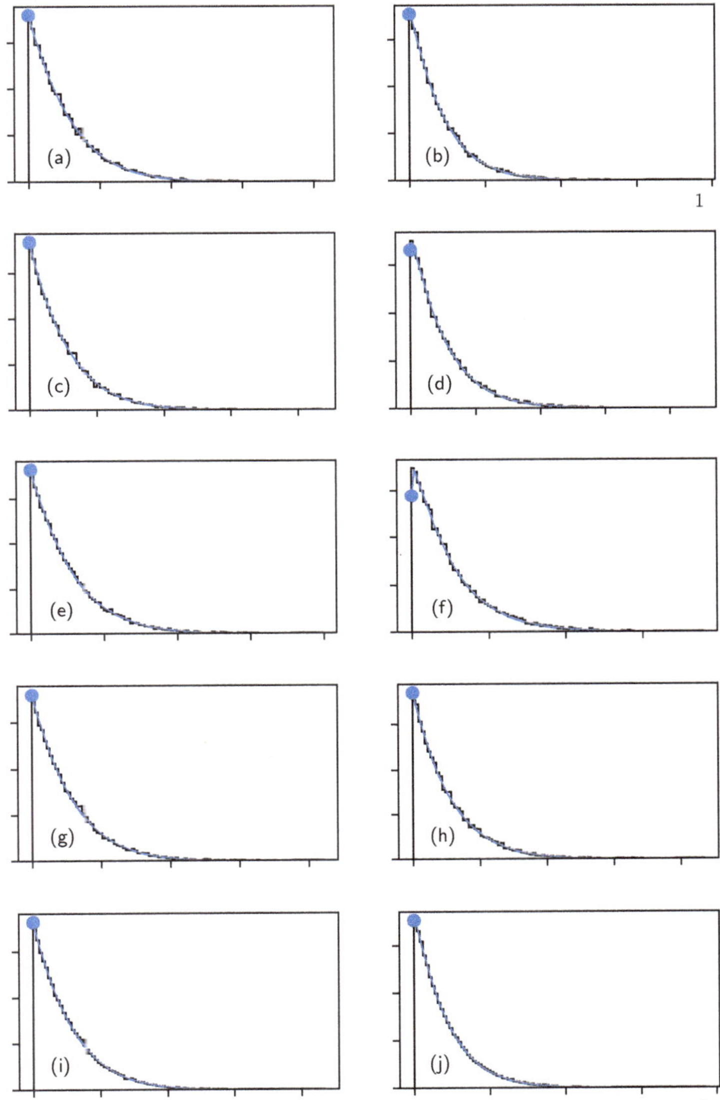

Figure 8. Level spacing distribution $P(S)$ for the two billiards: of $B = 0.1953$ (**left column**) and of $B = 0.083$ (**right column**), along with the best fitting Berry–Robnik–Brody curves. The abscissa represents S and the ordinate gives $P(S)$. The full circles represent the value of $P(S = 0)$. The parities are: even-even (**a**,**b**), even-odd (**c**,**d**), odd-even (**e**,**f**), odd-odd (**g**,**h**), and all parities together (**i**,**j**). For each particular parity, there are about 20,000 levels; for all parities together, there are about 80,000 levels. The values of (β, ρ_1) for the left column are (top to bottom): (0.798, 0.684), (1.000, 0.716), (1.000, 0.709), (1.000, 0.690), (1.000, 0.700). The classical value $\rho_1 = 0.7196$. The values of (β, ρ_1) for the right column are (top to bottom): (0.316, 0.666), (0.178, 0.588), (0.129, 0.474), (1.000, 0.738), (0.324, 0.660). The classical value $\rho_1 = 0.8383$.

5. Conclusions

In this paper, the statistical properties of the oscillations of the cumulative spectral staircase function (the integrated density of energy levels) around the corresponding mean value were studied, in order to compare the function obtained with the theoretical predictions of Steiner [11–14] for the fully chaotic and integrable (regular) systems. In billiards, the mean behavior is asymptotically exactly described by the Weyl formula (5). In the case of the integrable maximally irrational rectangle billiard, almost a Gaussian distribution of the fluctuations was surprisingly observed, in contrary to the expectations of Steiner's theory. Nevertheless, the standard deviation as a function of the wavenumber k was found to be proportional to $\sqrt{k}$, precisely in agreement with Steiner's prediction for all integrable systems.

In the two mixed-type lemon billiards, $B = 0.1953$ and 0.083, where regular and chaotic regions coexist in the classical phase space, the Gaussian distribution is found which is surprising, as the theory predicts a Gaussian distribution only for the fully chaotic (ergodic) systems. Thus, in these two cases, one observes that there is very little difference between the mixed-type systems and the fully chaotic systems. Moreover, this is even more surprising as in the two billiards studied here, the fraction of the chaotic part of the phase space was only 0.28 and 0.16, respectively. This implies that the distribution of the fluctuations of the spectral staircase functions around the mean behavior is not a very significant criterion for quantum chaos. This conclusion is corroborated by the result obtained for the integrable rectangle billiard.

The level spacing distribution of the energy spectra at high lying levels was also explored, starting at $k_0 = 2880$, and the Berry–Robnik–Brody distribution was found in both lemon billiard cases. In the case $B = 0.1953$, the results were in agreement with the Berry–Robnik picture [8,24], showing that the localization of the chaotic eigenstates is very weak, the level repulsion parameter β is close to one, and the quantum Berry–Robnik parameter ρ_2 is close to the classical one. On the other hand, in the lemon billiard $B = 0.083$ one observed relatively strong localization: β is substantially lower than one and varies widely over the four parities. Likewise, the quantum ρ_1 is substantially smaller than the classical value 0.8383 and varies widely over the four parities, which is certainly due to the complexity of the classical phase space, as the fine structure of the classical phase space is not yet resolved by the Poincaré–Husimi functions. This analysis will be the topics of forthcoming paper by us [9], using the approach of [4].

Author Contributions: Conceptualization and methodology, Č.L. and M.R.; investigation, Č.L., D.L. and M.R.; software package creation, Č.L.; numerical calculations and graphic design, Č.L. and D.L. writing and editing, M.R.; final editing in MDPI format, D.L. and M.R.; supervision M.R. All authors read and agreed to the published version of the manuscript.

Funding: This research was funded by Slovenian Research Agency (ARRS) Grant Number J1-9112 awarded to M.R.

Data Availability Statement: Not applicable.

Acknowledgments: This paper is dedicated to our friend Michael I. Tribelsky of M. V. Lomonosov Moscow State University, on the occasion of his 70th birthday, in thankfulness for his great scientific life opus over the past 50 years.

Conflicts of Interest: The authors declare no conflict of interest.The funders had no role in the design of the study; in the collection, analyses, or interpretation of data; in the writing of the manuscript, nor in the decision to publish the results.

References

1. Heller, E.J.; Tomsovic, S. Postmodern quantum mechanics. *Phys. Today* **1993**, *46*, 38–46. [CrossRef]
2. Lozej, Č. Stickiness in generic low-dimensional Hamiltonian systems: A recurrence-time statistics approach. *Phys. Rev. E* **2020**, *101*, 052204. [CrossRef] [PubMed]
3. Lozej, Č.; Lukman, D.; Robnik, M. Effects of stickiness in the classical and quantum ergodic lemon billiard. *Phys. Rev. E* **2021**, *103*, 012204. [CrossRef]

4. Lozej, Č.; Lukman, D.; Robnik, M. Classical and quantum mixed-type lemon billiards without stickiness. *Nonlinear Phenom. Complex Syst. (Minsk)* **2021**, *24*, 1–18. [CrossRef]
5. Stöckmann, H.-J. *Quantum Chaos—An Introduction*; Cambridge University Press: Cambridge, UK, 1999. doi: 10.1017/CBO9780511524622 [CrossRef]
6. Haake, F. *Quantum Signatures of Chaos*; Springer: Berlin, Germany, 2001. doi: 10.1007/978-3-642-05428-0 [CrossRef]
7. Robnik, M. Fundamental concepts of quantum chaos. *Eur. Phys. J. Spec. Top.* **2016**, *225*, 959–976. doi: 10.1140/epjst/e2016-02649-0 [CrossRef]
8. Robnik, M. A Brief introduction to stationary quantum chaos in generic systems *Nonlinear Phenom. Complex Syst. (Minsk)* **2020**, *23*, 172–191. doi: 10.33581/1561-4085-2020-23-2-172-191 [CrossRef]
9. Lozej, Č.; Lukman, D.; Robnik, M. Two mixed-type lemon billiards with complex phase space structure. **2021**, to be submitted.
10. Vergini, E.; Saraceno, M. Calculation by scaling of highly excited states of billiards. *Phys. Rev. E* **1995**, *52*, 2204–2207. [CrossRef]
11. Steiner, F. Quantum chaos; In *Universität Hamburg 1994: Schlaglichter der Forschung zum 75. Jahrestag*; Ansorge, R., Ed.; Reimer: Hamburg, Germany, 1994; p. 543.
12. Aurich, R.; Bolte, J.; Steiner, F. Universal signatures of quantum chaos. *Phys. Rev. Lett.* **1994**, *73*, 1356–1359. [CrossRef]
13. Backer, A.; Steiner, F.; Stifter, P. Spectral statistics in the quantized cardioid billiard. *Phys. Rev. E* **1995**, *52*, 2463–2472. [CrossRef]
14. Aurich, R.; Bäcker, A.; Steiner, F. Mode fluctuations as fingerprints of chaotic and non-chaotic systems. *Intl. J. Mod. Phys.* **1997**, *11*, 805–849. [CrossRef]
15. Alt, H.; Backer, A.; Dembowski, C.; Graf, H.D.; Hofferbert, R.; Rehfeld, H.; Richter, A. Mode fluctuation distribution for spectra of superconducting microwave billiards. *Phys. Rev. E* **1998**, *58*, 1737. [CrossRef]
16. Lopac, V.; Mrkonjić, I.; Radić, D. Classical and quantum chaos in the generalized parabolic lemon-shaped billiard. *Phys. Rev. E* **1999**, *59*, 303–311. [CrossRef]
17. Makino, H.; Harayama, T.; Aizawa, Y. Quantum-classical correspondences of the Berry-Robnik parameter through bifurcations in lemon billiard systems. *Phys. Rev. E* **2001**, *63*, 056203. [CrossRef] [PubMed]
18. Lopac, V.; Mrkonjić, I.; Radić, D. Chaotic behavior in lemon-shaped billiards with elliptical and hyperbolic boundary arcs. *Phys. Rev. E* **2001**, *64*, 016214. [CrossRef] [PubMed]
19. Chen, J.; Mohr, L.; Zhang, H.K.; Zhang, P. Ergodicity of the generalized lemon billiards. *Chaos* **2013**, *23*, 043137. [CrossRef] [PubMed]
20. Bunimovich, L.; Zhang, H.K.; Zhang, P. On another edge of defocusing: Hyperbolicity of asymmetric lemon billiards. *Commun. Math. Phys.* **2016**, *341*, 781–803. [CrossRef]
21. Bunimovich, L.A.; Casati, G.; Prosen, T.; Vidmar, G. Statistical properties of the localization measure of chaotic eigenstates and the spectral statistics in a mixed-type billiard *Exp. Math.* **2019**, *1*, 10.
22. Berry, M.V. Regularity and chaos in classical mechanics, illustrated by three deformations of a circular 'billiard'. *Eur. J. Phys.* **1981**, *2*, 91. [CrossRef]
23. Greene, J.M. A method for determining a stochastic transition. *J. Math. Phys.* **1979**, *20*, 1183–1201. [CrossRef]
24. Berry, M.V.; Robnik, M. Semiclassical level spacings when regular and chaotic orbits coexist. *J. Phys. A Math. Gen.* **1984**, *17*, 2413–2421. [CrossRef]
25. Lozej, Č. Tansport and Localization in Classical and Quantum Billiards Ph.D. Thesis, University of Maribor, Maribor, Slovenia, 2020. Available online: https://dk.um.si/Dokument.php?id=148000&lang=eng (accessed on 15 July 2021).
26. Brody, T.A. A statistical measure for the repulsion of energy levels. *Lett. Nuovo Cimento* **1973**, *7*, 482–484. doi: 10.1007/BF02727859 [CrossRef]
27. Brody, T.A.; Flores, J.; French, J.B.; Mello, P.A.; Pandey, A.; Wong, S.S.M. Random-matrix physics: Spectrum and strength fluctuations. *Rev. Mod. Phys.* **1981**, *53*, 385–479. [CrossRef]
28. Batistić, B.; Robnik, M. Semiempirical theory of level spacing distribution beyond the Berry-Robnik regime: modeling the localization and the tunneling effects. *J. Phys. A Math. Theor.* **2010**, *43*, 215101. [CrossRef]

Article

Gain-Assisted Optical Pulling Force on Plasmonic Graded Nano-Shell with Equivalent Medium Theory

Yamin Wu [1], Yang Huang [1,*], Pujuan Ma [2] and Lei Gao [3,*]

1 School of Science, Jiangnan University, Wuxi 214122, China; wxwym@jiangnan.edu.cn
2 Shandong Provincial Engineering and Technical Center of Light Manipulations & Shandong Provincial Key Laboratory of Optics and Photonic Device, School of Physics and Electronics, Shandong Normal University, Jinan 250014, China; pujuan-ma@sdnu.edu.cn
3 Department of Photoelectric Science and Energy Engineering, Suzhou City University, Suzhou 215104, China
* Correspondence: yanghuang@jiangnan.edu.cn (Y.H.); leigao@suda.edu.cn (L.G.)

Abstract: The tunable optical pulling force on a graded plasmonic core-shell nanoparticle consisting of a gain dielectric core and graded plasmonic shell is investigated in the illumination of a plane wave. In this paper, the electrostatic polarizability and the equivalent permittivity of the core-shell sphere are derived and the plasmonic enhanced optical pulling force in the antibonding and bonding dipole modes of the graded nanoparticle are demonstrated. Additionally, the resonant pulling force occurring on the dipole mode is shown to be dependent on the aspect ratio of the core-shell particle, which is illustrated by the obtained equivalent permittivity. This shows that the gradation of the graded shell will influence the plasmonic feature of the particle, thus further shifting the resonant optical force peaks and strengthening the pulling force. The obtained results provide an additional degree of freedom to manipulate nanoparticles and give a deep insight into light–matter interaction.

Keywords: optical force; graded plasmonic material; core-shell particle; optical gain

Citation: Wu, Y.; Huang, Y.; Ma, P.; Gao, L. Gain-Assisted Optical Pulling Force on Plasmonic Graded Nano-Shell with Equivalent Medium Theory. *Physics* **2021**, *3*, 955–967. https://doi.org/10.3390/physics3040060

Received: 2 September 2021
Accepted: 22 October 2021
Published: 3 November 2021

Publisher's Note: MDPI stays neutral with regard to jurisdictional claims in published maps and institutional affiliations.

1. Introduction

The change of the field gradients or linear momentum carried by photons will give rise to the optical force [1]. Radiation pressure induced by photon momentum exchange always pushes objects in the light flow direction, which is known as "optical pushing". In contrast, if the light–matter momentum transfer leads to the backward motion of the objects, this phenomenon is named "optical pulling". The optical pulling force is a more novel phenomenon than the pushing one because it requires many more critical conditions to realize it and has many more potential applications in nano-manipulation [2–8]. One possible way to obtain the optical pulling force to pull the object towards the light is by increasing the forward scattering via Gaussian beam [9,10], Bessel Beam [11], and other tractor beams. Recently, the optical pulling forces acting on a nano-object consisting of chiral [12–14], hyperbolic [15], and gain [16–18] materials have been widely investigated. The introduction of a gain material could provide additional forward scattering strength with the appropriate gain threshold to achieve the optical pulling force [17]; therefore, investigating the threshold gain for the pulling force in different gain-assist nanostructures constitutes another task for researchers. Actually, the threshold gain for the pulling force was analytically studied for nano-spheres, thin cylinders and thin slabs [18]. Moreover, the continuous modulation from the pushing to the pulling force was exhibited by controlling the incident angle of the interfering plane waves near the Fano resonance of the plasmonic nanoparticle [19]. The plasmonic enhanced optical force has been applied to quantum measurement, signal detection, and other fields [20–24]. Furthermore, tunable optical pulling forces originating from plasmon singularity and Fano resonance on plasmonic nanoparticles have been investigated in detail [25–27].

The approaches to realize and maximize the optical fulling force rely on designing a specific tractor beam [2,3,5,27–30], modifying the surroundings of the manipulated objects [31–33], or utilizing a gain-assist structure [15,18,25,34]. The resonant interplay between plasmonic structures and gain media and the coupling with a gain medium located in the core of a metallic nanoshell, when excited by means of an external pump, produces intense changes of the electromagnetic fields around the structure, thus producing novel features which can be useful for a variety of applications, such as photothermal therapy enhanced spectroscopy, and spaser [35–38]. Graded materials are the materials whose material properties can vary continuously in space with a gradient coefficient, and graded core-shell spheres show a widely tunable plasmonic response band [39–42]. The near field distribution [43], far field scattering as well as nonlinear response [44] enhanced by plasmon resonances could be adjustable by changing the gradient coefficient and aspect ratio of the inner and outer radii of the sphere. More recently, the nonlinear optical properties of graded magnetite nanoparticles in a colloid were investigated experimentally as a sample of the extension from the electrical field to the magnetic field [45]. The gradient coefficient provides us with a new freedom to control the plasmonic feature of the graded core-shell particle and could thus be further adopted to tailor the optical pulling force on the core-shell structure. Recently, the optical trapping force on a gain-enriched metallic nanoshell by a Gaussian beam was investigated [38], which opened perspectives for gain assisted optomechanics where nonlinear optical forces are finely tuned to efficiently trap manipulate, channel, and deliver an externally controlled nanophotonic system. The aim of this paper is to explore the optical pulling force on a gain-assist graded core-shell sphere at the nanoscale and to investigate the dependence of the pulling force on the gain threshold and the degree of gradation. Moreover, the equivalent permittivity of the graded core-shell is adopted to explain the pulling force that occurs in different plasmon resonant modes in the cases of low and high aspect ratios.

2. Models and Methods

We consider a core-shell sphere consisting of a dielectric gain core and graded plasmonic shell illuminated by a plane electric field in the host medium with a relative permittivity, $\varepsilon_{\rm h}$. Here, we make the assumption that the dielectric function of the graded materials varies along the radial direction, r, in the spherical coordinates (r, θ, φ) and can be written as $\varepsilon_{\rm s}(r)$. The inner and outer radii are a and b. For simplicity, the size of the particle is assumed to be much smaller than the incident wavelength, and therefore the retardation effect is neglected and a long-wavelength approximation could be adopted [46] The electric potentials in whole space satisfy the following equation within the quasi-static approximation [43]:

$$\nabla \cdot \left[\varepsilon_\beta(r) \nabla\, \phi_\beta\right] = 0, \tag{1}$$

where ϕ_β is the electric potential in each area ($\beta = {\rm c, s, h}$ indicates the core, the shell and the host media, respectively) and could derivate the local electric field by:

$$\mathbf{E}_\beta = -\nabla \phi_\beta. \tag{2}$$

In the illumination of an external uniform electric field along the z-direction, it can be written as:

$$\frac{1}{r^2}\frac{\partial}{\partial r}\left[r^2 \varepsilon_\beta(r)\frac{\partial \phi_\beta}{\partial r}\right] + \frac{1}{r\sin\theta}\frac{\partial}{\partial \theta}\left[\varepsilon_\beta(r)\frac{\sin\theta}{r}\frac{\partial \phi_\beta}{\partial \theta}\right] = 0, \tag{3}$$

The electric potentials in each region have the following general expressions:

$$\begin{aligned}
\phi_{\rm c}(r,\theta) &= -ArP_1(\cos\theta)E_0, && r \le a,\\
\phi_{\rm s}(r,\theta) &= [A_1 R_1^+(r) + B_1 R_1^-(r)]P_1(\cos\theta)E_0, && a < r \le b,\\
\phi_{\rm h}(r,\theta) &= (-r + b^3 B r^{-2})P_1(\cos\theta)E_0, && r > b
\end{aligned} \tag{4}$$

where A, B, A_1, B_1 are four unknown coefficients to be determined, $P_n(\cos\phi)$ is the n-th order Legendre polynomials, and the radial function, $R_n(r)$, in the shell region satisfies the following equation:

$$\frac{\partial}{\partial r}\left[r^2\varepsilon_s(r)\frac{\partial R_n(r)}{\partial r}\right] - n(n+1)[\varepsilon_s(r)R_n(r)] = 0, \tag{5}$$

where $R_n^+(r)$ and $R_n^-(r)$ are the two solutions that are regular at the origin and infinity, respectively, and are the key to achieving the polarizability of the graded core-shell particle.

Considering the boundary condition, i.e., the continuity of the potentials and normal electric displacements:

$$\begin{aligned}
\phi_c(r,\theta)|_{r=a} &= \phi_s(r,\theta)|_{r=a}, \\
\phi_s(r,\theta)|_{r=b} &= \phi_h(r,\theta)|_{r=b}, \\
-\varepsilon_c\frac{\partial\phi_c(r,\theta)}{\partial r}\Big|_{r=a} &= -\varepsilon_s(r)\frac{\partial\phi_s(r,\theta)}{\partial r}\Big|_{r=a}, \\
-\varepsilon_s(r)\frac{\partial\phi_s(r,\theta)}{\partial r}\Big|_{r=b} &= -\varepsilon_h\frac{\partial\phi_h(r,\theta)}{\partial r}\Big|_{r=b}.
\end{aligned} \tag{6}$$

The coefficients are obtained as:

$$\begin{aligned}
&A_1 = -3bT_1^-(a)/T(a,b), \\
&B_1 = 3bT_1^+(a)/T(a,b), \\
&A = -a^{-1}[A_1R_1^+(a) + B_1R_1^-(a)], \\
&B = [F(b) - \varepsilon_h]/[F(b) + 2\varepsilon_h], \\
&T(a,b) = T_2^+(b)T_1^-(a) - T_2^-(b)T_1^+(a), \\
&T_1^\pm(a) = R_1^\pm(a) - a\tfrac{\varepsilon_s(a)}{\varepsilon_c}\tfrac{\partial}{\partial r}R_1^\pm(a), \\
&T_2^\pm(b) = 2R_1^\pm(b) + b\tfrac{\varepsilon_s(b)}{\varepsilon_h}\tfrac{\partial}{\partial r}R_1^\pm(b), \\
&R_1(r) = A_1R_1^+(r) + B_1R_1^-(r), \\
&F(b) = \tfrac{b\varepsilon_s(b)}{R_1(b)}\tfrac{\partial R_1(b)}{\partial r}.
\end{aligned} \tag{7}$$

The spherical shell permittivity profile is given by the graded Drude model [43]:

$$\varepsilon_s(r,\omega) = \varepsilon_b - \frac{\omega_p^2(r)}{\omega(\omega + i\Gamma)}, \tag{8}$$

where $\omega_p(r)$ and Γ are the spatially varying plasmon frequency and the relaxation rate, respectively. r We introduce a graded plasmon frequency $\omega_p^2(r) = \omega_p^2(0)(1 - hr^k)$ in the graded Drude model [47], where h and k are two gradient coefficients denoting the gradation of the shell. Without loss of generality, we further normalize the external field frequencies ω and the relaxation rate Γ by $\omega_p(0)$. Consequently, Equation (8) reduces to $\varepsilon_s(r,\omega) = \varepsilon_b - \left(1 - hr^k\right)/[\omega(\omega + i\Gamma)]$. In the case of $1 - hr^k > 0$, the graded Drude model has a positive imaginary part, indicating that it is a lossy material.

The gain effect of the dielectric core can be realized by using a semiconductor or dye molecules with pumping [48]. In the theoretical part, we describe the gain core with a relative permittivity function whose imaginary part has a negative value, i.e., $\varepsilon_c = \varepsilon_{c0} + i\varepsilon_{cg}$ with $\varepsilon_{cg} < 0$ corresponding to the material gain. The electrostatic polarizability including the radiation reaction of the graded core-shell sphere can be written as:

$$\alpha = \frac{\alpha_0}{1 - i\frac{2}{3}\frac{k_h^3\alpha_0}{4\pi\varepsilon_0\varepsilon_h}} \tag{9}$$

with $\alpha_0 = 4\pi\varepsilon_0\varepsilon_h b^3 B$, $k_h = \frac{2\pi}{\lambda}\sqrt{\varepsilon_h}$ and ε_0 being the permittivity of vacuum. The time-averaged optical forces on the nonmagnetic Rayleigh nanoparticle for the incident wave are expressed as:

$$\langle F \rangle = \frac{1}{2} k_h E_0^2 \mathrm{Im}(\alpha). \tag{10}$$

We normalize the force with $F_0 = \pi b^2 S_{inc}/c$ (S_{inc}) is the power flow density of the incident wave and c is the speed of light). In the long-wavelength approximation, the dipole contribution dominates the electric response of the dielectric particle. If the dipolar factor B of the core-shell particle vanishes in Equation (4), this means that the core-shell particle and the host medium are the same in view of the dielectric response. Substituting ε_{eq} for ε_h in the dipolar factor B and implying $B = 0$ yield the self-consistency equation [49], which is solved to obtain the equivalent permittivity of the graded core-shell particle:

$$\varepsilon_{eq}(b) = \frac{b\varepsilon_s(b)}{R_1(b)} \frac{\partial R_1(b)}{\partial r}. \tag{11}$$

If one considers the nongraded case and the gradient coefficient is set to $h = 0$, then the equivalent permittivity of the graded core-shell particle is naturally reduced to the nongraded case:

$$\varepsilon_{eq}(b) = \varepsilon_s(0) \frac{2a^3[\varepsilon_c - \varepsilon_s(0)] + b^3[\varepsilon_c + 2\varepsilon_s(0)]}{a^3[\varepsilon_s(0) - \varepsilon_c] + b^3[\varepsilon_c + 2\varepsilon_s(0)]}. \tag{12}$$

3. Results and Discussion

The graded Drude model in Equation (8) shows that the gradient coefficient h plays a more important role than k within the present framework because the permittivity is independent of k when the radius is fixed, especially for the dipole moment, which we mainly focus on in the following. Figure 1 illustrates the permittivity of the graded shell as a function of the gradient coefficient h with the following parameter: $\omega_p(0) = 1.367 \times 10^{16}\ \mathrm{s}^{-1}$, $\Gamma = 2.733 \times 10^{13}\ \mathrm{s}^{-1}$, $\varepsilon_{c0} = 2.1$ and $\varepsilon_h = 1$. This shows that the graded shell can change from being dielectric-like to metallic-like with an increasing incident wavelength. On the other hand, the graded shell is more like a metal when h is low within the present spectrum. Actually, from Equation (8) one concludes that the dielectric-like or metallic-like profile of the graded Drude model is dependent on the choice of h and k, and for a plasmonic shell lower h and lower k values are needed. In what follows, the parameter range of h is used as being from 0 to 0.3 to achieve the plasmonic profile of the shell.

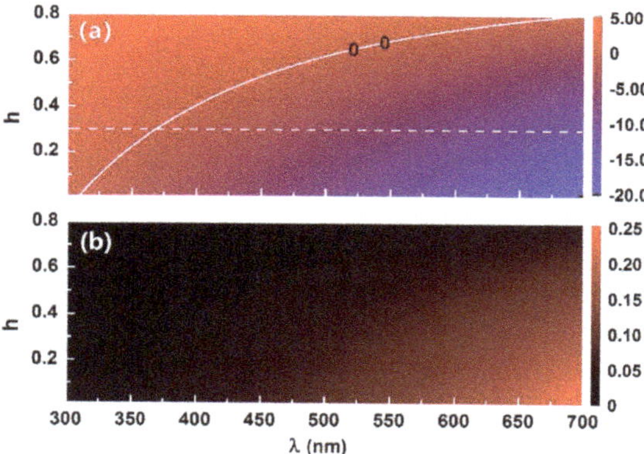

Figure 1. Real (**a**) and imaginary (**b**) parts of the graded shell permittivity $\varepsilon_s(b)$ as a function of the gradient coefficient, h, and the incident wavelength, λ. The white curve in (**a**) indicates a value of zero. Other parameters are the outer radius $b = 10$ nm and the gradient coefficient $k = 0.4$.

Figure 2 shows the normalized time-averaged optical force as a function of the incident wavelength with different gradient coefficients h and gain coefficients ε_{cg} for cases with a low aspect ratio ($a/b = 0.2$) and high aspect ratio ($a/b = 0.8$), respectively. The evolution of the optical force on the gradient coefficient h in the non-gain case ($\varepsilon_{cg} = 0$) is well seen in Figure 2a. As it is a graded plasmonic core-shell sphere, one expects two surface plasmon resonant modes, i.e., the bonding dipole mode at the long wavelength λ_+ and the antibonding dipole mode at the short wavelength λ_-. These two resonant modes lead to enhanced optical forces, and meanwhile the resonant wavelength is blue-shifted and the magnitudes of the peak value are decreased when the gradient coefficient h increases. It is easy to understand that both the real and imaginary parts of the permittivity of the graded shell are determined by h (see Figure 1), especially the real part which dramatically influences the resonant wavelength. When the gain level is increased, an enhanced negative optical force occurs in the antibonding dipole mode. There exists a critical level of gain in order to achieve the maximal optical pulling force, and a further increase of the gain level could not give rise to a stronger optical pulling force. On the other hand, the absolute value of the resonant peak is dramatically decreased with an increasing h for both the positive and negative forces. As for the optical force enhanced with the bonding dipole mode, it is not sensitive to the gain in the low aspect ratio case.

In contrast to the low aspect ratio case, the gain has a more dramatic influence on the bonding mode than on the antibonding mode in the high a/b case. An extremely strong optical pushing force (see Figure 2f) and pulling force (see Figure 2g) are achieved with the same level of gain as in the previous case. In addition, the gradient coefficient is found to dramatically enhance the optical pulling force in the bonding dipole mode, as shown in Figure 2g. Up to now, the gain-assisted optical pulling force has been found in both the antibonding mode and bonding mode in the graded core-shell particle. However, it should be mentioned that these two kinds of negative forces arise from different origins in terms of the equivalent medium, which is considered in detail below.

We now adopt the equivalent permittivity of the graded core-shell sphere, as shown in Equation (11), and plot the equivalent permittivity of the graded core-shell sphere for both low and high aspect ratio cases in Figures 3 and 4. It is found that $\text{Im}(\varepsilon_{eq})$ could exceed -7 in the equivalent permittivity spectrum of the low aspect ratio case due to the surface plasmon resonance. Note that ε_{cg} is merely -0.07 in the present model, which is not a high level of gain [50]; however, the equivalent sphere achieves an extremely high plasmonic enhanced effective gain level. As for the case of the high aspect ratio, the surface plasmon resonance leads to an extremely high loss peak in the equivalent permittivity spectrum. It should be mentioned that there exists only one resonant mode occurring in the inner surface of the core-shell sphere in the equivalent permittivity spectrum. This is because the equivalent permittivity obtained by the self-consistency method is independent of the host medium and this resonance is the intrinsic property of the core-shell sphere itself. Actually, the intrinsic plasmon resonance arises from the plasmonic singularity [25], and the resonant curve shapes vary with different ε_{cg}. If increases further, $\text{Im}(\varepsilon_{eq})$ in the high aspect ratio case will show a negative value peak.

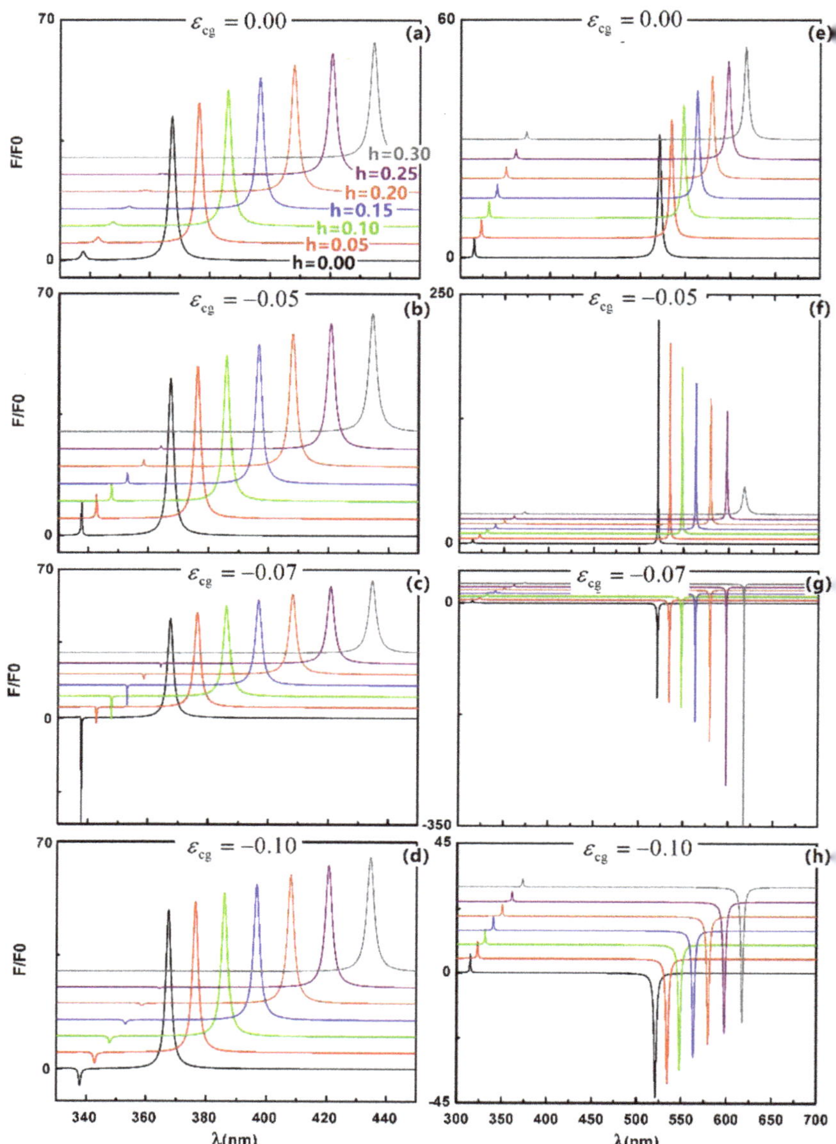

Figure 2. Normalized time-averaged optical force as a function of the incident wavelength λ for low aspect ratio, $a/b = 0.2$ (**a–d**), and high aspect ratio, $a/b = 0.8$ (**e–h**) with different gain coefficients ε_{cg} and gradient coefficients h. Other parameters are $b = 10$ nm and $k = 0.4$. For the normalization $F_0 = \pi b^2 S_{inc}/c$ (S_{inc} is the power flow density of the incident wave) is used.

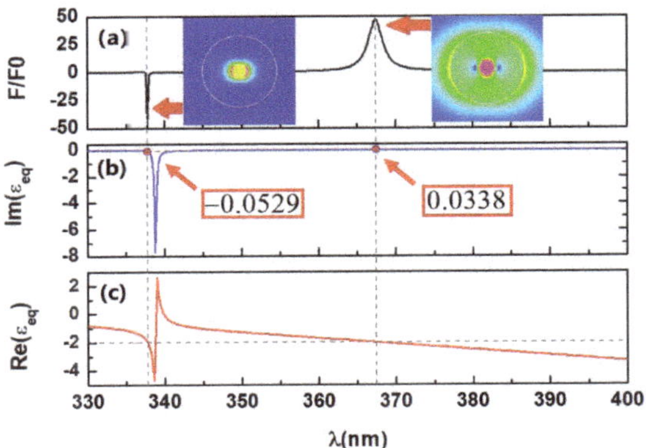

Figure 3. (**a**) Normalized optical force for the case of the low aspect ratio, as shown in Figure 2, with $\varepsilon_{cg} = -0.07$ and $h = 0$. (**b**) The imaginary and (**c**) real parts of the equivalent permittivity. The insets in (**a**) indicate the near field intensity distributions at the resonant wavelengths.

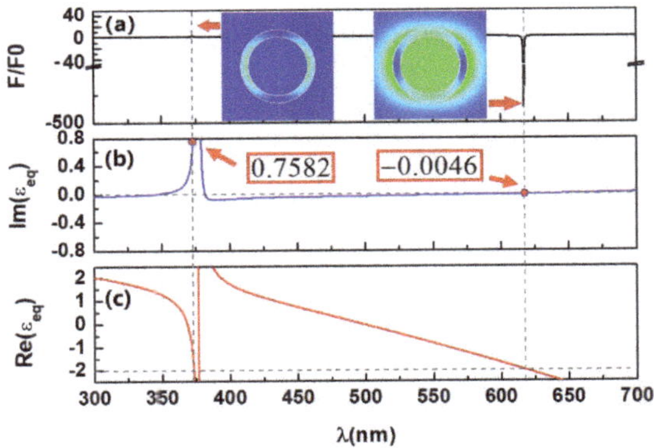

Figure 4. (**a**) Normalized optical force for the case of the high aspect ratio, as shown in Figure 2, with $\varepsilon_{cg} = -0.07$ and $h = 0.3$. (**b**) The imaginary and (**c**) real parts of the equivalent permittivity. The insets in (**a**) indicate the near field intensity distributions at the resonant wavelengths.

To give a simple approach for finding the resonant optical force and, on other hand, to mathematically explain why the resonant optical pulling force exists in the bonding (or anti-bonding) resonant modes corresponding to a high (or low) aspect ratio, we now introduce the equivalent permittivity of the core-shell sphere as obtained in Equation (12). When the core-shell particle is modeled as a sphere with an equivalent permittivity, the optical force on the corresponding equivalent sphere in the long-wavelength approximation is:

$$\langle F \rangle = 2\pi\varepsilon_0\varepsilon_h k_h b^3 E_0^2 \frac{3\mathrm{Im}(\varepsilon_{eq}) + 2(k_h b)^3 \left[\mathrm{Re}(\varepsilon_{eq}) - \varepsilon_h\right]^2/3}{\left[\mathrm{Re}(\varepsilon_{eq}) + 2\varepsilon_h\right]^2 + \left[\mathrm{Im}(\varepsilon_{eq})\right]^2 + 4(k_h b)^3 \mathrm{Im}(\varepsilon_{eq})\varepsilon_h}. \quad (13)$$

Equation (13) illustrates that the optical force can be enhanced by the surface plasmon resonances occurring when $\mathrm{Re}(\varepsilon_{eq}) = -2\varepsilon_h$, and the optical pulling force can be achieved when the numerator of Equation (13) has a negative value. Note that the second term of the numerator is always positive and that the potential negative optical force requires $\mathrm{Im}(\varepsilon_{eq})$ to be negative and its absolute value to be larger than the second term. On the other hand, by substituting $\mathrm{Re}(\varepsilon_{eq}) = -2\varepsilon_h$ into Equation (13) and ignoring the second term for a sufficiently small $k_h b$, Equation (13) reduces to a simple version as

$\langle F \rangle = 6\pi\varepsilon_0\varepsilon_h k_h b^3 E_0^2/\mathrm{Im}(\varepsilon_{eq})$, which indicates that the absolute value of the negative optical force is inversely proportional to $\mathrm{Im}(\varepsilon_{eq})$.

We highlight the values of $\mathrm{Im}(\varepsilon_{eq})$ at resonant wavelengths in Figures 3 and 4, and this explains why, mathematically, the optical pulling force occurs in different modes with the low and high aspect ratios in vacuum ($\varepsilon_h = 1$). In general, F/F_0 is negative when the equivalent sphere plays a role as the active gain medium with $\mathrm{Im}(\varepsilon_{eq}) < 0$, and F/F_0 is positive when the equivalent sphere is lossy with $\mathrm{Im}(\varepsilon_{eq}) > 0$. Moreover, it is concluded that the stronger optical pushing force can be achieved when the equivalent loss is low ($\mathrm{Im}(\varepsilon_{eq}) = 0.0338$) (see Figure 3b), and one can realize a stronger optical pulling force when the equivalent gain is low ($\mathrm{Im}(\varepsilon_{eq}) = -0.0046$) (see Figure 4b). With the equivalent permittivity, one can predict where the pulling force occurs in a simpler way. The inserts indicate the near field intensity distributions of the graded core-shell sphere in the different resonant modes for both low and high aspect ratio cases. It is well seen that the antibonding mode is dominated by the surface plasmon resonant on the inner surface. In contrast, the bonding mode is more influenced by the plasma on the outer surface. A dramatically high concentrated local field intensity in the dielectric core is found in the low aspect ratio case, especially in the bonding mode.

To further demonstrate this, the phase diagrams of the equivalent permittivity as a function of λ and h for both cases are plotted in Figure 5. The parameter space for $\mathrm{Im}(\varepsilon_{eq}) < 0$ is clearly plotted with a red–black color, and the pulling force peaks are located on the white lines that indicate the resonant conditions, $\mathrm{Re}(\varepsilon_{eq}) = -2\varepsilon_h$. Moreover, the pushing force peaks on the white lines lying in the gray region ($\mathrm{Im}(\varepsilon_{eq}) > 0$) are indicated in Figure 5 as well. The resonant pulling forces in the high aspect ratio case are generally stronger than those in the low aspect ratio (see Figure 2c,g), as soon as $|\mathrm{Im}(\varepsilon_{eq})|$ are much smaller on the pulling force line in Figure 5b, according to the previous analysis of Equation (13). Actually, in spite of the giant resonant pushing/pulling force on the white lines, in Figure 5b there exists a broader parameter space for a slight pulling force where $\mathrm{Im}(\varepsilon_{eq}) < 0$. Let us remark here that $\mathrm{Re}(\varepsilon_{eq})$ might not rigorously satisfy the resonant condition, i.e., $\mathrm{Re}(\varepsilon_{eq}) > -2\varepsilon_h$ with h increasing since the loss of the graded shell is inversely proportional to h, as illustrated in Figure 1b; consequently, there exists a cut-off for the resonant white line in the low aspect ratio case when h reaches ~ 0.24. Still, a tiny pulling force peak occurs in the bonding mode in Figure 2c with a large h.

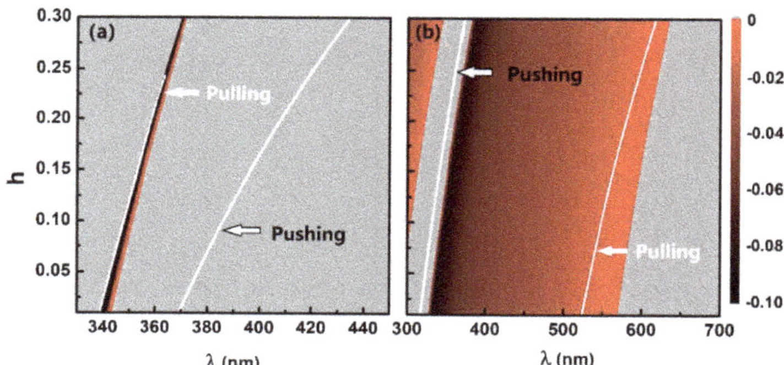

Figure 5. $\mathrm{Im}(\varepsilon_{eq})$ versus incident wavelength, λ, and gradient coefficient, h, for the cases of (**a**) low aspect ratio ($a/b = 0.2$) and (**b**) high aspect ratio ($a/b = 0.8$). Gray regions indicate the parameter space for a positive value. The white lines show the positions of $\mathrm{Re}(\varepsilon_{eq}) = -2\varepsilon_h$. The gain coefficient is set as $\varepsilon_{cg} = -0.07$ for both cases.

Finally, we investigate the dependence of the pulling force on the aspect ratio of the graded core-shell particle in Figure 6. The resonant pulling force tends to occur in the gain-assisted core-shell sphere with either a low aspect ratio or high aspect ratio (see the rectangle), and in contrast the resonant pushing force exists in the case of a wider moderate aspect ratio. On the other hand, the resonant wavelength together with the resonant pulling force peak varies with the graded coefficient h. This gives rise to a broader parameter space in order to realize the resonant pulling force in the antibonding mode and, on the contrary, a smaller space in the bonding mode with a high h.

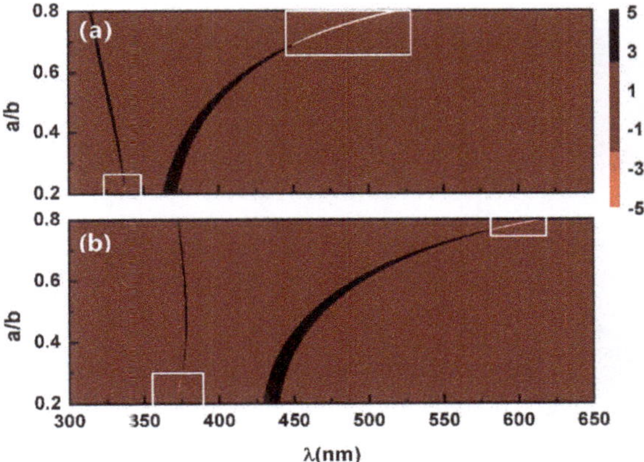

Figure 6. Normalized optical force, F/F_0, versus incident wavelength, λ, and aspect ratio, a/b, in the cases of (**a**) $h = 0$ and (**b**) $h = 0.3$ with $\varepsilon_{cg} = -0.07$. The white curves indicate a pulling force much stronger than -5, which is marked by a white rectangle.

It should be noted that, in this paper, we use a flat imaginary part of ε_c instead of a more realistic frequency-dependent one [25,35,38]. In what follows, a brief comparison between the results with the Lorentzian dependent permittivity and those with a flat constant permittivity for the gain media are given in order to certify the validity of this study. Thus, the gain media is described by the permittivity with a Lorentzian shape [38] as:

$$\varepsilon_c(\omega) = \varepsilon_{c0} - \frac{\varepsilon_{cg}\Delta}{2(\omega - \omega_g) + i\Delta}, \tag{14}$$

where ε_{c0} is the background permittivity of the gain core, ω_g is the emission centerline of the gain elements, and $\Delta = 2/\tau$ is the width of the Lorentzian shape where τ is the energy relaxation time of the gain. ε_{cg} is a dimensionless parameter measuring the amount of gain present in the system.

Figures 7 and 8 illustrate that when the gain lines shape is centered exactly on the plasmon resonant frequency, the optical pulling force obtained by $\varepsilon_c(\omega)$ in Equation (14) is the same as that with a flat constant ε_c. It is worth noting that $\varepsilon_c(\omega)$ at the plasmon resonant position has the same real and imaginary values as the flat one. One of the goals in this study is to demonstrate the different behaviors of the optical force in plasmonic resonant modes of a graded nanoshell with different aspect ratios (i.e., a/b) on the same level of gain. Thus, what makes sense is the central gain amount at the resonant frequency. From this point of view, the conclusions made with the flat constant model are the same as those made with the frequency-dependent one. However, it is found that the overall gain amount ($\varepsilon_{cg} = -0.07$) slightly exceeds the spaser threshold (see Figure 9) for both cases in Figures 7 and 8, which might lead to a spaser instability [35].

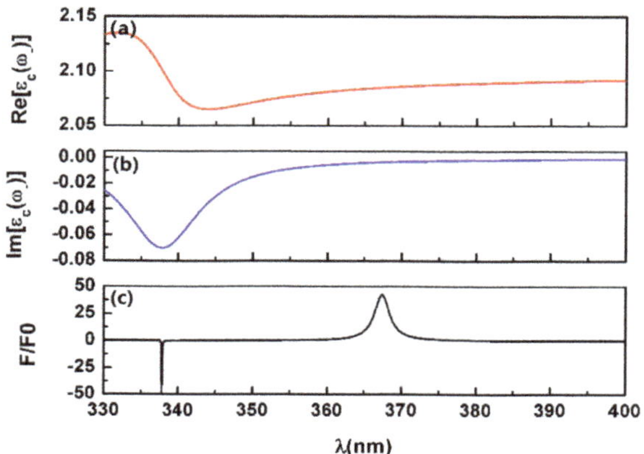

Figure 7. Real (**a**) and imaginary (**b**) parts of $\varepsilon_c(\omega)$ in the case of a low aspect ratio (a/b = 0.2) of Figure 2c with $\varepsilon_{cg} = -0.07$ and $h = 0$. (**c**) The same as Figure 3a but using the present $\varepsilon_c(\omega)$. Parameters: $\tau = 10^{-14}$ s, $\omega_g = \omega_- = 2\pi c/\lambda_-$, $\lambda_- = 337.8$ nm.

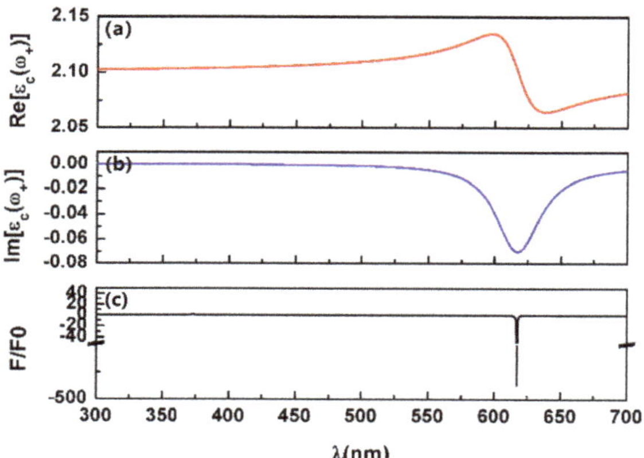

Figure 8. Real (**a**) and imaginary (**b**) parts of $\varepsilon_c(\omega)$ in the case of a high aspect ratio (a/b = 0.8) of Figure 2g with $\varepsilon_{cg} = -0.07$ and $h = 0.3$. (**c**) The same as Figure 4a but using the present $\varepsilon_c(\omega)$. Parameters: $\tau = 10^{-14}$ s, $\omega_g = \omega_+ = 2\pi c/\lambda_+$, $\lambda_+ = 617.4$ nm.

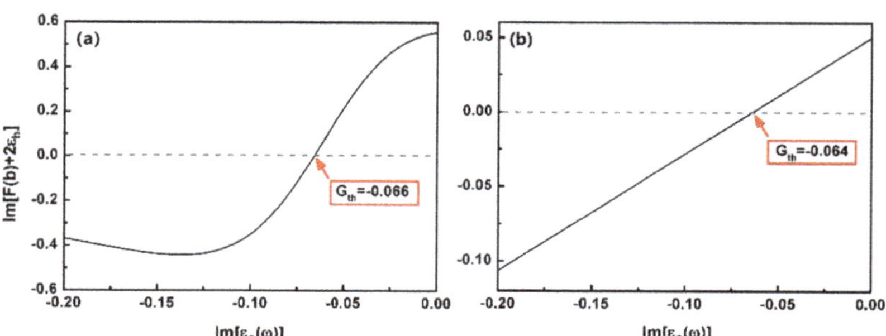

Figure 9. Finding the spaser threshold gain G_{th} for the cases of (**a**) Figure 7 and (**b**) Figure 8.

According to Ref. [35], a more complex model should be introduced. However, the analytical results for the coefficient B in Equation (4) contain the hyper-geometric function $F(\alpha, \beta, \gamma, z)$

in $R_1^+(r)$ and $R_1^-(r)$, i.e., $R_1^+(r) = rF(\alpha_1, \beta_1, \gamma_1, z)$ and $R_1^-(r) = r^{-2}F(\alpha_{-1}, \beta_{-1}, \gamma_{-1}, z)$, where $\alpha_{\pm 1} = \left[k \pm 3 - \sqrt{(k+1)^2 + 8}\right]/2k, \beta_{\pm 1} = \left[k \pm 3 + \sqrt{(k+1)^2 + 8}\right]/2k, \gamma_{\pm 1} = (\pm 3 + k)/k$ and $z = -hr^k/(\omega^2 + i\omega\Gamma - 1)$. Thus, it is not straight to analyze the temporal dynamic evolution of the dipole moments and study the condition of instability, especially for the case of the metallic nanoshell where more than one plasmonic resonant mode exists. The exact nature of this final state requires a thorough study dependent on the various model parameters, which is not within the scope of this paper.

4. Conclusions

In summary, in this paper, giant gain-assisted resonant pulling forces are demonstrated on a graded core-shell nanoparticle in a long-wavelength approximation. It is found that these plasmonic enhanced pulling forces can exist in either the antibonding or bonding modes based on the choice of the aspect ratio of the core-shell sphere. Generally, the antibonding mode in the low aspect ratio case and the bonding mode in the high aspect ratio would lead to the resonant pulling force, and this could be demonstrated by the obtained equivalent permittivity of the graded core-shell sphere. The gradation of the shell has a dramatic influence on the resonant wavelength of the pulling force and could strengthen the resonant pulling force with the same level of gain as in the nongraded case. Moreover, the parameter space for realizing the pulling force is broadened with a higher gradient coefficient. The present study may give a deep insight into the mechanism of the pulling force in gain systems and offer an effective way to obtain large negative forces for nano-manipulation.

Author Contributions: Conceptualization, Y.H.; methodology, Y.H. and L.G.; formal analysis, Y.W. and Y.H.; investigation, Y.W. and Y.H.; writing—original draft preparation, Y.W. and Y.H.; writing—review and editing, Y.H., P.M. and L.G. All authors have read and agreed to the published version of the manuscript.

Funding: This research was funded by the National Natural Science Foundation of China, grant number 11704158, 11774252, 92050104 and 12004225, the Natural Science Foundation of Jiangsu Province, grant number BK20170170, and the Suzhou Prospective Application Research Project, Grant number SYG202039.

Data Availability Statement: This study does not report any data.

Conflicts of Interest: The authors declare no conflict of interest.

References

1. Xin, H.; Li, Y.; Liu, Y.; Zhang, Y.; Xiao, Y.; Li, B. Optical Forces: From Fundamental to Biological Applications. *Adv. Mater.* **2020**, *32*, 2001994. [CrossRef] [PubMed]
2. Chen, J.; Ng, J.; Lin, Z.; Chan, C.T. Optical Pulling Force. *Nat. Photonics* **2011**, *5*, 531–534. [CrossRef]
3. Novitsky, A.; Qiu, C.-W.; Wang, H. Single Gradientless Light Beam Drags Particles as Tractor Beams. *Phys. Rev. Lett.* **2011**, *107*, 203601. [CrossRef] [PubMed]
4. Dogariu, A.; Sukhov, S.; Sáenz, J. Optically Induced "Negative Forces". *Nat. Photonics* **2013**, *7*, 24–27. [CrossRef]
5. Gao, D.; Novitsky, A.; Zhang, T.; Cheong, F.C.; Gao, L.; Lim, C.T.; Luk'yanchuk, B.; Qiu, C.-W. Unveiling the Correlation between Non-Diffracting Tractor Beam and Its Singularity in Poynting Vector. *Laser Photonics Rev.* **2015**, *9*, 75–82. [CrossRef]
6. Zhang, L.; Qiu, X.; Zeng, L.; Chen, L. Multiple Trapping Using a Focused Hybrid Vector Beam. *Chin. Phys. B* **2019**, *28*, 094202. [CrossRef]
7. Ling, L.; Guo, H.-L.; Huang, L.; Qu, E.; Li, Z.-L.; Li, Z.-Y. The Measurement of Displacement and Optical Force in Multi-Optical Tweezers. *Chin. Phys. Lett.* **2012**, *29*, 014214. [CrossRef]
8. Lepeshov, S.; Krasnok, A. Virtual Optical Pulling Force. *Optica* **2020**, *7*, 1024. [CrossRef]
9. Guo, G.; Feng, T.; Xu, Y. Tunable Optical Pulling Force Mediated by Resonant Electromagnetic Coupling. *Opt. Lett.* **2018**, *43*, 4961. [CrossRef]
10. Lee, E.; Huang, D.; Luo, T. Ballistic Supercavitating Nanoparticles Driven by Single Gaussian Beam Optical Pushing and Pulling Forces. *Nat. Commun.* **2020**, *11*, 2404. [CrossRef] [PubMed]
11. Novitsky, A.; Qiu, C.-W. Pulling Extremely Anisotropic Lossy Particles Using Light without Intensity Gradient. *Phys. Rev. A* **2014**, *90*, 053815. [CrossRef]

12. Ding, K.; Ng, J.; Zhou, L.; Chan, C.T. Realization of Optical Pulling Forces Using Chirality. *Phys. Rev. A* **2014**, *89*, 063825 [CrossRef]
13. Li, G.; Wang, M.; Li, H.; Yu, M.; Dong, Y.; Xu, J. Wave Propagation and Lorentz Force Density in Gain Chiral Structures. *Opt Mater. Express* **2016**, *6*, 388. [CrossRef]
14. Wang, M.; Li, H.; Gao, D.; Gao, L.; Xu, J.; Qiu, C.-W. Radiation Pressure of Active Dispersive Chiral Slabs. *Opt. Express* **2015**, *23* 16546. [CrossRef] [PubMed]
15. Shalin, A.S.; Sukhov, S.V.; Bogdanov, A.A.; Belov, P.A.; Ginzburg, P. Optical Pulling Forces in Hyperbolic Metamaterials. *Phys Rev. A* **2015**, *91*, 063830. [CrossRef]
16. Chen, H.; Gao, L.; Zhong, C.; Yuan, G.; Huang, Y.; Yu, Z.; Cao, M.; Wang, M. Optical Pulling Force on Nonlinear Nanoparticles with Gain. *AIP Adv.* **2020**, *10*, 015131. [CrossRef]
17. Duan, X.-Y.; Wang, Z.-G. Fano Resonances in the Optical Scattering Force upon a High-Index Dielectric Nanoparticle. *Phys. Rev. A* **2017**, *96*, 053811. [CrossRef]
18. Mizrahi, A.; Fainman, Y. Negative Radiation Pressure on Gain Medium Structures. *Opt. Lett.* **2010**, *35*, 3405. [CrossRef]
19. Chen, H.; Ye, Q.; Zhang, Y.; Shi, L.; Liu, S.; Jian, Z.; Lin, Z. Reconfigurable Lateral Optical Force Achieved by Selectively Exciting Plasmonic Dark Modes near Fano Resonance. *Phys. Rev. A* **2017**, *96*, 023809. [CrossRef]
20. Song, C.; Yang, S.; Li, X.M.; Li, X.; Feng, J.; Pan, A.; Wang, W.; Xu, Z.; Bai, X. Optically Manipulated Nanomechanics of Semiconductor Nanowires. *Chin. Phys. B* **2019**, *28*, 054204. [CrossRef]
21. Wang, H.-C.; Li, Z.-P. Advances in Surface-Enhanced Optical Forces and Optical Manipulations. *Acta Phys. Sin.* **2019**, *68*, 144101 [CrossRef]
22. Li, S.; Li, H.-Z.; Xu, J.-P.; Zhu, C.-J.; Yang, Y.-P. Squeezed Property of Optical Transistor Based on Cavity Optomechanical System *Acta Phys. Sin.* **2019**, *68*, 174202. [CrossRef]
23. Gu, K.-H.; Yan, D.; Zhang, M.-L.; Yin, J.-Z.; Fu, C.-B. Quantum Control of Fast/Slow Light in Atom-Assisted Optomechanical Cavity. *Acta Phys. Sin.* **2019**, *68*, 054201. [CrossRef]
24. Zhang, X.-L.; Bao, Q.-Q.; Yang, M.-Z.; Tian, X.-S. Entanglement Characteristics of Output Optical Fields in Double-Cavity Optomechanics. *Acta Phys. Sin.* **2018**, *67*, 104203. [CrossRef]
25. Gao, D.; Shi, R.; Huang, Y.; Gao, L. Fano-Enhanced Pulling and Pushing Optical Force on Active Plasmonic Nanoparticles. *Phys. Rev. A* **2017**, *96*, 043826. [CrossRef]
26. Bian, X.; Gao, D.L.; Gao, L. Tailoring Optical Pulling Force on Gain Coated Nanoparticles with Nonlocal Effective Medium Theory. *Opt. Express* **2017**, *25*, 24566. [CrossRef]
27. Chen, H.; Liu, S.; Zi, J.; Lin, Z. Fano Resonance-Induced Negative Optical Scattering Force on Plasmonic Nanoparticles. *ACS Nano* **2015**, *9*, 1926–1935. [CrossRef]
28. Sukhov, S.; Dogariu, A. Negative Nonconservative Forces: Optical "Tractor Beams" for Arbitrary Objects. *Phys. Rev. Lett.* **2011**, *107*, 203602. [CrossRef]
29. Novitsky, A.; Qiu, C.-W.; Lavrinenko, A. Material-Independent and Size-Independent Tractor Beams for Dipole Objects. *Phys. Rev. Lett.* **2012**, *109*, 023902. [CrossRef]
30. Brzobohatý, O.; Karásek, V.; Šiler, M.; Chvátal, L.; Čižmár, T.; Zemánek, P. Experimental Demonstration of Optical Transport, Sorting and Self-Arrangement Using a 'Tractor Beam'. *Nat. Photonics* **2013**, *7*, 123–127. [CrossRef]
31. Salandrino, A.; Christodoulides, D.N. Reverse Optical Forces in Negative Index Dielectric Waveguide Arrays. *Opt. Lett.* **2011**, *36*, 3103. [CrossRef] [PubMed]
32. Kajorndejnukul, V.; Ding, W.; Sukhov, S.; Qiu, C.-W.; Dogariu, A. Linear Momentum Increase and Negative Optical Forces at Dielectric Interface. *Nat. Photonics* **2013**, *7*, 787–790. [CrossRef]
33. Petrov, M.I.; Sukhov, S.V.; Bogdanov, A.A.; Shalin, A.S.; Dogariu, A. Surface Plasmon Polariton Assisted Optical Pulling Force: Surface Plasmon Polariton Assisted Optical Pulling Force. *Laser Photonics Rev.* **2016**, *10*, 116–122. [CrossRef]
34. Alaee, R.; Christensen, J.; Kadic, M. Optical Pulling and Pushing Forces in Bilayer *PT*-Symmetric Structures. *Phys. Rev. Appl.* **2018**, *9*, 014007. [CrossRef]
35. Veltri, A.; Chipouline, A.; Aradian, A. Multipolar, Time-Dynamical Model for the Loss Compensation and Lasing of a Spherical Plasmonic Nanoparticle Spaser Immersed in an Active Gain Medium. *Sci. Rep.* **2016**, *6*, 33018. [CrossRef] [PubMed]
36. Caligiuri, V.; Pezzi, L.; Veltri, A.; De Luca, A. Resonant Gain Singularities in 1D and 3D Metal/Dielectric Multilayered Nanostructures. *ACS Nano* **2017**, *11*, 1012–1025. [CrossRef]
37. Pezzi, L.; Iatì, M.A.; Saija, R.; De Luca, A.; Maragò, O.M. Resonant Coupling and Gain Singularities in Metal/Dielectric Multishells: Quasi-Static Versus T-Matrix Calculations. *J. Phys. Chem. C* **2019**, *123*, 29291–29297. [CrossRef]
38. Polimeno, P.; Patti, F.; Infusino, M.; Sánchez, J.; Iatì, M.A.; Saija, R.; Volpe, G.; Maragò, O.M.; Veltri, A. Gain-Assisted Optomechanical Position Locking of Metal/Dielectric Nanoshells in Optical Potentials. *ACS Photonics* **2020**, *7*, 1262–1270. [CrossRef]
39. Huang, J.P.; Yu, K.W.; Gu, G.Q.; Karttunen, M. Electrorotation in Graded Colloidal Suspensions. *Phys. Rev. E* **2003**, *67*, 051405. [CrossRef] [PubMed]
40. Gao, L.; Huang, J.P.; Yu, K.W. Effective Nonlinear Optical Properties of Composite Media of Graded Spherical Particles. *Phys. Rev. B* **2004**, *69*, 075105. [CrossRef]
41. Gao, L.; Yu, K.W. Second- and Third-Harmonic Generation in Random Composites of Graded Spherical Particles. *Phys. Rev. B* **2005**, *72*, 075111. [CrossRef]

42. Huang, J.P.; Yu, K.W. Effective Nonlinear Optical Properties of Graded Metal-Dielectric Composite Films of Anisotropic Particles. *J. Opt. Soc. Am. B* **2005**, *22*, 1640. [CrossRef]
43. Wei, E.-B.; Sun, L.; Yu, K.-W. Controlling Electric Field Distribution by Graded Spherical Core-Shell Metamaterials. *Chin. Phys. B* **2010**, *19*, 107802. [CrossRef]
44. Huang, J.P.; Hui, P.M.; Yu, K.W. Second-Harmonic Generation in Graded Metal-Dielectric Films of Anisotropic Particles. *Phys. Lett. A* **2005**, *342*, 484–490. [CrossRef]
45. Espinosa, D.H.G.; Oliveira, C.L.P.; Figueiredo Neto, A.M. Influence of an External Magnetic Field in the Two-Photon Absorption Coefficient of Magnetite Nanoparticles in Colloids and Thin Films. *J. Opt. Soc. Am. B* **2018**, *35*, 346. [CrossRef]
46. Veltri, A.; Aradian, A. Optical Response of a Metallic Nanoparticle Immersed in a Medium with Optical Gain. *Phys. Rev. B* **2012**, *85*, 115429. [CrossRef]
47. Huang, J.; Yu, K. Enhanced Nonlinear Optical Responses of Materials: Composite Effects. *Phys. Rep.* **2006**, *431*, 87–172. [CrossRef]
48. Strangi, G.; De Luca, A.; Ravaine, S.; Ferrie, M.; Bartolino, R. Gain Induced Optical Transparency in Metamaterials. *Appl. Phys. Lett.* **2011**, *98*, 251912. [CrossRef]
49. Gao, L.; Wan, J.T.K.; Yu, K.W.; Li, Z.Y. Effects of Highly Conducting Interface and Particle Size Distribution on Optical Nonlinearity in Granular Composites. *J. Appl. Phys.* **2000**, *88*, 1893–1899. [CrossRef]
50. Huang, Y.; Xiao, J.J.; Gao, L. Antiboding and Bonding Lasing Modes with Low Gain Threshold in Nonlocal Metallic Nanoshell. *Opt. Express* **2015**, *23*, 8818–8828. [CrossRef]

 physics

MDPI

Article

Polygon-Based Hierarchical Planar Networks Based on Generalized Apollonian Construction

Mikhail V. Tamm [1,2,*], Dmitry G. Koval [1] and Vladimir I. Stadnichuk [1]

[1] Faculty of Physics, Moscow State University, 119992 Moscow, Russia; dimkov97@mail.ru (D.G.K.); stadn@polly.phys.msu.ru (V.I.S.)
[2] CUDAN Open Lab, Tallinn University, 10120 Tallinn, Estonia
* Correspondence: tamm@polly.phys.msu.ru

Abstract: Experimentally observed complex networks are often scale-free, small-world and have an unexpectedly large number of small cycles. An Apollonian network is one notable example of a model network simultaneously having all three of these properties. This network is constructed by a deterministic procedure of consequentially splitting a triangle into smaller and smaller triangles. In this paper, a similar construction based on the consequential splitting of tetragons and other polygons with an even number of edges is presented. The suggested procedure is stochastic and results in the ensemble of planar scale-free graphs. In the limit of a large number of splittings, the degree distribution of the graph converges to a true power law with an exponent, which is smaller than three in the case of tetragons and larger than three for polygons with a larger number of edges. It is shown that it is possible to stochastically mix tetragon-based and hexagon-based constructions to obtain an ensemble of graphs with a tunable exponent of degree distribution. Other possible planar generalizations of the Apollonian procedure are also briefly discussed.

Keywords: scale-free networks; Apollonian network; random planar graphs; generating functions

Citation: Tamm, M.V.; Koval, D.G.; Stadnichuk, V.I. Polygon-Based Hierarchical Planar Networks Based on Generalized Apollonian Construction. *Physics* **2021**, 3, 998–1014. https://doi.org/10.3390/physics3040063

Received: 1 July 2021
Accepted: 11 October 2021
Published: 8 November 2021

Publisher's Note: MDPI stays neutral with regard to jurisdictional claims in published maps and institutional affiliations.

1. Introduction

It is often convenient to present big volumes of data as a graph, i.e., as a set of objects and binary relations (bonds) between them. This approach naturally arises in numerous contexts ranging from physics of disordered systems [1] and biology [2] to sociology [3] and linguistics (see, e.g., [4–6]). The rapid growth in information technology ensures that larger and larger datasets of this type are becoming available. This naturally stimulates interest in the tools to analyze these datasets and simple (or not so simple) reference mathematical models, which can be used to probe their properties. Thus, a rapid development in the last 20 years of a new interdisciplinary field on the boundary of the random graph theory, the data analysis and the statistical physics, known as complex network theory [7–9], has occurred.

Among the structural characteristics typical for many experimentally observed networks, there are three especially common and striking (see, e.g., [7]): (i) the small-world property (a very small average node-to-node distance measured along the network), (ii) extremely wide, approximately a power-law distribution of the node degrees (the networks with this property are often called 'scale-free'), and (iii) large, as compared to referent randomized networks, the concentration of the short circles (e.g., triangles). It is reasonably easy to construct a model network that has one or two of these characteristics, e.g., Erdős–Rényi graphs [10,11] are small-world, random geometrical graphs [12] that have a large clustering coefficient. The Barabási–Albert model [13] generates small-world scale-free networks. The Watts–Strogatts model [14] generates small-world graphs with a large clustering coefficient, etc. Generating all three properties simultaneously is much harder. Random geometric networks in a hyperbolic space [15–17] constitute one example of networks with these properties. Another one is the Apollonian network.

The Apollonian network [18,19] is a planar graph that arises naturally as a network representation of the Apollonian gasket, a remarkable object, which is, apparently, the first known fractal (interestingly, its exact fractal dimension is still unknown) [20,21]. The construction of this network can be explained recursively as follows; see Figure 1. Take a triangle, pick a point inside it and connect it to the three corners of the triangle. As a result, one obtains a set of 3 adjacent triangles that form the first-generation Apollonian network Now, pick a point inside each of the three triangles, and connect it to its corners, this gives a second-generation Apollonian network, then repeat ad infinitum. The resulting network has been studied extensively in recent years, and it has been shown to have many beautiful properties. For example, the degree distribution and the clustering coefficient have been calculated [18], as well as the average path length [22]. Notably, there is an interesting non-planar interpretation of the Apollonian network. Namely, it can be thought of as a simplicial complex in the following way [23]. A first-generation Apollonian network is a tetrahedron (3-simplex). A second-generation Apollonian network consists of four tetrahedra: the original one and another three, each having a common two-face with the original one. A third-generation Apollonian network consists of 13 tetrahedra: one produced in the first generation, three produced in the second generation and nine new ones attached to each free face of the three second-generation tetrahedra, etc. Thus, one can think of an Apollonian network as a regular rooted tree of tetrahedra touching each other by common faces. This construction is easy to visualize in a 3-dimensional (3D) space (see Figure 1b,c), and it makes the Apollonian network a natural discretization of the 3D hyperbolic space in the same way as a regular tree is a natural discretization of the hyperbolic plane. Many properties of the Apollonian network can be calculated exactly, which makes it a nice toy model for the study of various properties of real scale-free networks. As a result, there have been a significant number of papers in recent years studying percolation [23], spin models [24], signal spreading [25], synchronization [26], traffic [27], random walks [28,29], etc., on the Apollonian network.

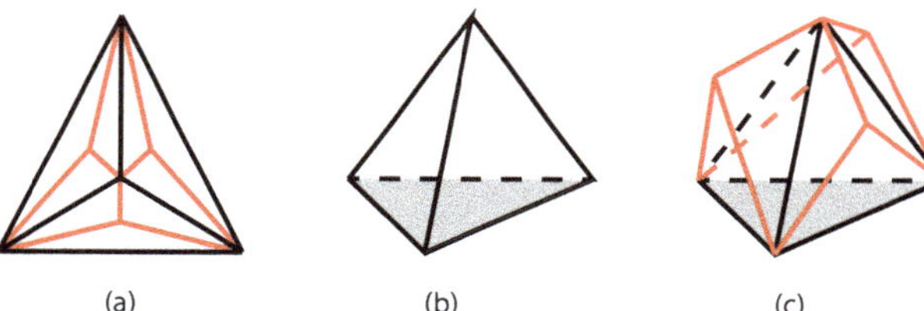

Figure 1. Apollonian network: (**a**) first (black) and second (black and red) generations of the Apollonian network; (**b**) first generation of the same Apollonian network represented as a tetrahedron in 3-dimensional space; (**c**) Apollonian networks of higher generations can be thought of as rooted trees constructed from adjacent tetrahedra. Here, second generation is shown; the shaded face functions as a root of the tree.

Despite being such a beautiful and well-studied object, the Apollonian network has certain drawbacks as a model of real networks. Most importantly, it is a single deterministic object with certain fixed properties, e.g., a fixed degree distribution with a fixed power law exponent $\gamma = \ln 3/\ln 2$. Importantly, that degree distribution is not a true power law but rather a log-periodic distribution consisting of a sequence of atoms at points 3×2^n and a power-law envelope. This means that the network is scale-invariant only with respect to certain discrete renormalizations and thus do not have the full set of properties of a true power law distribution; see [30] for a recent discussion. One natural generalization is a random Apollonian network [31–33], which is constructed, instead of a regular generation-by-generation process, by sequential partitioning of arbitrarily chosen triangles. The average degree distribution in such network is a true power law with exponent $\gamma_R = 3$ [32]. Notably, to the best of our knowledge, random Apollonian graphs remain the only scale-free planar graph model with a continuously growing size for which the exact degree distribution exponent is known. Another way of generalizing the network is to consider the k-simpliceswith $k > 3$ as building blocks of the network construction procedure. This gives rise to multidimensional Apollonian networks [34,35].

In this paper, the authors suggest another way of generalizing the Apollonian network construction. As a result, a novel famility of small-world, scale-free planar networks is obtained. The main idea is to construct an Apollonian-style iteration procedure based on polygons with different numbers of edges. The paper is organized as follows. In Section 2, a tetragon-based Apollonian-style network is constructed and the corresponding degree distribution is explicitely calculated. Then, the suggested procedure is generalized to polygons with an arbitrary even number of edges. In Section 4, it is shown that it is possible to construct a continuous one-parametric family of models interpolating between the tetragon- and hexagon-based models and demonstrate that the models in this family have a power low degree distribution with an exponent depending on the parameter, so it is possible to adjust it to fit the desired degree distribution (note that the adjustable exponent of the degree distribution can be obtained by different means in the so-called Evolving Apollonian networks [26,33]). Last section, summarizes the results of the paper and discusses further open questions and possible generalizations.

2. Tetragon-Based Network

2.1. Definition

Among several possible ways of generalizing the procedure described above to the case of polygons, consider the following procedure defined here for tetragons but easy to generalize for any polygon with an even number of edges. Note that given that we make such a generalization in further sections, we prefer to use the term 'tetragon' rather than 'quadrilateral' for a polygon with 4 sides in order to make the terminology more uniform.

Take a tetragon and pick a point inside it; then choose (at random) a pair of non-adjacent vertices of the original tetragon and connect them with a polyline with one new intermediate point. One now has two adjacent tetragons, for which one can repeat this construction, as shown in Figure 2. Importantly, contrary to the standard Apollonian network, which is a deterministic object, the network resulting from this procedure is stochastic. Indeed, already in the second generation, there are three topologically different realizations of the network, see Figure 2B. Notably, at any generation, this network has no triangles and is, in fact, bipartite.

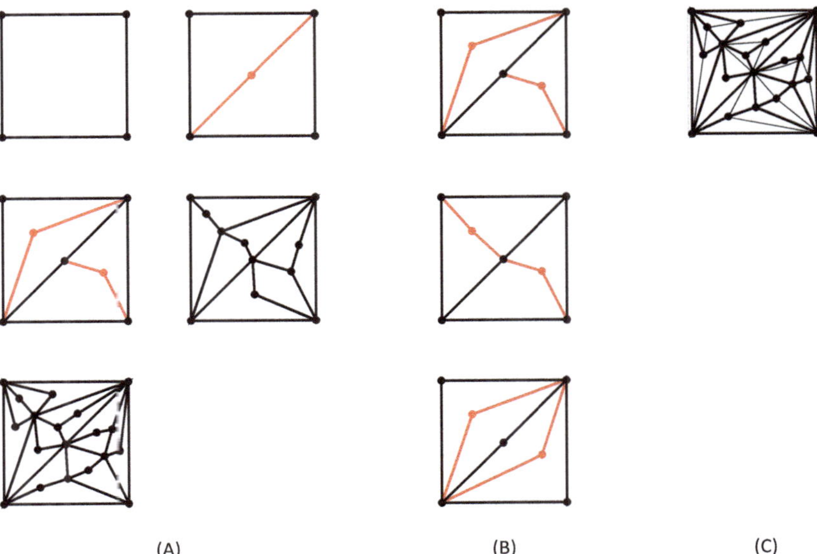

Figure 2. Constructionof a tetragon-based network: (**A**) representatives of the tetragon-networks up to the 4th generation; (**B**) three possible topologically different realizations of the second-generation network; (**C**) random triangulation of the fourth-generation network.

2.2. Degree Distribution

The natural question to ask about this newly introduced class of planar Apollonian like networks is what is the degree distribution of the nodes $G_n(k)$ in the n-th generation of the network and what distribution $\mathcal{G}(k)$ it converges to for $n \to \infty$ (here and in what follows the term "degree distribution" is used to mean the probability density function, i.e., the probability for the node to have a degree equal to k, as opposed to the cumulative distribution function, i.e., the probability for the node to have a degree larger or equal to k). By analogy with the Apollonian networks, one expects $\mathcal{G}(k)$ to be scale-free, i.e.,

$$\mathcal{G}(k) \simeq Ck^{-\alpha}, \quad k \gg 1, \tag{1}$$

with some yet unknown constants C and α.

To calculate the degree distribution $G_n(k)$, note that the degree of any given node is a random variable, whose distribution $F_{n-m}(k)$ for all nodes except the four initial ones depends only on the number of generations between the generation m at which it was created and current generation n. Indeed, each node with degree k has exactly k adjacent tetragons ($k-1$ for the four initial nodes), and at every step of the recurrent procedure each of these tetragons is split in two, which results in the creation of a new edge adjacent to the node with the probability $1/2$ (in the other half of the cases, the splitting path does not go through the given node). These splitting events happen independently for all tetragons. The overall degree distribution is therefore calculated by averaging over degree distributions of different generations:

$$G_n^{\text{all}}(k) = \frac{4F_n^{(0)}(k) + \sum_{m=1}^{n} Q_m F_{n-m}(k)}{4 + \sum_{m=1}^{n} Q_m} = \frac{4}{2^n+3} F_n^{(0)}(k) + \frac{2^n-1}{2^n+3} G_n(k), \quad G_n(k) = \frac{\sum_{m=1}^{n} Q_m F_{n-m}(k)}{\sum_{m=1}^{n} Q_m}, \tag{2}$$

where $Q_m = 2^{m-1}$ is the number of nodes created in the m-th generation, $F_n^{(0)}(k)$ is the degree distribution of the four initial nodes, and it is convenient to introduce $G_n(k)$, the degree distribution of all nodes except four initial ones.

2.3. Recurrence Relation for $F_n(K)$

To construct the recurrence relation for $F_n(k)$ proceed as follows. Let l be the degree of a node in the $(n-1)$-th generation. This means that this node has l tetragons adjacent to it and when constructing the n-th generation of the network, each of them will be split in half and with probability $1/2$, the splitting path will go through the node under consideration. Every such path increases the degree of the node by one. Thus, the overall degree may increase by l', $0 \le l' \le l$, with the probability $2^{-l}\binom{l}{l'}$, leading to

$$F_n(k) = \sum_{l=\lfloor (k+1)/2 \rfloor}^{k} 2^{-l} \binom{l}{k-l} F_{n-1}(l) \text{ for } n \ge 1, \quad F_0(k) = \delta_{k,2}, \tag{3}$$

where the fact that all nodes are created with degree 2 is taken into account and the notation $\lfloor x \rfloor$ is introduced for the integer part of x (i.e., greatest integer less or equal to x). This equation can be written down in a simpler form in terms of a generating function,

$$f_n(\lambda) = \sum_{k=2}^{\infty} \lambda^k F_n(k). \tag{4}$$

Indeed, after substituting Equation (3), one obtains $f_0(\lambda) = \lambda^2$ and

$$\begin{aligned} f_n(\lambda) &= \sum_{k=2}^{\infty} \sum_{l=\lfloor (k+1)/2 \rfloor}^{k} \lambda^k 2^{-l} \binom{l}{k-l} F_{n-1}(l) \\ &= \sum_{l=2}^{\infty} \sum_{m=0}^{l} (\lambda/2)^l \binom{l}{m} \lambda^m F_{n-1}(l) = \sum_{l=2}^{\infty} \left(\frac{\lambda(1+\lambda)}{2}\right)^l F_{n-1}(l) = f_{n-1}\left(\frac{\lambda(1+\lambda)}{2}\right), \end{aligned} \tag{5}$$

where the order of summation is changed, $m = k - l$ is introduced, and the binomial formula,

$$(1+\lambda)^l = \sum_{m=0}^{l} \binom{l}{m} \lambda^m, \tag{6}$$

is used. The recurrence relation for the four initial nodes is a bit different because in their case, the node of degree l has only $l-1$ adjacent tetragons:

$$F_n^{(0)}(k) = \sum_{l=\lfloor k/2 \rfloor+1}^{k} 2^{-l+1} \binom{l-1}{k-l+1} F_{n-1}^{(0)}(l) \text{ for } n \geq 1, \;\; F_0(k) = \delta_{k,2}, \tag{7}$$

which leads to the following equation for the generating function

$$f_n^{(0)}(\lambda) = \sum_{k=2}^{\infty} \lambda^k F_n^{(0)}(k) = \frac{2}{1+\lambda} f_{n-1}^{(0)}\left(\frac{\lambda(1+\lambda)}{2}\right), \tag{8}$$

In the $n \to \infty$ limit, both $f_n(\lambda)$ and $f_n^{(0)}(\lambda)$ converge to zero for all $|\lambda| < 1$. Indeed, the probability to have any finite degree many generations after the creation of a node is exponentially small.

2.4. Generating Function of the Degree Distribution

Combining Equation (2) for $G_n(k), G_n^{\text{all}}(k)$ and the equations for the generating functions (5) and (8), one gets the recurrence relation for the full degree distributions in terms of generating functions:

$$g_n(\lambda) = \sum_k G_n(k)\lambda^k, \;\; g_n^{\text{all}}(\lambda) = \sum_k G_n^{\text{all}}(k)\lambda^k. \tag{9}$$

For $g_n(\lambda)$, one gets:

$$(2^{n+1}-1)g_{n+1}(\lambda) = 2^n\lambda^2 + (2^n - 1)g_n\left(\frac{\lambda(1+\lambda)}{2}\right) \text{ for } n \geq 0; \;\; g_0(\lambda) = \lambda^2, \tag{10}$$

which in the limit of large n reduces to

$$g_{n+1}(\lambda) = \frac{1}{2}\lambda^2 + \frac{1}{2} g_n\left(\frac{\lambda(1+\lambda)}{2}\right). \tag{11}$$

Contrary to Equation (5) and (8), Equation (10) has a non-trivial limiting solution for $n \to \infty$. Indeed, if $G_n(k)$ converges to a limiting form $\mathcal{G}(k)$, then

$$\bar{g}(\lambda) = \sum \mathcal{G}(k)\lambda^k = \sum \lim_{n\to\infty} G_n(k)\lambda^k = \lim_{n\to\infty} \sum G_n(k)\lambda^k = \lim_{n\to\infty} g_n(\lambda). \tag{12}$$

where the summation and the limit are transposed, as one can do for convergent positive series. Thus, $\bar{g}(\lambda)$ is a solution of the functional equation,

$$2\bar{g}(\lambda) = \lambda^2 + \bar{g}\left(\frac{\lambda(1+\lambda)}{2}\right), \;\; \bar{g}(\lambda) = \sum_k \mathcal{G}(k)\lambda^k. \tag{13}$$

It seems impossible to solve this equation for all λ; however, it is possible to extract most important information about $\mathcal{G}(k)$ directly from the equation. Indeed, the behavior of the distribution for the small and large k is controlled by the behavior of the generating function in the vicinity of $\lambda = 0$ and $\lambda = 1$, respectively. For small λ, substituting

$$\bar{g}(\lambda) = p_2\lambda^2 + p_3\lambda^2 + p_4\lambda^4 + \dots \tag{14}$$

into Equation (13), one obtains:

$$p_2 = \frac{4}{7},\ p_3 = \frac{16}{105},\ p_4 = \frac{16}{155},\ p_5 = \frac{64}{1519},\ \text{etc.} \tag{15}$$

for the limiting probabilities of having a node degree equal to 2, 3, 4, 5, ...

In turn, from the behavior of $\bar{g}(\lambda)$ in the vicinity of $\lambda = 1$, one can extract both the value of α defined in Equation (1) and the values of existing moments of $\mathcal{G}(k)$. Indeed, define

$$\mathcal{G}_{\text{reg}}(k) = \mathcal{G}(k) - Ck^{-\alpha}, \tag{16}$$

and, accordingly,

$$g_{\text{reg}}(\lambda) = \bar{g}(\lambda) - C\text{Li}_{\alpha}(\lambda). \tag{17}$$

Equation (1) implies that $\mathcal{G}_{\text{reg}}(k)$ converges to zero with growing k faster than $k^{-\alpha}$, which guarantees that $g_{\text{reg}}(\lambda)$ is less singular then the first one in the vicinity of $\lambda = 1$. Thus, at $\lambda \to 1$, the function $\bar{g}(\lambda)$ has a singularity of the type $(1-\lambda)^{\alpha-1}$ and has smooth derivatives up to the order $\lfloor \alpha - 1 \rfloor$. Thus, in the lowest orders in $\epsilon = 1 - \lambda$,

$$\bar{g}(\lambda) = \sum_{i=0}^{\lfloor \alpha-1 \rfloor} a_i \epsilon^i + C\Gamma(1-\alpha)\epsilon^{\alpha-1} + o(\epsilon^{\alpha-1}), \tag{18}$$

where values of a_i depend on the small-k behavior of $\mathcal{G}(k)$ and contain information about the momenta of the distribution:

$$a_0 = \sum_k \mathcal{G}(k);\ \ a_1 = -\sum_k k\mathcal{G}(k) = -\langle k \rangle_\infty,\ \text{etc.} \tag{19}$$

Now, substituting $\lambda(\lambda+1)/2 = 1 - 3\epsilon/2 + \epsilon^2/2$ into Equation (13) and equating coefficients in front of different powers of ϵ, one obtains:

$$\begin{aligned} 2a_0 &= 1 + a_0,\ \ a_0 = 1, \\ 2a_1 &= -2 + 3a_1/2,\ \ a_1 = -4, \\ 2C\Gamma(1-\alpha) &= C\Gamma(1-\alpha)(3/2)^{\alpha-1},\ \ \alpha = 1 + \tfrac{\ln 2}{\ln(3/2)} = \tfrac{\ln 3}{\ln 3 - \ln 2} \approx 2.70951\ldots \end{aligned} \tag{20}$$

Thus, $\lfloor \alpha - 1 \rfloor = 1$, and only zeroth and first moments of the distribution converge:

$$\sum_k \mathcal{G}(k) = a_0 = 1;\ \ \langle k \rangle_\infty = -a_1 = 4, \tag{21}$$

while all the higher moments, starting from $\langle k^2 \rangle$, diverge with growing n.

It is instructive to calculate the exact values of moments $\langle k \rangle_n, \langle k^2 \rangle_n$ for all finite n. To do this, note that

$$\langle k \rangle_n = \left.\frac{dg_n(\lambda)}{d\lambda}\right|_{\lambda=1};\ \ \langle k^2 \rangle_n = \langle k \rangle_n + \left.\frac{d^2 g_n(\lambda)}{d\lambda^2}\right|_{\lambda=1}. \tag{22}$$

Equation (10) implies

$$(2^{n+1}-1)g'_{n+1}(\lambda) = 2^{n+1}\lambda + (2^n - 1)\left(\frac{2\lambda+1}{2}\right)g'_n\left(\frac{\lambda(1+\lambda)}{2}\right), \tag{23}$$

which, for $\lambda = 1$, leads to

$$\left(1 - 2^{-n-1}\right)\langle k \rangle_{n+1} = 1 + \frac{3}{4}(1 - 2^{-n})\langle k \rangle_n. \tag{24}$$

Substituting

$$b_n = 4 - (1 - 2^{-n})\langle k \rangle_n, \tag{25}$$

and allowing for the initial condition $\langle k\rangle_1 = 2, b_1 = 3$, one obtains:

$$b_n = \frac{3}{4}b_{n-1} = 4\left(\frac{3}{4}\right)^n \tag{26}$$

and, thus,

$$\langle k\rangle_n = 4\frac{1-(3/4)^n}{1-(1/2)^n}. \tag{27}$$

This is the average degree of all nodes except the original four at the n-th step of the network-generation process. Given Equation (2), one obtains the average degree of *all* nodes:

$$\langle k\rangle_n^{\text{all}} = \sum k G_n^{\text{all}}(k) = \frac{(2^n-1)\langle k\rangle_n + 4(1+(3/2)^n)}{2^n+3} = 4\frac{(1-(3/4)^n)2^n + 1 + (3/2)^n}{2^n+3} = 4\frac{2^n+1}{2^n+3}. \tag{28}$$

2.5. Second Moment of the Finite-Generation Distribution

To calculate the second moment, take the second derivative of Equation (10):

$$(2^{n+1}-1)g''_{n+1}(\lambda) = 2^{n+1} + (2^n-1)g'_n\left(\frac{\lambda(1+\lambda)}{2}\right) + (2^n-1)\left(\frac{2\lambda+1}{2}\right)^2 g''_n\left(\frac{\lambda(1+\lambda)}{2}\right) \tag{29}$$

and take into account Equation (55). Substituting $\lambda = 1$ and allowing for the fact that

$$g'_n(1) = \langle k\rangle_n;\ g''_n(1) = \langle k^2\rangle_n - \langle k\rangle_n \tag{30}$$

leads to

$$(2^{n+1}-1)(\langle k^2\rangle_{n+1} - \langle k\rangle_{n+1}) = 2^{n+1} + (2^n-1)\langle k\rangle_n + \frac{9}{4}(2^n-1)(\langle k^2\rangle_n - \langle k\rangle_n) \tag{31}$$

or

$$(2^{n+1}-1)\langle k^2\rangle_{n+1} - \frac{9}{4}(2^n-1)\langle k^2\rangle_n = 2^{n+1} + (2^{n+1}-1)\langle k\rangle_{n+1} - \frac{5}{4}(2^n-1)\langle k\rangle_n, \tag{32}$$

which, after substituting Equation (27), simplifies to

$$(2^{n+1}-1)\langle k^2\rangle_{n+1} - \frac{9}{4}(2^n-1)\langle k^2\rangle_n = 2^n\left(5-\left(\frac{3}{4}\right)^n\right). \tag{33}$$

Now define the sequence

$$a_n = \langle k^2\rangle_n\frac{2^n-1}{2^n} \tag{34}$$

and its generating function $F(s) = \sum a_n s^n$. The recurrency for a_n reads:

$$a_{n+1} = \frac{9}{8}a_n + \frac{5}{2} - \frac{1}{2}\left(\frac{3}{4}\right)^n \tag{35}$$

for $n \geq 1$, and $a_1 = 2$. Then:

$$F(s)\left(1-\frac{9}{8}s\right) = 2s + \frac{5}{2}\frac{s^2}{1-s} - \frac{3}{8}\frac{s^2}{1-3s/4}. \tag{36}$$

In order to proceed further, note that

$$\begin{aligned}\frac{2s}{1-9s/8} &= -\frac{16}{9}+\frac{16}{9}\frac{1}{1-9s/8}, \\ \frac{s^2}{(1-s)(1-9s/8)} &= \frac{8}{9}-8\frac{1}{1-s}+\frac{64}{9}\frac{1}{1-9s/8}, \\ \frac{s^2}{(1-3s/4)(1-9s/8)} &= \frac{32}{27}-\frac{32}{9}\frac{1}{1-3s/4}+\frac{64}{27}\frac{1}{1-9s/8}.\end{aligned} \tag{37}$$

Thus,

$$F(s)=\frac{56}{3}\frac{1}{1-9s/8}-20\frac{1}{1-s}+\frac{4}{3}\frac{1}{1-3s/4};\ a_n=\frac{56}{3}\left(\frac{9}{8}\right)^n-20+\left(\frac{3}{4}\right)^{n-1}, \tag{38}$$

and

$$\langle k^2\rangle=(1-2^{-n})^{-1}\left[\frac{56}{3}\left(\frac{9}{8}\right)^n-20+\left(\frac{3}{4}\right)^{n-1}\right]. \tag{39}$$

Proceeding in the same way, one obtains:

$$\langle k^2\rangle^{(0)}=\frac{4}{3}\left(\frac{9}{4}\right)^n+\frac{5}{3}\left(\frac{3}{2}\right)^n+1. \tag{40}$$

Thus, the total average degree is:

$$\begin{aligned}\langle k^2\rangle_{\text{all}} &= \frac{2^n-1}{2^n+3}\langle k^2\rangle+\frac{4}{2^n+3}\langle k^2\rangle^{(0)} \\ &= \left(1+\frac{3}{2^n}\right)^{-1}\left[24\left(\frac{9}{8}\right)^n-20+8\left(\frac{3}{4}\right)^n+4\left(\frac{1}{2}\right)^n\right]\approx 24\left(\frac{9}{8}\right)^n-20,\end{aligned} \tag{41}$$

where the approximal equality holds for $n \gg 1$. Comparing Equations (39) and (41) shows that, interestingly, the four initial nodes contribute a finite fraction to the overall value of $\langle k^2\rangle_{\text{all}}$, which converges to $2/9$ for large n.

2.6. Scaling Form of the Degree Distribution

For large n, the degree distribution $G_n(k)$ converges to $\mathcal{G}(k)$. Typically (see, e.g., [36]) one expects the ratio of these functions $\Phi_n(k)=G_n(k)/\mathcal{G}(k)$ to attain a universal shape for large n. More precisely, it means that there exists a scaling function $\phi(x)$ and a sequence K_n for which

$$\frac{\Phi_n(k)}{\phi(k/K_n)}\to 1 \text{ for } n\to\infty. \tag{42}$$

Here, K_n is the characteristic scale of the distribution of the n-th generation graph, and it diverges as $n\to\infty$. One must demand $\phi(0)=1$ in order for $G_n(k)$ to converge to $\mathcal{G}(k)$ and $\phi(\infty)=0$ in order for k to be bound from above for any finite n. There are various ways of obtaining the scaling factor K_n, e.g., one can use the large-n behavior of $\langle k^2\rangle_n$:

$$\langle k^2\rangle_n=\sum k^2G_n(k)\sim K_n^{(3-\alpha)}, \tag{43}$$

where it is taken into account that (contrary to the lower moments) $\langle k^2\rangle_n$ is controlled by the tail of the distribution. Substituting Equation (41), one obtains:

$$K_n^{3-\alpha}=(9/8)^n;\ \to\ K_n=(3/2)^n, \tag{44}$$

which is to be expected since the typical maximal degree of the network increases by a factor $3/2$ on each step.

In order to check the predictions of the model, 2×10^5 realizations of the networks of up to the 14th generation were generated. Figure 3a shows the resulting degree distributions for sequential generations of the network. It can be seen from Figure 3b that after renormalization of the abscisse and ordinate axes by the factors $(3/2)^n$ and $k^{-\alpha}$, respectively, the data collapse perfectly on a single scaling curve $\phi(x)$.

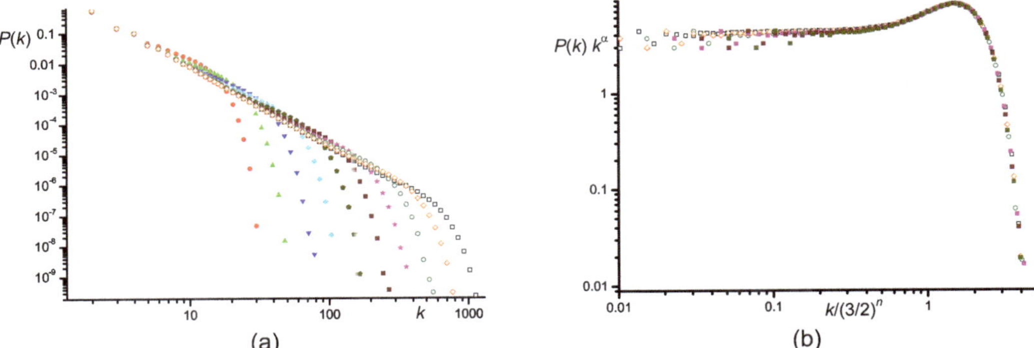

Figure 3. (**a**) Degree distributions of the tetragon-based networks of generations 5–14 as indicated. The results shown are obtained after averaging over 2×10^5 realizations and logarithmic binning with step 1.1. (**b**) Same distributions but rescaled as the axes show.

3. Polygon-Based Networks for Polygons with Any Even Number of Edges

The procedure suggested in Section 2 can be easily generalized for any even number of edges $2m$ ($m \geq 2$) in the generating polygon; see generalization for $m = 3$ in Figure 4. This procedure results in a sequence of planar scale-free network models with degree distributions converging to

$$\mathcal{G}_m(k) \simeq C_m k^{-\alpha_m}, \quad k \gg 1, \tag{45}$$

with m-dependent exponents α_m. At each generation, each polygon is split by a path connecting directly opposite nodes. There are m different ways of such a splitting, so each node of a polygon participates in the splitting with probability $1/m$. This allows generalizing the recurrence relations (3) and (7) for the degree distribution of a node n generations after its creation in the following way:

$$F^{(0)}_{n,m}(k) = \sum_{l=\lfloor k/2\rfloor+1}^{k} \binom{l-1}{k-l+1}\left(\frac{1}{m}\right)^{k-l+1}\left(\frac{m-1}{m}\right)^{2l-k-2} F^{(0)}_{n-1,m}(l) \quad \text{for } n \geq 1 \;\; F^{(0)}_0(k) = \delta_{k,2}, \tag{46}$$

for the original $2m$ nodes, and

$$F_{n,m}(k) = \sum_{l=\lfloor (k+1)/2\rfloor}^{k} \binom{l}{k-l}\left(\frac{1}{m}\right)^{k-l}\left(\frac{m-1}{m}\right)^{2l-k} F_{n-1,m}(l) \quad \text{for } n \geq 1; \;\; F_{0,m}(k) = \delta_{k,2}; \tag{47}$$

for all the rest. The number of nodes created at n-th generation ($n \geq 1$) is $(m-1)2^n$. Proceeding in exactly the same way as before, gives

$$f^{(0)}_{n,m}(\lambda) = \sum_{k=2}^{\infty} \lambda^k F^{(0)}_n(k) = \frac{m}{m-1+\lambda} f^{(0)}_{n-1}\left(\frac{\lambda(m-1+\lambda)}{m}\right), \tag{48}$$

$$\begin{aligned} g_{n,m}(\lambda) &= \frac{1}{(m-1)(2^n-1)} \sum_{l=1}^{n}\sum_{k=2}^{\infty}(m-1)2^l F_{n-l,m}(k)\lambda^k \\ &= \frac{2^{n-1}}{2^n-1}\lambda^2 + \frac{2^{n-1}-1}{2^n-1} g_{n-1,m}\left(\frac{\lambda(m-1+\lambda)}{m}\right) \text{ for } n \geq 1; \;\; g_{1,m}(\lambda) = \lambda^2, \end{aligned} \tag{49}$$

and

$$g_{n,m}^{\text{all}}(k) = \frac{2m}{(m-1)2^n + m + 1} f_{n,m}^{(0)}(\lambda) + \frac{(m-1)(2^n - 1)}{(m-1)2^n + m + 1} g_{n,m}(\lambda) \tag{50}$$

for the generating functions of the degree distributions of the original, newly created and all nodes of the network, $f_{n,m}^{(0)}(\lambda)$, $g_{n,m}(\lambda)$ and $g_{n,m}^{\text{all}}(\lambda)$, respectively. The limiting function,

$$\bar{g}_m(\lambda) = \lim_{n \to \infty} g_{n,m}(\lambda), \tag{51}$$

satisfies

$$\bar{g}_m(\lambda) = \frac{\lambda^2}{2} + \frac{1}{2} \bar{g}_m\left(\frac{\lambda(m-1+\lambda)}{m}\right), \tag{52}$$

and its behavior is easy to analyze both in the vicinity of $\lambda = 0$ and $\lambda = 1$. Expanding Equation (52) for small λ, one obtains:

$$\begin{aligned} p_2^m &= \frac{m^2}{K_{2,m}}, \; p_3^m = \frac{2m^3(m-1)}{K_{2,m}K_{3,m}}, \; p_4^m = \frac{m^3(2m^3 + 5(m-1)^3)}{\prod_{l=2}^{4} K_{l,m}}, \\ p_5^m &= \frac{m^5(m-1)^2(12m^4 + 8m^3(m-1) + 14(m-1)^4)}{\prod_{l=2}^{5} K_{l,m}}, \; \text{etc.,} \end{aligned} \tag{53}$$

where a short-hand notation, $K_{l,m} = 2m^l - (m-1)^l$, is introduced.

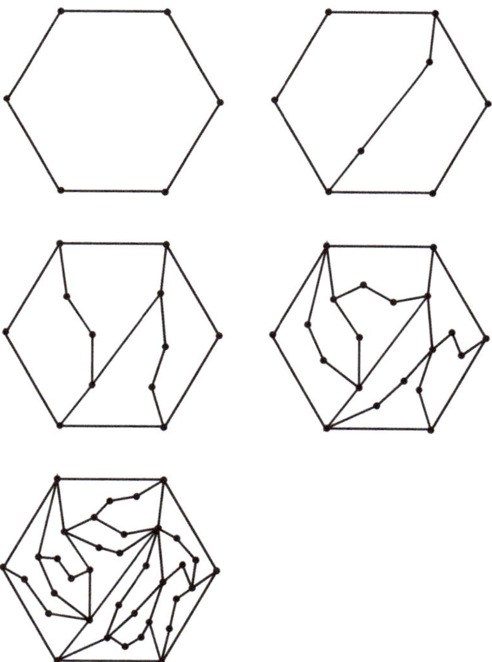

Figure 4. Construction of the hexagon-based network (up to 4th generation).

In turn, in the vicinity of $\lambda = 1$ $\bar{g}_m(\lambda)$, it takes the form of Equation (18). Substituting the ansatz (18) into Equation (52), one obtains:

$$\begin{aligned} 2a_0 &= 1 + a_0, \ a_0 = 1, \\ 2a_1 &= -2 + (m+1)a_1/m, \ a_1 = -2m/(m-1), \\ 2a_2 &= 1 - a_1/m + (m+1)^2 a_2/m^2, \ a_2 = \frac{(m+1)m^2}{(m-1)(m^2-2m-1)}, \\ 2C\Gamma(1-\alpha_m) &= C\Gamma(1-\alpha_m)(m+1/m)^{\alpha_m - 1}, \ \alpha_m = 1 + \frac{\ln 2}{\ln(m+1) - \ln m}. \end{aligned} \tag{54}$$

Thus, for any $m \geq 3$ $\lfloor \alpha_m - 1 \rfloor \geq 2$, the second moment of $\mathcal{G}_m(k)$ converges. The moments are controlled by the coefficients a_i:

$$\sum_k \mathcal{G}(k) = a_0 = 1, \ \sum_k k\mathcal{G}(k) = -a_1 = \frac{2m}{m-1}, \ \sum_k k^2 \mathcal{G}(k) = 2a_2 - a_1 = \frac{2m(2m^2 - m - 1)}{(m-1)(m^2 - 2m - 1)}. \tag{55}$$

Since the second moment of $\mathcal{G}_m(k)$ is now controlled by the values of distribution at small k, the initial $2m$ nodes do not contribute to the second moment.

Once again, in order to check the predictions 2×10^5 realizations of the networks of up to the 12th generation were generated. The results are shown in Figure 5a, and in Figure 5b the plot in the renormalized coordinates is shown. The collapse of the data onto a single master curve is apparent, although it is somewhat worse than in Figure 3b. Presumably, this happens because the typical degrees in the hexagon-based network are much smaller than in the tetragon-network of the same size, and finite size effects are therefore more important.

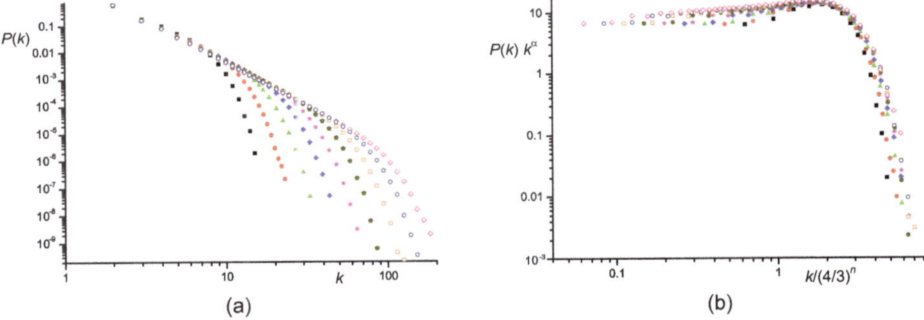

Figure 5. (**a**) Degree distributions of the hexagon-based networks of generations 4–12 as indicated. The results shown are obtained after averaging over 2×10^5 realizations and logarithmic binning with step 1.1. (**b**) Same distributions but rescaled as the axes show.

4. Polygon-Based Networks with Smoothly Changing Exponent of the Degree Distribution

As a result of Section 3, one now has a sequence of Apollonian-like models that generate planar scale-free networks with a discrete sequence of degree-distribution exponents $\alpha_m = 1 + \ln 2/(\ln(m+1) - \ln m)$, $m = 2, 3, \ldots$ Is it possible to further generalize the model to make α change continuously and take any intermediate values, including, for example, $\alpha = 3$, corresponding to the point where the second moment of the degree distributions diverges for the first time?

It turns out that this is indeed possible. One way to do that is as follows. Assume that when introducing a new shortcut dividing a polygon in two, one makes the resulting partition to be a pair of tetragons with probability p and a pair of hexagons with probability $1-p$. That is to say, if the original polygon is a tetragon, then with probability p introduce a 2-step path connecting opposite vertices, and with probability $q = 1 - p$, a 4-step path;

if the original polygon is a hexagon, the new path connecting two opposite vertices is a 1-step path with probability p and a 3-step path with probability q.

Here, we restrict ourselves to this simplest construction, although it is possible to create more complicated rules. For example, one can introduce correlations between generations in a Markovian way so that there is a matrix p_{ij} of probabilities for a tetragon to give birth to a couple of tetragons, a hexagon to give birth to a couple of tetragons, etc. As a result, it might be possible to construct a network that is, for example, tetragon-dominated at large scales (early generations) and hexagon-dominated at small scales (later generations).

Once again , consider a node, which is created at generation n_0, and let us study its degree distribution at generation $n_0 + n$. This degree distribution depends only on n and on the number of edges of the initial two faces adjacent to the node where tetragons or hexagons.

Let the average fraction of tetragons at a given generation be p and the fraction of hexagons be $q = 1 - p$. Then, for each face adjacent to a given node, the probability that this face is a tetragon is

$$\pi(p) = \frac{4p}{4p + 6q} = \frac{2p}{3 - p}. \tag{56}$$

Assume now that different faces adjacent to a node are tetragons (hexagons) independently from each other. Generally speaking, that is not true: when a new edge is created the two faces on the sides of it have a similar number of edges. However, one might expect that as the degree of the node grows the correlations become less and less relevant. In this approximation, the probability for a node of degree k to have exactly l adjacent tetragons is

$$\binom{k}{l} \pi^l (1 - \pi)^{k-l}. \tag{57}$$

when the next generation is created, a new edge adjacent to the node under consideration is created with probability 1/2 for each tetragon face and with probability 1/3 for each hexagon face. Therefore, one can write down the following approximate equation for the probability $P(k + r|k, p)$ of the node that has degree $k + r$ at the next generation given that it had degree k in the previous one:

$$P(k + r|k, p) = \sum_{l=0}^{k} \sum_{s=0}^{r} \binom{k}{l} \pi^l (1 - \pi)^{k-l} \binom{l}{s} \left(\frac{1}{2}\right)^l \binom{k - l}{r - s} \left(\frac{1}{3}\right)^{r-s} \left(\frac{2}{3}\right)^{k-l-r+s}, \tag{58}$$

where the binomial coefficients $\binom{m}{n}$ are assumed to be zeros if $n > m$ or $n < 0$. Now introduce the probability $F_n(k|p)$ for a node to have degree k n generations after its creation and the corresponding generation function,

$$f_n(\lambda|p) = \sum_{k=2}^{\infty} F_n(k)\lambda^k. \tag{59}$$

Then, $f_0(\lambda|p) = \lambda^2$,

$$F_n(k) = \sum_{k'=\lfloor \frac{k+1}{2} \rfloor}^{k} P(k|k', p)F_{n-1}(k'), \tag{60}$$

and

$$f_n(\lambda|p) = \sum_{k=2}^{\infty} \sum_{k'=\lfloor \frac{k+1}{2} \rfloor}^{k} \lambda^k P(k|k', p)F_{n-1}(k') = \sum_{k'=2}^{\infty} F_{n-1}(k')\lambda^{k'} \sum_{k=k'}^{2k'} \lambda^{k-k'} P(k|k', p). \tag{61}$$

Using Equation (58), it is easy to calculate the second sum on the right-hand side:

$$\begin{aligned}
\sum_{r=0}^{k'} \lambda^r P(k'+r|k',p) &= \sum_{r=0}^{k'} \sum_{l=C}^{k'} \sum_{s=0}^{r} \frac{k'!}{s!(l-s)!(r-s)!(k'-l-r+s)!} \pi^l (1-\pi)^{k'-l} \left(\tfrac{1}{2}\right)^l \left(\tfrac{1}{3}\right)^{r-s} \left(\tfrac{2}{3}\right)^{k'-l-r+s} \lambda^r \\
&= \sum_{r=0}^{k'} \sum_{l=C}^{k'} \sum_{s=0}^{r} \frac{k'!}{s!(r-s)!(l-s)!(k'-l-r+s)!} \left(\tfrac{\pi\lambda}{2}\right)^s \left(\tfrac{\pi}{2}\right)^{l-s} \left(\tfrac{(1-\pi)\lambda}{3}\right)^{r-s} \left(\tfrac{2(1-\pi)}{3}\right)^{k'-l-r+s} \\
&= \left(\tfrac{\pi\lambda}{2} + \tfrac{\pi}{2} + \tfrac{(1-\pi)\lambda}{3} + \tfrac{2(1-\pi)}{3}\right)^{k'},
\end{aligned} \tag{62}$$

which leads to the following equation for the generating function:

$$f_n(\lambda) = f_{n-1}\left(\lambda\left(\frac{4-\pi}{6} + \frac{2+\pi}{6}\lambda\right)\right) = f_{n-1}\left(\frac{2-p+\lambda}{3-p}\lambda\right), \tag{63}$$

where Equation (56) is taken into account to obtain to the last expression. Proceeding as before, one gets the equation for the generating function of the full limiting degree distribution $g_\infty(\lambda)$ (except for the initial set of nodes):

$$g_\infty(\lambda) = \frac{\lambda^2}{2} + \frac{1}{2} g_\infty\left(\frac{2-p+\lambda}{3-p}\lambda\right). \tag{64}$$

Expanding $g_\infty(\lambda)$ for $\lambda = 1-\epsilon$, $\epsilon \ll 1$ in the form (compare Equation (18)):

$$\bar{g}(\lambda) = \sum_{i=0}^{\lfloor \alpha(p)-1 \rfloor} a_i \epsilon^i + C\Gamma(1-\alpha(p))\epsilon^{\alpha(p)-1} + o(\epsilon^{\alpha(p)-1}), \tag{65}$$

and equating the coefficients term by term exactly in the same way as in Section 2, one gets the following equation for the degree distribution exponent $\alpha(p)$:

$$2 = \left(\frac{4-p}{3-p}\right)^{\alpha(p)-1}, \quad \alpha(p) = 1 + \frac{\ln 2}{\ln(4-p) - \ln(3-p)}. \tag{66}$$

Thus, for example, the interesting case $\alpha(p) = 3$ when the second moment of the degree distribution diverges for the first time corresponds to

$$p|_{\alpha=3} = 2 - \sqrt{2} \approx 0.58579\ldots \tag{67}$$

In Figure 6, the numerical data for the degree distribution of the mixed networks are shown. One can clearly see that the slope of the distribution gradually changes with changing p. Moreover, after rescaling the degree distribution with the power law prescribed by Equation (66), all curves are approximately flat for small k, validating the approximation of independent phases.

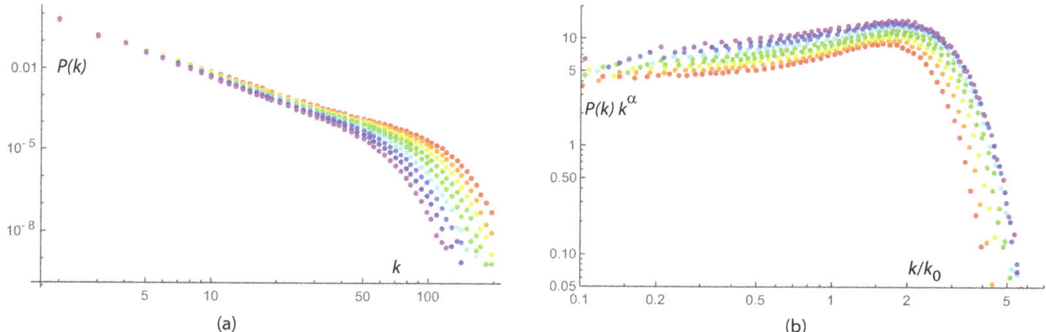

Figure 6. (**a**) Degree distributions of n = 10th-generation mixed tetragon–hexagon networks with varying p, rainbow changing color from $p = 0.9$ (red) to $p = 0.1$ (violet). The results shown are obtained after averaging over 10^5 realizations and logarithmic binning with step 1.05. (**b**) Same distributions but rescaled: $P(k)$ is rescaled by its theoretical behavior $k^{-\alpha}$, with α given by Equation (66), k is rescaled by a factor $k_0(p,n) = ((4-p)/(3-p))^n$, which approximates the growth of the maximal accessible degree with the number of generations n.

5. Concluding Remarks and Open Questions

This paper presents one possible class of planar random graphs constructed from polygons with an even number of edges using a procedure similar to the construction of Apollonian graphs [18]. The $2m$-polygon-based graphs have a limiting power law degree distribution with the exponent,

$$\alpha_m = 1 + \frac{\ln 2}{\ln(m-1) - \ln m}, \tag{68}$$

and the moments of the degree distribution are given by Equations (21) and (55). The second moment of the degree distribution diverges as $(9/8)^n$ with the number of generations n in the case of tetragon-based graphs (see Equation (41)) and converges to a finite value in Equation (55) for the polygons with a larger number of edges. Moreover, as described in Section 4, it is possible to construct a mixed model based on two different polygons (tetragons and hexagons in our example) so that on all stages of construction, tetragons are formed with probability p and hexagons with probability $1-p$. By varying p, one can adjust the slope of the degree distribution in order to achieve a desired value in a way reminiscent of evolving Apollonian networks [26].

Clearly, all graph classes presented here are small world. Indeed, the diameter of the graphs grows at most linearly with the number of generations:

$$d_{n+1} \le d_n + 2\lfloor m/2 \rfloor, \tag{69}$$

where d_n is the diameter of the n-generation graph. In turn, the total number of nodes grows exponentially with the number of generations; thus, the diameter is, at most, proportional to the logarithm of the number of nodes.

The shortest cycles in the graphs presented here are $2m$, and, in particular, there are no triangles in them, so, generally speaking, the clustering coefficient is zero. However, this should not obscure the fact that there is actually a huge number of short cycles in these graphs. Indeed, consider the following auxiliary construction: let the polygon-based construction be exactly as presented above up to n-th generation, but then connect all the nodes belonging to the same face on the last generation of the procedure, so the smallest faces (i.e., faces constructed on the last step) are considered to be complete graphs K_{2m} ($(2m-1)$-simplices). The large-scale structure of the resulting graph (including, e.g., the slope of the degree distribution) will be the same as in the original polygon-based procedure, but a finite fraction of nodes (those created in the n-th generation of the construction) will have clustering coefficient 1, guaranteeing that the average clustering coefficient of the whole graph remains finite as $n \to \infty$. In order to use polygon-based graphs as a toy model for experimental systems, it might be reasonable to add a random fraction of them in order to fit the observed clustering coefficient instead of adding all possible links connecting the vertices in the smallest faces.

Interestingly, polygon-based graphs with even m are bipartite, see Figure 7a. In that sense, the tetragon-based graph seems to be a natural generalization of the Apollonian construction for the case of bipartite graphs. We expect that there might be a connection between bipartite polygon-based graphs and space-filling bearings, which allow only cycles of even lengths [37] in a way similar to the connection between original Apollonian networks and space-filling systems of embedded disks. Exploring this question goes, however, beyond the scope of this paper.

This paper restricts itself to just one class of possible generalizations of the Apollonian construction based on polygons of arbitrary sizes. It is quite easy to suggest various other generalizations. The most obvious example is, probably, the random polygon construction where new graphs are constructed not generation-by-generation by splitting all the polygons of the previous generation at once, but rather by randomly choosing on each step a face to split. Figure 7b presents an example realization of such a tetragon-based random graph. In the standard Apollonian case, it is known that the exponent of the degree distribution is different for the regular and random constructions. Calculating the degree distribution of random Apollonian-like polygon-based graphs remains an interesting open question.

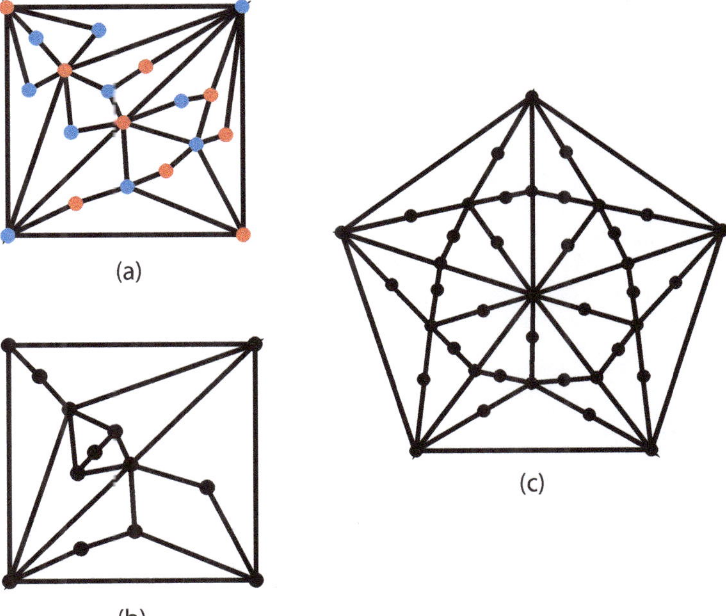

Figure 7. (**a**) Tetragon-based networks are bipartite. (**b**) An example of a particular realization of a randon tetragon-based graph. (**c**) An example of a 2nd-generation deterministic pentagon-based graph.

Another, this time a completely deterministic generalization, is as follows. Consider a polygon with an odd number of edges $2m+1, m \geq 1$. Put a point inside the polygon and connect it with all vertices of the polygon by chains of m edges and $(m-1)$ nodes. This splits a polygon into $2m+1$ faces, each having exactly $2m+1$ edges. On the next step, repeat this procedure for each of the faces and proceed ad infinitum. Figure 7c shows the second-generation pentagon-based graph obtained via such procedure. Clearly, this construction is an even more direct generalization of the Apollonian graph construction (indeed, $m = 1$ case is just the Apollonian graph itself). However, it means that it has standard drawbacks of the Apollonian graph in a sense that it is a single deterministic object rather than a stochastic ensemble of graphs and that its limiting degree distribution is not a power law but rather a log-periodic function with a power law envelope.

We think that the classes of graphs presented here are a useful addition to the toolkit of toy models to model scale-free graphs. Indeed, while having the main advantages of the Apollonian networks, they have additional flexibility in a sense that one might regulate the slope of the degree distribution and the average clustering coefficient in the way described above. In particular, such graphs might be, in our opinion, useful in the applications where graph planarity is essential [38], for example, in quantitative geography, such as the study of the formation of the systems of interconnected cities. On the other side, studying percolation, spectral properties, diffusion, synchronization, epidemic spreading, etc., on these generalized graphs might allow to systematically study the influence of varying degree exponents on these phenomena, which, to the best of our knowledge, have not yet been done for the scale-free planar networks.

Author Contributions: Conceptualization, M.V.T.; Investigation, M.V.T., D.G.K. and V.I.S.; Methodology, M.V.T.; Software, V.I.S.; Supervision, M.V.T.; Visualization, M.V.T.; Writing—original draft M.V.T.; Writing—revised version, M.V.T. All authors have read and agreed to the published version of the manuscript.

Funding: This research was funded by RSF, grant number 21-11-00215.

Acknowledgments: The authors are grateful to H. Hermann for encouraging comments on the idea of this work and to G. Bianconi, S. Nechaev and P. Krapivsky for many interesting discussions. The research presented here was undertaken in the Laboratory of Nonlinear, Nonequilibrium and Complex Systems at Moscow State University, which, for the last 10 years, has been headed by Professor M.I. Tribelsky. We are very grateful for his valuable scientific and personal advice, kindness and support over all these years and would like to dedicate this work to him on the occasion of his 70th birthday.

Conflicts of Interest: The authors declare no conflict of interest.

References

1. Tikhonov, K.S.; Mirlin, A.D. From Anderson localization on random regular graphs to many-body localization. *Ann. Phys.* **2021** in press. [CrossRef]
2. Sneppen, K.; Zocchi, G. *Physics in Molecular Biology*; Cambridge University Press: Cambridge, UK, 2005. [CrossRef]
3. Jackson, M.O. *Social and Economic Networks*; Princeton University Press: Princeton, NJ, USA, 2010.
4. Kenett, Y.N.; Anaki, D.; Faust, M. Investigating the structure of semantic networks in low and high creative persons. *Front. Hum Neurosci.* **2014**, *8*, 407. [CrossRef] [PubMed]
5. Stella, M.; Beckage, N.M.; Brede, M.; Domenico, M.D. Multiplex model of mental lexicon reveals explosive learning in humans. *Sci. Rep.* **2018**, *8*, 2259. [CrossRef] [PubMed]
6. Valba, O.V.; Gorsky, A.S.; Nechaev, S.K.; Tamm, M.V. Analysis of English free association network reveals mechanisms of efficient solution of Remote Association Tests. *PLoS ONE* **2021**, *16*, e0248986.
7. Newman, M. *Networks*; Oxford University Press: Oxford, UK, 2018. [CrossRef]
8. Barabasi, A.L. *Network Science*; Cambridge University Press: Cambridge, UK, 2016.
9. Dorogovtsev, S. *Complex Networks*; Oxford University Press: Oxford, UK, 2010. [CrossRef]
10. Erdős, P.; Rényi, A. On Random Graphs I. *Publ. Math.* **1959**, *6*, 290–297.
11. Krapivsky, P.L.; Redner, S.; Ben-Naim, E. *A Kinetic View of Statisrical Physics*; Cambridge University Press: Cambridge, UK, 2010. [CrossRef]
12. Balister, P.; Bollobás, B.; Sarkar, A. Percolation, connectivity, coverage and colouring of random geometric graphs. In *Handbook of Large-Scale Random Networks*; János Bolyai Mathematics Society Studies; Bollobás, B., Kozma, R., Miklós, D., Eds.; Springer Berlin/Heidelberg, Germany. [CrossRef]
13. Barabási, A.L.; Albert, R. Emergence of scaling in random networks. *Science* **1999**, *286*, 509–512. [CrossRef]
14. Watts, D.; Strogatz, S. Collective dynamics of 'small-world' networks. *Nature* **1998**, *393*, 440–442. [CrossRef]
15. Krioukov, D.; Papadopoulos, F.; Kitsak, M.; Vahdat, A.; Boguna, M. Hyperbolic geometry of complex networks. *Phys. Rev. E* **2010**, *82*, 036106. [CrossRef]
16. Boguna, M.; Papadopoulos, F.; Krioukov, D. Sustaining the internet with hyperbolic mapping. *Nat. Commun.* **2010**, *1*, 62 [CrossRef]
17. Papadopoulos, F.; Kitsak, M.; Serrano, M.Á.; Boguná, M.; Krioukov, D. Popularity versus similarity in growing networks. *Nature* **2012**, *489*, 537–540. [CrossRef]
18. Andrade, J.S., Jr.; Herrmann, H.J.; Andrade, R.F.S.; da Silva, L. Apollonian networks: Simultaneously scale-free, small world, Euclidean, space filling, and with matching graphs. *Phys. Rev. Lett.* **2005**, *94*, 018702. [CrossRef]

19. Doye, J.P.K.; Massen, C.P. Self-similar disk packings as model spatial scale-free networks. *Phys. Rev. E* **2005**, *71*, 016128. [CrossRef] [PubMed]
20. Kasner, E.; Supnick, F. The Apollonian packing of circles. *Proc. Natl. Acad. Sci. USA* **1943**, *29*, 378–384. [CrossRef] [PubMed]
21. Boyd, D.W. The residual set dimension of the Apollonian packing. *Mathematika* **1973**, *20*, 170–174. [CrossRef]
22. Zhang, Z.; Chen, L.; Zhou, S.; Fang, L.; Guan, J.; Zou, T. Analytical solution of average path length for Apollonian networks. *Phys. Rev. E* **2008**, *77*, 017102. [CrossRef] [PubMed]
23. Bianconi, G.; Ziff, R.M. Topological percolation on hyperbolic simplicial complexes. *Phys. Rev. E* **2018**, *98*, 052308. [CrossRef]
24. Andrade, R.F.S.; Hermann, H.J. Magnetic models on Apollonian networks. *Phys. Rev. E* **2005**, *71*, 056131. [CrossRef]
25. Lind, P.G.; da Silva, L.R.; Andrade, J.S., Jr.; Hermann, H.J. Spreading gossip in social networks. *Phys. Rev. E* **2007**, *76*, 036117. [CrossRef]
26. Zhang, Z.; Rong, L.; Zhou, S. Evolving Apollonian networks with small-world scale-free topologies. *Phys. Rev. E* **2006**, *74*, 046105. [CrossRef]
27. Mendes, G.A.; da Silva, L.R.; Hermann, H.J. Traffic gridlock on complex networks. *Phys. A* **2012**, *391*, 362. [CrossRef]
28. Huang, Z.-G.; Xu, X.-J.; Wu, Z.-X.; Wang, Y.-H. Walks on Apollonian networks. *Eur. Phys. J. B* **2006**, *51*, 549–553. [CrossRef]
29. Zhang, Z.; Guan, J.; Xie, W.; Qi, Y.; Zhou, S. Random walks on the Apollonian network with a single trap. *Europhys. Lett.* **2009**, *86*, 10006. [CrossRef]
30. Newberry, M.G.; Savage, V.M. Self-similar processes follow a power law in discrete logarithmic space. *Phys. Rev. Lett.* **2019**, *122*, 158303. [CrossRef] [PubMed]
31. Zhou, T.; Yan, G.; Zhou, P.-L.; Fu, Z.-Q.; Wang, B.-H. Random apollonian networks. *arXiv* **2004**, arXiv:0409414.
32. Zhou, T.; Yan, G.; Wang, B.-H. Maximal planar networks with large clustering coefficient and power-law degree distribution. *Phys. Rev. E* **2005**, *71*, 046141. [CrossRef] [PubMed]
33. Kolossvary, I.; Komjathy, J.; Vago, L. Degrees and distances in random and evolving Apollonian networks. *Adv. Appl. Prob.* **2016**, *48*, 865–902. [CrossRef]
34. Zhang, Z.; Comellas, F.; Fertin, G.; Rong, L. High-dimensional Apollonian networks. *J. Phys. A* **2006**, *39*, 1811. [CrossRef]
35. Zhang, Z.; Rong, L.; Comellas, F. High-dimensional random Apollonian networks. *Phys. A* **2006**, *364*, 610–618. [CrossRef]
36. Setna, J.P. *Statistical Mechanics: Entropy, Order Parameters, and Complexity*; Oxford University Press: Cambridge, UK, 2006. [CrossRef]
37. Oron, G.; Herrmann, H.J. Generalization of space-filling bearings to arbitrary loop size. *J. Phys. A* **2000**, 33, 1417–1434. [CrossRef]
38. Barthelemy, M. Spatial networks. *Phys. Rep.* **2011**, *499*, 1–101. [CrossRef]

Article

Bounds on Energies and Dissipation Rates in Forced Dynamos

Michael Proctor †

Department of Applied Mathematics and Theoretical Physics (DAMTP), University of Cambridge, Clarkson Road, Cambridge CB3 0WA, UK; mrep@cam.ac.uk
† Current address: King's College, University of Cambridge, King's Parade, Cambridge CB2 1ST, UK.

Abstract: This paper is concerned with limits on kinetic and magnetic energies and dissipation rates in forced flows that lead to dynamo action and a finite amplitude magnetic field. Rigorous results are presented giving upper and lower limits on the values of these quantities, in a simple cubic geometry with periodic boundary conditions, using standard inequalities. In addition to the general case, results in the special case of the Archontis dynamo are presented, in which fields and flows are closely similar in much of the domain.

Keywords: magnetohydrodynamics; dynamo theory; rigorous bounds

1. Introduction

This paper sets out some rigorous bounds on statistically steady dynamos driven by steady forcing. Such dynamos have been widely investigated by a number of authors; for a review with references see [1]. Of particular interest is the so-called Archontis dynamo [2,3] in which the magnetic field is almost aligned with the velocity field over much of the domain. In the latter case the velocity field and magnetic field can be almost steady with steady forcing while in other cases the fields and flows are disordered and aperiodic in time, so that averages over both space and time have to be assumed constant to make progress. This permits the construction of both upper and lower bounds on the time-averaged magnetic energy or dissipation, and the ratios of these quantities to their kinetic equivalents. A number of these results have been given elsewhere (see references in Section 3), but others are apparently novel. Some general results are given, and also more restrictive bounds that obtain (approximately) when the fields and flows are of Archontis type. The analysis shows results for the simple case where the magnetic Prandtl is unity and also where it is small, as found in laboratory fluids.

The plan of this paper is as follows: After an introduction in Sections 1 and 2, the mathematical problem is presented in Section 3.

Citation: Proctor, M. Bounds on Energies and Dissipation Rates in Forced Dynamos. *Physics* **2022**, *4*, 933–939. https://doi.org/10.3390/physics4030061

Received: 11 July 2022
Accepted: 16 August 2022
Published: 23 August 2022

2. Methods

The analysis is performed using a number of known inequalities (Poincaré, Cauchy–Schwartz, Hölder, and others) that have a long history in the literature; while the bounds that are derived are only approximate in the sense that results for real flows will not be close to the extreme limits derived, they have the merit of being universal and are useful as a consistency check on any numerical computations. There have been a number of rigorous results derived for magnetic field growth due to prescribed flow; see, e.g., [4]. Here, flows that are driven by a steady space-periodic forcing are examined. Results are presented both for the flows and fields themselves, and for the Elsasser variables, defined as the sum and difference of the velocity and magnetic fields.

3. Governing Equations

We consider forced incompressible flow $\mathbf{u}^*$ and dynamo generated magnetic field $\mathbf{b}^*$ in a periodic box off side L. The forcing $\mathbf{f}^*(\mathbf{x}^*)$ is also taken to be incompressible. As well as the velocity forcing we have the Lorenz force due to the magnetic field. The momentum Equation for $\mathbf{u}^*$ and the induction equation for $\mathbf{b}^*$ can be written in the dimensionless forms

$$\frac{\partial \mathbf{u}}{\partial t} + R\mathbf{u}\cdot\boldsymbol{\nabla}\mathbf{u} = -\nabla P + R\mathbf{b}\cdot\boldsymbol{\nabla}\mathbf{b} + \nabla^2\mathbf{u} + \mathbf{f}, \tag{1}$$

$$\frac{\partial \mathbf{b}}{\partial t} = R\boldsymbol{\nabla}\times(\mathbf{u}\times\mathbf{b}) + P_m^{-1}\nabla^2\mathbf{b}, \tag{2}$$

where $\mathbf{x}^* = \mathbf{x}/d$, $\mathbf{u}^* = R\nu\mathbf{u}/d$, $\mathbf{b}^*$ (measured in Alfvén velocity units) $= R\nu\mathbf{b}/d$, $\mathbf{f}^* = \nu^2 R d^{-3}\mathbf{f}$ and $P_m = \nu/\eta$ (the magnetic Prandtl number), where ν is the kinematic viscosity, $d = L/2\pi$ where L is the length scale of the system, and η the magnetic diffusivity. The parameter R is defined such that $\langle|\mathbf{f}|^2\rangle_S = 1$, where $\langle\cdot\rangle_S$ is defined as an average over the periodic box. Thus, R plays the role of a Reynolds number or inverse viscosity. For simplicity, the assumption is made that the forcing $\mathbf{f}$ is independent of time, and it is further supposed that all quantities are periodic over a box of side L. It is also supposed that $\boldsymbol{\nabla}\cdot\mathbf{u} = \boldsymbol{\nabla}\cdot\mathbf{f} = \boldsymbol{\nabla}\cdot\mathbf{b} = 0$, and that $\langle\mathbf{b}\rangle_S = \langle\mathbf{u}\rangle_S = \langle\mathbf{f}\rangle_S = 0$. (It may be checked that these averages do not change during the evolution of the system.) We are interested in a system in a statistically steady state, and define $\langle\cdot\rangle$ to be the time average of $\langle\cdot\rangle_S$ over time.

4. Results

4.1. No Magnetic Field

Equations (1) and (2) have solutions with $\mathbf{b} \equiv 0$. In that case we can can find bounds on the kinetic energy and viscous dissipation in terms of properties of $\mathbf{f}$. Defining $\mathcal{E}^2 \equiv \langle|\mathbf{u}|\rangle^2$ $\mathcal{D}^2 \equiv \langle|\nabla\mathbf{u}|\rangle^2$ we have the following results, noting that $\mathcal{E}^2 \le \mathcal{D}^2$ in the chosen geometry.

4.1.1. Upper Bound

Multiplying Equation (1) by $\mathbf{u}$ and averaging and use of the Cauchy and Poincaré inequalities gives

$$\mathcal{E}^2 \le \mathcal{D}^2 = \langle\mathbf{u}\cdot\mathbf{f}\rangle \le \mathcal{D}\left\langle|\nabla\mathbf{g}|^2\right\rangle^{\frac{1}{2}}, \tag{3}$$

where $\mathbf{g}$ is defined by $\mathbf{f} = -\nabla^2\mathbf{g}$, $\langle\mathbf{g}\rangle_S = 0$, so that one has the upper bound,

$$\mathcal{E} \le \mathcal{D} \le \Gamma, \quad \text{where} \quad \Gamma^2 = \left\langle|\nabla\mathbf{g}|^2\right\rangle \le 1. \tag{4}$$

The last of these inequalities is shown, recalling that $\langle|\mathbf{f}|^2\rangle = 1$, by noting that $\Gamma^2 = \langle\mathbf{f}\cdot\mathbf{g}\rangle \le \langle|\mathbf{g}|^2\rangle^{\frac{1}{2}} \le \Gamma$.

4.1.2. Lower Bound

Multiplying Equation (1) by $\mathbf{g}$ and averaging, recalling that $\mathbf{g}$ is independent of time, one obtains:

$$-R\langle\mathbf{u}\cdot(\mathbf{u}\cdot\nabla\mathbf{g})\rangle - \left\langle\mathbf{u}\cdot\nabla^2\mathbf{g}\right\rangle = -R\langle\mathbf{u}\cdot(\mathbf{u}\cdot\nabla\mathbf{g})\rangle + \mathcal{D}^2 = \langle\mathbf{f}\cdot\mathbf{g}\rangle = \Gamma^2. \tag{5}$$

One can bound the terms on the left hand side using the above inequalities and the Hölder inequality to get (defining $\max_{i,j,\mathbf{x}}(|\partial g_i/\partial x_j|) = A$, where A is a constant):

$$(RA+1)\mathcal{D}^2 \ge RA\mathcal{E}^2 + \mathcal{D}^2 \ge \Gamma^2. \tag{6}$$

Results similar to both these upper and lower bounds have been obtained earlier, e.g., [5,6].

4.2. Dynamc Inequalities

We now introduce the induction equation and the Lorentz force term and try to generalise the above results. These are now framed as inequalities on the magnetic energy and magnetic dissipation; we define $\mathcal{E}_m^2 \equiv \langle |\mathbf{b}|^2 \rangle$, $\mathcal{D}_m^2 \equiv \langle |\nabla \mathbf{b}|^2 \rangle$.

4.2.1. Upper Bound

Multiplying Equation (1) by $\mathbf{u}$ and Equation (2) by $\mathbf{b}$, averaging and adding, one finds that the cubic integrals cancel and that

$$(P_m^{-1}\mathcal{D}_m^2 + \mathcal{D}^2) = \langle \mathbf{u} \cdot \mathbf{f} \rangle \leq \Gamma \mathcal{D}, \quad \text{so that} \tag{7}$$

$$\mathcal{D}_m^2 \leq P_m \left(\Gamma \mathcal{D} - \mathcal{D}^2 \right) \leq \frac{P_m \Gamma^2}{4}. \tag{8}$$

4.2.2. Lower Bound

More surprising perhaps is the fact that the magnetic dissipation is bounded below, for given values of R and $\mathcal{D}$. Multiplying Equation (1) by $\mathbf{g}$, using Equation (8) and averaging, one finds:

$$-R\langle(\mathbf{u} \cdot (\mathbf{u} \cdot \nabla \mathbf{g}) - \mathbf{b} \cdot (\mathbf{b} \cdot \nabla \mathbf{g})\rangle + (P_m^{-1}\mathcal{D}_m^2 + \mathcal{D}^2) = \Gamma^2. \tag{9}$$

Then by analogous methods to the non-magnetic case one finally obtains:

$$\begin{aligned} RA(\mathcal{D}_m^2 + \mathcal{D}^2) + (P_m^{-1}\mathcal{D}_m^2 + \mathcal{D}^2) &\geq RA(\mathcal{E}_m^2 + \mathcal{E}^2) + (P_m^{-1}\mathcal{D}_m^2 + \mathcal{D}^2) \geq \Gamma^2, \quad \text{or} \\ \mathcal{D}_m^2 \geq \frac{\Gamma^2 - RA\mathcal{E}^2 - \mathcal{D}^2}{RA + P_m^{-1}} &\geq \frac{\Gamma^2 - (RA+1)\mathcal{D}^2}{RA + P_m^{-1}}. \end{aligned} \tag{10}$$

These results are related to those shown by Tilgner [7] in the context of the G.O. Roberts dynamo.

4.2.3. Dissipation Ratio

One can write these results alternately in terms of the ratio of magnetic to kinetic dissipation. Let $\mathcal{D}_m^2 = k^2\mathcal{D}^2$; then (here $\mathcal{G} = 1/\mathcal{D}$):

$$\frac{\Gamma^2\mathcal{G}^2 - (1 + RA)}{RA + P_m^{-1}} \leq k^2 \leq P_m(\Gamma\mathcal{G} - 1). \tag{11}$$

The largest possible value of k_{max}^2 of k^2 is given by the intersection of the boundary curves of the inequalities in Equation (11). Eliminating $X = \Gamma\mathcal{G}$ at the intersection, one finds:

$$k_{\text{max}}^2 = \tfrac{1}{2}\left[P_m(Z-1) + \sqrt{P_m^2(1-Z)^2 + 4P_m(2Z+1)} \right], \quad \text{where } Z = P_m RA. \tag{12}$$

k_{max}^2 appears to increase monotonically with R. For many liquid metals $P_m \ll 1$; in that case, if $k_{\text{max}}^2 = O(1)$, one must have $R = O(P_m^{-2})$.

4.3. The Archontis Dynamo

Now these results can be applied to the Archontis dynamo to get what are probably rather conservative estimates on the relation between R and P_m necessary to sustain such a dynamo.

The original forcing function used by Archontis takes the form,

$$\mathbf{f} = \sqrt{\frac{2}{3}}(\sin z, \sin x, \sin y). \tag{13}$$

Then one can see that $\mathbf{g} = \mathbf{f}$, $A = \sqrt{2/3}$, $\Gamma = 1$. For the Archontis dynamo, $k^2 \approx 1$ and so the boundary of the region in (R, P_m) space, where this is possible, is given by putting both the expressions in Equation (11) equal to unity, giving the approximate inequality for such a dynamo:

$$R \geq \sqrt{\frac{1}{6} P_m^{-2}}. \tag{14}$$

Since, as previously noted, $P_m \ll 1$ in many physical systems, it can be seen that Archontis type dynamos can occur only for large values of R for such systems.

4.4. Use of Elsasser Variables

One can instead use the Elsasser variables $\mathbf{z}_\pm = \mathbf{u} \pm \mathbf{b}$. The equations can be written

$$\frac{\partial \mathbf{z}_\pm}{\partial t} + R\mathbf{z}_\mp \cdot \boldsymbol{\nabla} \mathbf{z}_\pm = -\nabla P + \frac{1}{2}(1 + P_m^{-1})\nabla^2 \mathbf{z}_\pm + \frac{1}{2}(1 - P_m^{-1})\nabla^2 \mathbf{z}_\mp + \mathbf{f}. \tag{15}$$

Then multiplying each equation by $\mathbf{z}_\pm$, one finds, in the statistically steady state:

$$\langle \mathbf{z}_\pm \cdot \mathbf{f} \rangle = \frac{1}{2}(1 + P_m^{-1})\left\langle |\nabla \mathbf{z}_\pm|^2 \right\rangle + \frac{1}{2}(1 - P_m^{-1})\langle \nabla \mathbf{z}_\pm \cdot \nabla \mathbf{z}_\mp \rangle. \tag{16}$$

Now defining $\mathcal{D}_\pm^2 = \left\langle |\nabla \mathbf{z}_\pm|^2 \right\rangle$ and using similar inequalities to those in previous Sections, one finds:

$$\Gamma \mathcal{D}_\pm + \frac{1}{2}|1 - P_m^{-1}|\mathcal{D}_\pm \mathcal{D}_\mp \geq \frac{1}{2}(1 + P_m^{-1})\mathcal{D}_\pm^2, \tag{17}$$

or, dividing through,

$$\Gamma \geq \frac{1}{2}(1 + P_m^{-1})\mathcal{D}_\pm - \frac{1}{2}|1 - P_m^{-1}|\mathcal{D}_\mp. \tag{18}$$

As a check consider the non-magnetic case for which $\mathbf{z}_+ = \mathbf{z}_-$. For $P_m < 1$ we reproduce an inequality similar to Equation (4). When $P_m > 1$ one obtains a weaker result. One can find relations corresponding to Equation (10) by multiplying Equation (15) by $\mathbf{g}$ and averaging, leading to

$$VR\mathcal{D}_\pm \mathcal{D}_\mp + \frac{1}{2}(1 + P_m^{-1})\Gamma \mathcal{D}_\pm + \frac{1}{2}|1 - P_m^{-1}|\Gamma \mathcal{D}_\mp \geq \Gamma^2. \tag{19}$$

where $\max_{\mathbf{x}}(|\mathbf{g}|) = V$. To understand what these inequalities imply consider the special case $P_m = 1$, and the Archontis forcing, where $V = \sqrt{2}, \Gamma = 1$. Then Equations (18) and (19) become:

$$\mathcal{D}_\pm \leq 1, \quad \sqrt{2}R\mathcal{D}_\pm \mathcal{D}_\mp + \mathcal{D}_\pm \geq 1. \tag{20}$$

For the Archontis dynamo, we have that $\mathcal{D}_- \ll \mathcal{D}_+$, say. It is straightforward to check, using the lower sign in Equation (20), that the smallest value of the ratio $r = \mathcal{D}_-/\mathcal{D}_+$ compatible with the above inequalities is $1/(\sqrt{2}R + 1)$, and so R has to be large for an Archontis type dynamo, as indeed is found in the experiments; see for example the detailed calculations for large R , described for $P_m = 1$ in [8], (with large R corresponding to small ϵ in their notation).

Similar results hold for general P_m. Consider the interesting case $P_m < 1$. Then a similar calculation gives, for $\mathcal{H} = VR$,

$$2\mathcal{H}r \geq 1 + r + \frac{1}{P_m}(1 - 2r - r^2) + \frac{r}{P_m^2}(r - 1). \tag{21}$$

How big should R - or $\mathcal{H}$ - be to get close to the Archontis configuration? If $r \ll 1$ then $\mathcal{H} \approx (1 + P_m^{-1})/2r \gg 1$. If $r \sim P_m \ll 1$ then $r \approx 1/\left((P_m(2\mathcal{H} + P_m^{-2})\right)$ and so $\mathcal{H}$ has to be very large, of order P_m^{-2}, to reduce r significantly below P_m.

Figure 1 shows the relation between $\mathcal{H}$ and r for three values of $P_m \leq 1$. It can immediately be seen that as P_m decreases (i.e., η increases for fixed ν) it becomes harder to make r small, i.e., to approach an Archontis state.

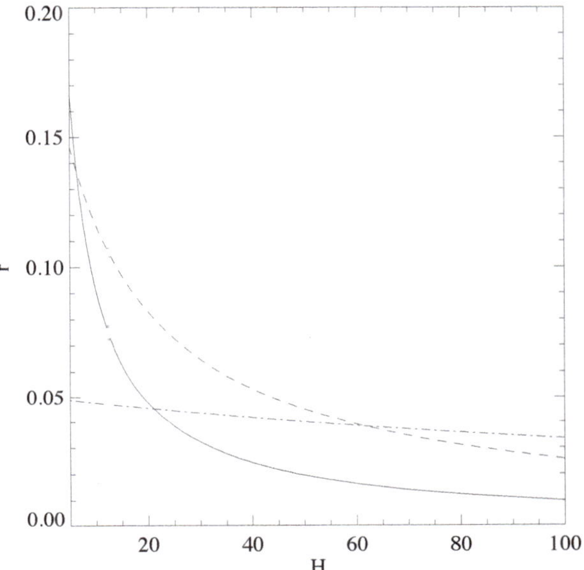

Figure 1. The minimum value of $r = \mathcal{D}_-/\mathcal{D}_+$ as a function of $\mathcal{H} = R\sqrt{2}$ for $P_m = 1$ (solid line); $P_m = 0.2$ (dashed line); $P_m = 0.05$ (dash-dotted line). See text for details.

The values of k and r are not straightforwardly related. By use of the Schwartz inequality it can be established that as long as $k, r < 1$ as envisaged one must have $(r+1)(k+1) \geq 2$, so that, for instance, if r is close to zero, then k must be close to unity but, since r is bounded below rather than above for fixed $\mathcal{H}$, then k is not constrained.

5. Discussion

In this short paper, I explore the rigorous results that constrain the magnetic and kinetic energies and dissipation rates in a forced, fully nonlinear dynamo. It is possible to obtain both lower and upper bounds on the magnetic dissipation for a given forcing. In particular, the parameter R, which measures the amplitude of the forcing, has to be very large when the magnetic Prandtl number is small. Similar bounds can be found for the Elsasser variables, $\mathbf{z}_\pm = \mathbf{u} \pm \mathbf{z}$, and these lead to the same conclusion in the case of the Archontis dynamo where $\mathbf{u}$ and $\mathbf{b}$ are closely aligned in much of the domain, while some of the bounds are likely to be rather loose, they are at least rigorous, and provide guidance as to where in parameter space to look for exceptional configurations such as those found by Archontis. Work is in progress with D.J.Galloway to examine the accuracy of the bounds through direct numerical simulations. In Appendix A, the author of this paper presents his brief reminiscent about contacts with M. Tribelsky to whose honour this issue is dedicated.

Funding: This research was supported by a visiting Fellowship at the University of Sydney.

Data Availability Statement: Not applicable.

Acknowledgments: I am grateful to D.J. Galloway for hosting me at the University of Sydney and for helpful discussions.

Conflicts of Interest: The author declares no conflict of interest.

Appendix A

I am delighted to dedicate this paper to my friend and colleague Michael Tribelsky. I have known him since meeting in Bayreuth in 1989 at a conference and we had valuable discussions during his visit to the Isaac Newton Institute, Cambridge, UK in 2005.

Figure A1. Michael Tribelsky (**left**) and Michael Proctor (**right**): Trinity College, Cambridge, UK in 2005.

References

1. Galloway, D.J. Nonlinear dynamos. In *Proceedings of the IAU Symposium No. 271: Astrophysical Dynamics: From Stars to Galaxies*; Brummell, N.H., Brun, A.S., Miesch M.S., Ponty, Y., Eds.; International Astronomical Union: Paris, France, 2011; pp. 297–303. [CrossRef]
2. Archontis, V.; Dorch, S.B.F.; Nordlund, A. Nonlinear MHD dynamo operating at equipartition. *Astron. Astrophys.* **2007**, *472*, 715–726. [CrossRef]
3. Dorch, S.B.F.; Archontis, V. On the saturation of astrophysical dynamos: Numerical experiments with the no-cosines flow. *Sol. Phys.* **2004**, *224*, 171–178. [CrossRef]
4. Proctor, M.R.E. Introduction to dynamo theory. In *Mathematical Aspects of Natural Dynamos*; Dormy, E., Soward, A.M., Eds.; Chapman and Hall/CRC: New York, NY, USA, 2007; pp. 20–41. [CrossRef]
5. Doering, C.; Foias, C. Energy dissipation in body-forced turbulence. *J. Fluid Mech.* **2002**, *467*, 289–306. [CrossRef]

6. Doering, C.; Eckhardt, B.; Schumacher, J. Energy dissipation in body-forced plane shear flow. *J. Fluid Mech.* **2003**, *494*, 275–284. [CrossRef]
7. Tilgner, A. Scaling laws and bounds for the turbulent G.O. Roberts dynamo. *Phys. Rev. Fluids* **2017**, *2*, 024606. [CrossRef]
8. Gilbert, A.D.; Ponty, Y.; Zheligovsky, V. Dissipative structures in a nonlinear dynamo. *Geophys. Astrophys. Fluid Dyn.* **2011**, *105*, 629–653. [CrossRef]

MDPI
St. Alban-Anlage 66
4052 Basel
Switzerland
Tel. +41 61 683 77 34
Fax +41 61 302 89 18
www.mdpi.com

Physics Editorial Office
E-mail: physics@mdpi.com
www.mdpi.com/journal/physics

www.ingramcontent.com/pod-product-compliance
Lightning Source LLC
LaVergne TN
LVHW071005170726
843515LV00006B/1079